People, Land and Water in the Arab Middle East

Studies in Environmental Anthropology
edited by Roy Ellen, University of Kent at Canterbury, UK

This series is a vehicle for publishing up-to-date monograph studies on particular issues in particular places which are sensitive to both socio-cultural and ecological factors (i.e. sea level rise and rain forest depletion). Emphasis will be placed on the perception of the environment, indigenous knowledge and the ethnography of environmental issues. While basically anthropological, the series will consider works from authors working in adjacent fields.

Volume 1
A Place Against Time
Land and Environment in the Papua New Guinea Highlands
Paul Sillitoe

Volume 2
People, Land and Water in the Arab Middle East
Environments and Landscapes in the Bilâd ash-Shâm
William and Fidelity Lancaster

Volumes in Production

Volume 3
Protecting the Arctic
Indigenous Peoples and Cultural Survival
Mark Nuttall

Volume 4
Transforming the Indonesian Uplands
Marginality, Power and Production
Tania Murray Li

This book is part of a series. The publisher will accept continuation orders which may be cancelled at any time and which provide for automatic billing and shipping of each title in the series upon publication. Please write for details.

People, Land and Water in the Arab Middle East

Environments and Landscapes in the Bilâd ash-Shâm

William Lancaster

former Director of the
British Institute at Amman for Archæology and History,
Senior Associate Member, St. Antony's College, Oxford

and

Fidelity Lancaster

harwood academic publishers
Australia • Canada • China • France • Germany • India
Japan • Luxembourg • Malaysia • The Netherlands
Russia • Singapore • Switzerland

Copyright © 1999 OPA (Overseas Publishers Association) N.V.
Published by license under the Harwood Academic Publishers
imprint, part of The Gordon and Breach Publishing Group.

All rights reserved.

No part of this book may be reproduced or utilized in any form or
by any means, electronic or mechanical, including photocopying
and recording, or by any information storage or retrieval system,
without permission in writing from the publisher. Printed in
Singapore.

Amsteldijk 166
1st Floor
1079 LH Amsterdam
The Netherlands

British Library Cataloguing in Publication Data

Lancaster, William
  People, land and water in the Arab Middle East :
  environments and landscapes in the Bilâd Ash-Shâm. –
  (Studies in environmental anthropology ; v. 2)
  1. Tribes – Jordan – Social conditions 2. Land use, Rural –
  Jordan 3. Water supply, Rural – Jordan – Management 4. Bilâd
  Ash-Shâm (Jordan) – Social life and customs
  I. Title II. Lancaster, Fidelity
  333.7'6'095695

ISBN 90-5702-322-9
ISSN 1025-5869

CONTENTS

List of Maps — vii
List of Figures — ix
List of Plates — xi
Preface — xiii

Chapter 1 Aims and Arguments; Methods of Working — 7
Chapter 2 The Framework to Social Practice — 53
Chapter 3 Physical Environments, Landscapes and Nature — 97
Chapter 4 Water — 129
Chapter 5 Land Use; The Practices of Production Systems using Land as a Primary Resource — 167
Chapter 6 Buildings and Other Structures — 239
Chapter 7 Productivity, Distribution and Consumption — 289
Chapter 8 Integration into Modernity — 343

Appendix — 397
Bibliography — 417
Index — 437

LIST OF MAPS

1	Bilâd ash-Shâm	xv
2	Qalamûn	1
3	North Kerak Plateau	2
4	Fainân/Dana/Qâdisîyya	3
5	Sakaka/Jawf Area	4
6	Harra/Hamad – Water and Grazing	5

LIST OF FIGURES

1	Diagram of a Ghadîr	134
2	Diagram of a Mahfûr	136
3	Anqa Mahafîr	137
4	Diagram of a Cistern	139
5	Diagram of a Thumaila	140
6	Shâfi's Farm – Irrigation System	148
7	Diagram of Irrigation System at W. Ibn Hammâd	149
8	Diagram of a Foggara	156
9	Tent of Mobile Herder	244
10	Tent – Permanent Herding Base	245
11	Tent at Annual Summer Site	246
12	Diagram of "Progression" – Tent to "Villa"	257
13	Diagram of a Khâna	268

LIST OF PLATES

1	Household vegetable garden – Sakaka	81
2	Date market – Sakaka	82
3	Hard grazed for years, plants re-appear with suitable rains – ar-Risha, eastern hamad, Jordan	83
4	Villa in Sakaka: note water-tower decorated as an incense burner	84
5	Mechanised large scale farming – near al-Jauf	85
6	Water filled *mahafir* in the *hamad* – east Jordan	86
7	Camels going to water in Wadi Ghwair: old gardens on right bank – southern Ghor, Jordan	87
8	Bronze age water storage still in use – the harra in north Jordan	88
9	A ghadir in the eastern *hamad* – Jordan	89
10	Gathering and milking sheep – eastern *hamad* of Jordan	90
11	Threshing machine – Karak plateau, Jordan	91
12	Vet injecting sheep, helped by the owner's daughter – Southern Jordan	92
13	*Jabban* making cheese – eastern *badia*, Jordan	93
14	Yarded dairy herd, central Jordan: bales of straw from Saudi Arabia	94
15	Herder's tent, *haush* and pick-up: water seep nearby – Karak plateau, Jordan	95
16	Spring-fed gardens at Dana – Southern Jordan	96

PREFACE

This book is not primarily for Arabists, nor are we Arabists. In addition, we work from speech. For these reasons, therefore, names of people, places and things have been transliterated as we heard them, and in a style to be accessible to the general reader. We have not made a distinction between *'ain* and *hamza* nor used the *ta marbuta*, nor included the 'heavy' letters except in quotations where the style adopted by each author has been retained. Nor have we used Arab adjectival forms or plurals. While recognising that these simplifications will irritate specialists, it is hoped that they will make reading easier for non-specialists.

It has not been possible to include all the information we have been given. And there are many aspects of land and water management of which we have remained unaware. Our information comes from observations, questions and discussions on specific occasions, so some of what we find out arises from chance, and alternatively, we miss other information that would, in retrospect, have been available because we had no idea that a particular specialist was present, or that certain techniques were in use. To quote a Rwala neighbour, "You can only see what you see, and hear what you hear. You don't notice what you don't know about, and you don't always listen to what your ears hear. It is not possible to know everything, that is why there is never complete agreement between people, even those who were present, as to what exactly happened in a fight. Even if people agree on what happened, each one will interpret its meaning slightly differently, because each is an individual. Only God knows the truth."

We are deeply grateful to the many bodies and individuals who have supported this work and made it possible. In Jordan, the Department of Antiquities and its Directors; His Excellency Sharif Fawaz, Head of the Badia (Desert) Police Force; and Shaikh Faisal bin Fawaz al-Sha'alan have all eased our paths. We wish to thank His Excellency the Ambassador of the Kingdom of Saudi Arabia in London, and the Governor of al-Jawf District, His Excellency al Amir Sultan bin Abdul Rahman al-Sudairy, for permission to revisit old friends and ask new questions. In Syria, the Ministry of Culture and the Department of Antiquities were helpful.

The British Institute at Amman for Archaeology and History sponsored much of our research. We are indebted to the Burqu'-Ruwaishid Project of Dr Alison Betts of Edinburgh University (and now of Sydney University), and to the Khirbet Faris Project of Dr Jeremy Johns of the Oriental Institute, University of Oxford, and Miss Alison McQuitty, British Institute at Amman for Archaeology and History, for asking us to participate. The World Bank and the Jordanian Royal Society for the Conservation of Nature asked us to report for their Dana Project.

The research would have been the poorer if we had not had access to the libraries of the British Institute at Amman for Archaeology and History, the American Centre for Oriental Research in Amman, and the Abdul Hamid Shoman Foundation Library, also in Amman; the library of the Oriental Institute, University of Oxford; and the Kirkwall Public Library, Orkney, whose staff were so helpful and patient in getting books and articles for us through the Inter-Library Loan Service.

Dr Clive Agnew of the Department of Geography, University College London, read Chapter IV and Dr Tom Carnie, formerly of the University of Glasgow, read Chapter VII. We are deeply appreciative of their help and comments.

Our deepest thanks go to all our hosts and friends in the towns and countryside of the Bilâd ash-Shâm. They are far too numerous to thank individually as their numbers, after twenty-five years, must run into thousands. Without their good sense, sympathy, and hospitality this book could never have been written. It is our hope that it will, in some small measure, honour their knowledge and generosity.

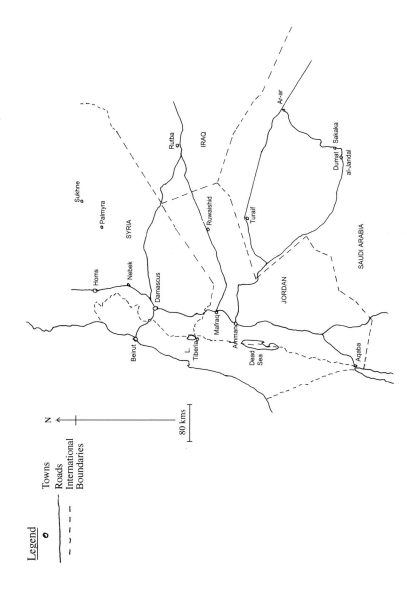

Map 1 Bilâd ash-Shâm

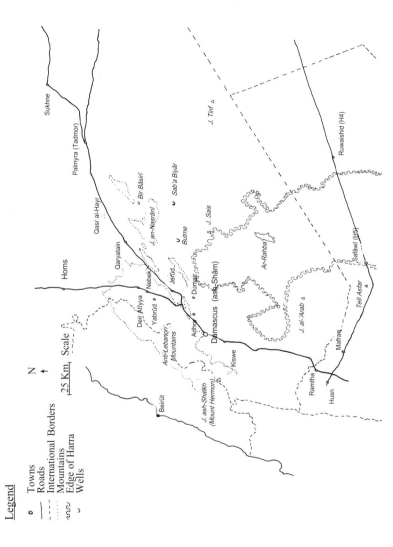

Map 2 Qalamûn

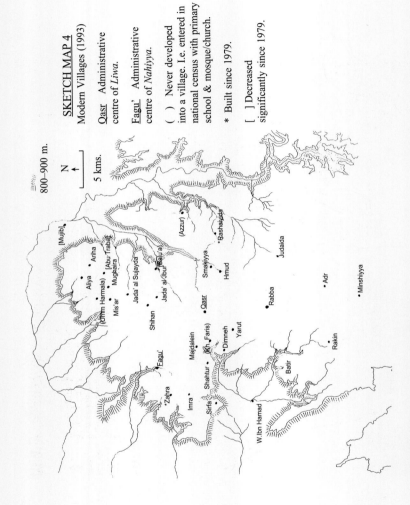

Map 3 North Kerak Plateau

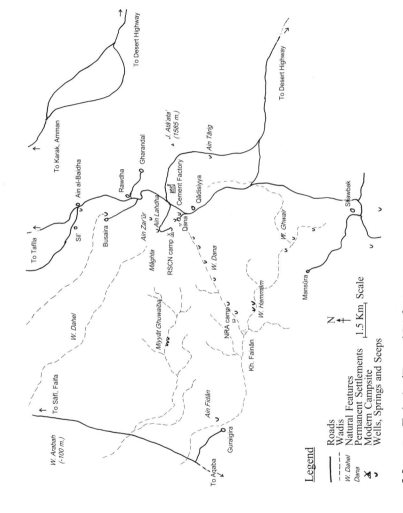

Map 4 Fainân/Dana/Qâdisîyya

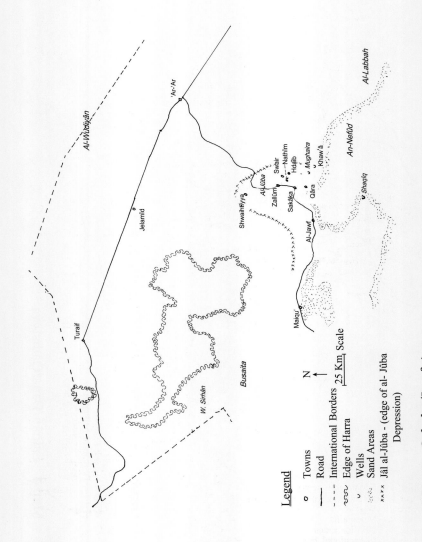

Map 5 Sakaka/Jawf Area

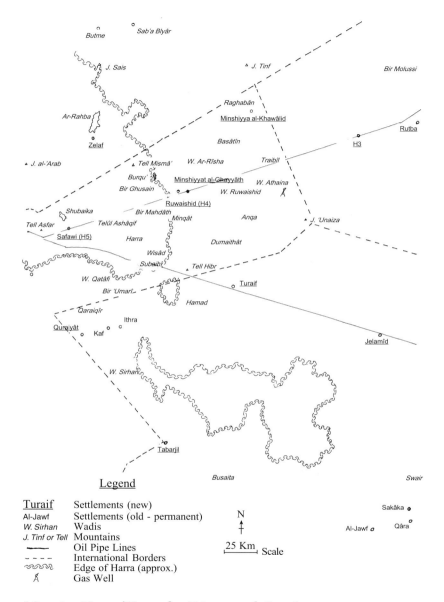

Map 6 Harra/Hamad – Water and Grazing

CHAPTER 1

AIMS AND ARGUMENTS; METHODS OF WORKING

The aim is to demonstrate local systems of land and water use in the Bilâd ash-Shâm in the present and recent past. These systems of use rest on local perceptions of the physical environments, and these perceptions are effected by the moral premises underlying culture and society. Local economic and political processes are, to a considerable extent, concerned with land and water use. Political and economic factors at work in the wider region affect particular manifestations of land and water systems. Transformations in these systems are seen as resulting from changes in such economic and political factors and technologies, perceived by local participants to be outside their control and to come from external sources. This is not to say that local people see themselves as within a stereotype of an unchanging rural East, they do not. They perceive transformations as negotiable between themselves as individuals representing particular economic and political strategies, and external agents bringing new demands, requirements and technologies. The capacity to transform is inherent in local social practice. Participants consider local land use systems as sustainable in themselves, with an integral flexibility of response to unpredictable variations and unforeseen factors. The compatibility of local systems with the economic and political demands of the nation state and the global economy are examined through local concepts of production, distribution and consumption.

The observations, conversations, and discussions from which our information and analysis are drawn, come from twenty five years of intermittently continual research with Rwala of Syria, Jordan and northern Saudi Arabia; Ahl al-Jabal of eastern Jordan; Ahl al-Karak, farmers and herders in central mountain Jordan; and 'Ata'ata, Amarin, Sa'idiyin, Rashaiyida and Azazme around Fainan and Dana in the Shera mountains in southern Jordan. Our various hosts have always informed themselves of our reasons for being there, and who our paymasters were. We have been fortunate in being able, largely, to fund ourselves from our own resources or to have been supported for subsistence rather than profit. An earlier book made no financial profits in relation to its research costs, nor has any post occupied as a result of a research reputation paid enough to be considered as profiteering from local knowledge. Our education and knowledge, language and literacy skills, income and resources have not marked us off significantly from our hosts in our view or theirs; any expertise in one area has been limited by a deficit in another; we are better off financially than some and poorer than others; our speech marks us as having learnt Arabic in the *badia*; and so on. We are assigned a place as relatively familiar but transient guests, part of the community or incorporated into family to ensure decent behaviour in a society where asking direct questions is inappropriate. As we 'use' our hosts as a resource, we have been 'used' by them. We are grateful for their hospitality, sponsorship, and generosity in sharing their knowledge. This book, like the earlier, aims to honour those who have shared with us; not by presenting only their abstracted ideal, the presentation of self, but, by including the particular practice of individuals and their self knowledge, to portray the diversity of individuals, relations within the countryside, and their uses of and negotiations between their theoretical generalisations with individual actions and events.

The argument is that local land use systems, which vary across the natural environments of the Bilâd ash-Shâm, have existed over time independently of centralised political and economic systems, although they may well co-exist. Developmental models, or models using centre and periphery, or world systems, do not adequately account for the observed local systems and the perception of them that we hear discussed by their users. Nor are the systems a simple response to particular environmental conditions and determined solely by them. The varied ecological regions are constructed as landscapes with specific potentialities and limitations through the seasons and over the years. Local systems are considered through the considerations of their participants, so that local theories and practice may be seen as primary information with less interpretative overlay from external concerns and interests. Ideas and practices of land and water use are integrated in local theories of social, political and economic action.

Researches into land and water use in different environments of the Arabian peninsula have led to an appreciation of what appear to be underlying and long-standing themes of social practice based on relatively similar premises. Similar physical environments do not support identical land use systems, although there are observable similarities; not all groups living in or using one environment have the same subsistence strategy. This might be explained by cultural backgrounds, demographic pressures or, say, a difference in permitted strategies from political events. However, all those with whom we discussed such matters emphasised similar processes and premises for their particular land use system, and for their perception of those of others. People emphasise the construction of a series of landscapes, of possible livelihoods compatible with those landscapes and with ways of living compatible with moral premises. The common factor in their appraisal of physical environments is that no one environment of itself can provide a secure living throughout a year and over the years. Two corollaries follow. Firstly, all livelihoods will be multiresource. Whatever the main subsistence activity is – herding, hunting and gathering, farming, or fishing – people have additional sources of income from supplying services or crafts made possible from their primary resource, or by operating with a second or third resource. Secondly, mobility between resource areas is implicit; this is usually the movement of people through the yearly progression of the seasons to different areas, together with the movement of people and/or goods between areas of surplus and deficit.

There is therefore an explicit and general concern with social relations and the modes of identification and processes of verification that enable mobility of persons and goods. These are predicated on common moral premises, articulated in a variety of ways, but all based on jural equality before God; from this follows a concern with individual autonomy and responsibility for one's actions. Land and water use systems are founded on the ideas people have about their environment and their society, as if society and its practices were unchanging. In one way, society changes all the time with members being born and dying, and the consequent changes in the personnel and composition of groups, together with losses and gains in capacities for knowledge and action. In another, its members adjust to, cope with and use repercussions from external factors with consequent probable internal changes. The development of particular productive or distributive strategies depends on what local people consider their markets are doing, and whether their personal networks have a surplus of the necessary inputs (cash, labour, time, knowledge, skills) to improve or extend technologies, production or distribution. Increases in wealth reaching an area, for example from more active markets or a greater demand for services, are considered to come from externally generated economic activities. Since thought is as important as action in the existence of society, possibilities for change and continuity are integral.

Groups holding the premises of jural equality before God and individual autonomy live in and/or use the countryside of the Bilâd ash-Shâm, follow many occupations, and would describe themselves and be described by others in different ways, depending on contexts. Such social groups of 'tribes', 'peasants', 'traders', 'nomads', 'villagers', 'townsmen' are often discussed as if these were discrete, bounded, bodies and as if such groups have 'rules' imposed upon them, derived from deterministic associations between environment and livelihood, or occupation and political action. Individuals of such groupings, on the contrary, speak of underlying similarities of social practice and of the premises on which practice is founded, since there is an awareness both of the generative construction of social practice by individuals and of deep continuities of such practice. Close similarities in particular practices may occur in widely separated and apparently disparate areas.[1]

Most rural inhabitants identify themselves in tribal terms; the remainder do so in family terms, considered to be like those of tribe on a smaller scale. Academic discussions of tribe in the Arab Middle East have changed over the years from the exotic and the social evolutionary towards a concept of tribe as a constructed abstraction by a variety of interested parties (Eickelman 1989: 127–8). A generalised 'the tribe' has given way to an acknowledgement of individual tribespeople as constructors of transmitted knowledge and actors in constantly renegotiated practice. Ethnographers must generalise from the particular, but in recognising process and practice, and the slippage between local theoretical generalisations and individual actions and events, are happier to present indeterminate information. The tribally varied information we had from longterm discussions about identity and access to resources gave rise to an idea of a 'conceptual infrastructure' (1992a), where tribal identity with its concomitant practice gives its holders privileged access to economic, political and social life. Johns (1994: 3)

---

[1] See Zubaida's discussion of boys' games of middle class Baghdad and as described by Musil (1928MC: 256–8) for Rwala from the early years of the century, and seen by us in the 1970s: (personal communication).

mis-reads the paper and wrongly understands it to say that a tribal conceptual infrastructure "determines" economic and social organisation, and that changes in, for example, material culture can be explained only by recourse to external factors. The basic idea of a conceptual infrastructure is that rather than an infrastructure of social institutions, manifested as physical structures and material evidences, it is possible to have a processual infrastructure for social activities, based on shared moral concepts, and manifested through social practice. Access to the jurally binding processes that constitute much of economic and political life between tribesmen, between individuals and groups from different tribes, and between tribesmen and non-tribal individuals and groups, is a major function of tribal identity, while such identity is presented by its holders in genealogical idioms and political metaphors and metonymies.

Local people do present changes in materials or techniques, or political and economic manifestations, as coming from outside. Responses to such new factors are generated from within current social practice, aligning changes with the moral premises underlying identities. These interactions between external introductions and internal formations comprise transformation. Since a community presents itself as a moral community, change that results in a transformation of social practice 'must' come from outside, since for change to arise internally implies the existing moral order to be untenable. Revelation is the plausible exception, although it is worth noting that Shaban (1976: 14–15) says Muhammad emphasised that he was not an innovator but a prophet who wanted to restore the application of the principles of eternal truth as had all the prophets from Abraham of which he was the last. The transmission of knowledge and its successive re-interpretations by transmitters and audience permits negotiation between past knowledge and changing situations so that bodies of knowledge become differentiated within society. In parallel, a community conscious of operating an economy from its own resources would logically expect economic change to be effected through external factors. Since local economies, whatever their main perceived strategy, encompass a variety of productive and service activities, their members assume that all could achieve some sort of livelihood anywhere in the peninsula, and beyond. It is assumed that livelihood would be achieved through help from dispersed family members, by adapting existing techniques to new areas, by developing new sources of income, learning and applying new technologies, using new political and economic opportunities, or by using landscapes out of favour with others.

As well as change and transformation, there is the alternative to tribally based identity. Individuals can become not tribal. Being 'not tribal' is seen as behaving not in accordance with the moral premises; duties and obligations to fellow tribal members are ignored, or the individual becomes unacceptable in other ways to his tribal fellows. Such behaviour may be less satisfactory and sustainable over time, although often more immediately profitable in terms of material wealth. Which forms of social practice individuals choose is that person's prerogative. Individuals and their families may pursue one identity and its practice in one locality and a second in another; or families may have some members following tribal social practice while others pursue individual self-interest to the exclusion of familial and neighbourly concerns. Inherent tension within 'tribal' families, like others, focuses around interpretations of members' behaviour in accordance with the underlying social premises and the opportunities afforded by individuals' perceptions of resource management under differing conditions of political action.

The inhabitants of the rural Bilâd ash-Shâm also describe themselves in terms of their main sources of livelihood; herders of camels, goats, sheep and formerly cows or farmers (Lancasters 1988). Each occupation was and is part of a wider multiresource economy with market and service components. Tribal and family groupings also participate in the political and economic activities of the (nation) state/s in which they live or to which they have access. These different groupings, with varied occupations and disparate environments, see their social identity to come from common and shared moral premises which predicates social practice, and from a shared past. Family or tribal groups can both split, usually from irresolvable differences, and accrete others, often through a reputation for mediation, hospitality and honourable political action. Tribal and family groups, of whatever occupational pursuit, are enmeshed in a variety of alliance networks, which carry potentials for rather than dictates of action. The use of 'tribal group' may jar but its use conveys an association of people using an area identifying themselves by tribal or family names, and behaving in accordance with tribal identity. 'Tribe' could be used if it were read as a metonymy, the name of the whole standing for a part, but a further gloss that a part is an entire named section would then be needed. Using locality names has the problem that a locality, as we were able to cover the wider region (because of state borders), is inconsistent with local use; nor are tribal sections and families discrete localised groups, although people talk as if they were. For these reasons, 'tribal group or grouping' presents realities more accurately.

A consideration of land and water systems entails awareness of past history. Members of all local groups account for their presence in a locality by past action. The depth of time may range from an explanation of movement into an area in the last few years because of internal disputes in a former locality, climatic variations, or political or economic pressures; from movement three or four centuries ago for similar reasons; or they may claim to have 'always' been present. Such explanations may rely on metaphors of cause and time, and conceal information rather than present it. To what extent use by mobile groups affected the sustainability of land and water resources is uncertain. Members of local groupings in various areas assume they could make a living from the resources available in other localities, even if such areas did not have precisely similar environments. This assumption reflects the varied and varying environments used in the livelihood strategies of any group, and the multiresource nature of modes of livelihood by an exponent of one occupation. Movement between known resource areas is axiomatic. As individuals from named groups range widely outside core areas, movement by a group or its parts to other regions is not seen as difficult or unusual. To establish claims of preferential access in a new area may be difficult. Initial use is achieved by claims of need, relationships through women or former neighbours, or by payment, while longer term claims to the right of residence come through social action. Local land use systems were not unchanging before the current situation of nation states and incorporation in a world economy, since the Bilâd ash-Shâm is part of wider economic structures which have affected local economies at various times.

The wider Bilâd ash-Shâm (see Map 1), in the views of local users, is the region bounded on the west by the mountains of the Anti-Lebanon which extend south as the hills either side of the Jordan valley, Dead Sea and Wadi Araba, and the eastern coast of the Red Sea. The northern edge is formed by the ranges of hills extending north-east from Damascus to Palmyra and Dair az-Zor.

Eastwards, the region falls away to the Euphrates and the Wadi Hauran. To the south, its limits are the northern slopes of the Nefud and the Hijaz. The mountain spine and its plateaux have the most rainfall, and the plateaux offer the best conditions for rainfed agriculture. The mountain slopes and valleys have springs; mountain slopes often provide opportunities for tree cultivation and rough grazings, while valley floors may be irrigated from springs and groundwater, as are other basins such as the Ghutas of Damascus and Qalamoun, Tadmur/Palmyra, and Jauf. The eastern *badia* regions comprise the gravel *hamad*, the black basalt *harra*s, the flint *sawwân* and the sand dunes of the Nefud. Drainage systems, or wadis, themselves the remains of earlier geological periods and associated freshwater lakes, provide temporary water holding systems in the *badia*. Larger groundwater basins feed oasis wells.

Local commentators assume a long continuity for their social practice and its structures, and place social activity in physical environments that are considered to be relatively similar to those of the present. Such an assumption does not commit local societies to the determinism of the '*longue durée*' since members acknowledge change in technologies and political and economic factors, while many tribal groups in the Bilâd ash-Shâm and its environs are known from before Islam. This is a good place to consider the questions of sources and materials. If we were repeating local sources, we would provide a variety of transmissions on a certain number of themes, which would be without chronological order. The methods and concerns of local historical traditions are comparable to those of Arab historical writings in that they rely on transmission and interpretation for their continuation and are driven by the concerns of local groupings. History may be presented as traditions, or as the necessary interests of the well-rounded respected man, or as wisdom, or as the necessary adjunct to ruling; local histories have been heard in all these presentations on various occasions. There is little chronology or linkage into specific rather than generalised outside events; the idiom is of 'the ruler of Mecca', 'when the Ottomans', 'the ibn Sa'ud', or 'when Glubb Pasha' – reminiscent of the dating of certain Safaïtic inscriptions by 'in the year when the Romans sacked X' and the epigrapher can come up with a choice of three different and widely separated years (Macdonald 1995a). Local and tribal histories are frequently concerned with the re-creation of a moral order and space broken by man's greed, weakness or selfishness, put in the discourse of violent political action and honour disgraced and defended, recalling pre-Islamic tribal traditions. Indeed, the concern for 'just rule' or moral order and the constant threat to this condition is a feature of mediaeval and later Islamic history and its historians (Chamberlain 1994; Khalidi 1994). Local and tribal histories[2] are largely without time; time is often seen either as all pervasive, or as continually circulating. This is not useful when there is a wish to establish chronologies for evaluating local commentators' view of the long continuities of their social practice and landscapes.

Material from archaeology, epigraphy and histories from other historiographies help in this. Archaeological evidence permits the validity of assuming long continuities in the physical environments, vegetation and fauna, and the corpus of productive strategies, as will be discussed later. Epigraphic remains confirm both agricultural works and seasonal herding locations, and establish the building of structures by groups for communal purposes. Written historical materials are of various types. There are the histories written by Islamic scholars over the centuries to transmit, interpret and order the various bodies of historical knowledge of concern to their societies, whose historiography is brilliantly illuminated by Khalidi (1994). From the sixteenth century there are histories written by "barbers, farmers, minor state officials, obscure military commanders" that "record the encounter between oral and popular culture and the high literary tradition", whose "mentalities still await scrutiny by modern anthropologists and historians of ideas" (Khalidi 1994: 233). There are formidable compilations of government documents from some periods, written to record or aid administration and the collection of taxation. For these to be informative it is necessary to know the reason for their recording and the methods of recording. Such documents were not, unlike scholarly histories, written to transmit learning but to aid administrative officials. A third class of documents are those left by travellers; some are immensely informed, others superficial. Often, they are particular, or generalise heavily from the particular, and vary in the quality of observation and analysis, reflecting both the concerns of the traveller and his time. The best are invaluable, since the depth of their observations often lead a Burckhardt, Doughty or Musil to record alternative

---

[2] Similarly Dresch (1989: 179) describes mountain Yemen histories as "where a great deal happens, ... but little is conceived to change".

versions, contradictory information and ambiguous analyses. Sources and their information are necessarily selective and selected. The nature of historical truth is an unending topic of fascination in the *badia* and the countryside, where interpretation is a recognised part of the transmission of knowledge, and where each individual is seen as an autonomous selector in transmitting and interpreting knowledge. The statement of al-Ma'arri (d.449/1058), quoted by Khalidi (1994: 186) "In every age there are myths in which men believe. Has any age ever monopolised truth?" would be, if not already known, appreciated in those discussions.

The differences between urban towns and cities, centres of political and economic life as well as learning and religion, and their rural hinterlands and the *badia* are often portrayed to be large. They may be so quantatively, but less so in the social processes and practice of the units of social life, the families and their networks. Local people describe their social landscapes through families and their histories, with implications of honour defended and reputation. Such descriptions lead to the presentation of a decentralised political arenas, familiar from the poems of the pre-Islamic *jâhaliyya*, mediaeval Damascus as analysed by Chamberlain (1994), and a late eighteenth century Jabal Nablus poem described by Doumani (1995: 19–20) to advance "an alternative framework to centralisation: unity through cultural solidarity and local identification, not through political hegemony." The aim is comparable to those of the Rwala and other bedouin tribes, recorded in Musil's works and Wallin (1854: 122–3), and Lancaster (1981). Networks of a variety of loose alliances, which had the potential for action by autonomous individuals as members of groups present on the ground, linked "settled peasant populations and the wild desert Arabs" (Finn 1878: i, 316). The setting out of this alternative attitude to political power is relevant since it is the position of many local groups, and contrary to the assumptions of many writings on political power among rural and tribal groupings in the Arab Middle East.

These tribes and families are not generalised 'tribes' and 'families', but particular. Their members use a genealogical idiom of descent through males to describe the various levels of inclusion, each of which is recruited by specific rather than replicated means. Men contract marriages preferentially with a *bint 'amm*, literally the daughter of the father's brother but for many groups with a second or third cousin rather than a first. This brings some families closer, and distances others. Closeness and distance are the modes of setting contexts in analyses of social action within and between groups. As there are no senior or junior branches, which would be contrary to the public claim of equality, and since women are always jurally members of their natal family, closeness and distance achieved through women allow internal group dynamics. Honour defended deflates to individual autonomy, with each responsible for his actions; and conflates to earned and defended group reputation. Providing for one's defence is a crucial part of honour, as is the protection of one who requests it. Action between individuals is achieved through formal or informal contracts, sponsorships and shares, backed by guarantees of recompense and witnesses. Action between those from social groups where one is non-tribal is achieved similarly, and 'makes each participant equal for that occasion' by recourse to a principle of structural equality derived from segmentary[3] practice within tribal and family discourse. Local ideas of power are concerned with the 'power to' act of all members, rather than with the 'power over' others by a few. These local ideas of power are embedded in the concept that all are equal before God, and therefore it follows that all are equidistant from God. The general construction of asymmetries of power between persons is thus inappropriate, although the construction of asymmetry between particular parties where one takes over the defence of the other for benefit is permissible but valid only for that situation. Ideas of power and ownership are more dependent on particular situations between individuals as prescriptive persons and moral equals, than on generalities between classes of persons, although using metonymy they are often described in general terms.

Hierarchy in this social logic is located less between persons or ranks but rather in contexts and situations where autonomy can be maintained and action achieved through alliance. Participants

---

[3] Herzfeld (1987: 164–9) emphasises segmentation as a mode of relative differentiation. From his Greek ethnographies, segmentation is concerned with rules (how you face outsiders) and strategies (how you describe your activities to insiders). Arab rural populations use the idiom of segmentation to differentiate relatively through metaphors of context and location. Dresch (1989: 70, n.5) comments revealingly on the classical coupling of 'honour and shame', established from southern European contexts. Although Arab tribal societies talk about 'honour' and 'shame', the application of the models of 'honour and shame' is inappropriate since "the relation of individual persons to society is rather differently conceived."

use idioms of closeness and distance, official and local, public and private, and formal and informal to describe or explicate action in particular contexts. These idioms do not define and are not presented as analytical structures. They are rather devices around which particular actions and motivations can be built and deconstructed, and between which actors and audience may shift within a situation. Actual behaviour arises from negotiations between the moral premises of the wider society at one level of thought and the need for daily decisions at a more immediate level. Lewis (1987: 154) gives a neat example in his analysis of the dealings of two of the Mhaid shaikhs of the Feda'an tribe. One supported the Turks and the other the French in 1920 and '21; the supporter of the French commented "If the Turks capture the territory my cousin will mediate for and protect me, and if the French occupy the area I will mediate for and protect my cousin." Shaikhs and other political leaders and representatives are accommodated to concepts of individual autonomy by a rearrangement of ideas of social function and representation of groups to similar and different others. Individual tribesmen, like people from other social groups, transmit and interpret variations on these general themes. One group might emphasise 'power to' and themselves as exemplifiers of this moral position, while their neighbours might emphasise their achievements of 'power over' through a variety of contested claims and negotiated events but based on the ability of 'power to'. The pragmatic character of local theoretical generalisations and analyses of actions negotiate different forms of social knowledge.

Such current and recent past tribal social practice does not of itself validate those of the long continuities and historical traditions. Ethnography may inform theoretical constructs of social activity, but can not provide solid evidence of past activity or transformation. That itself is a construction, whether by scholars of western or Islamic traditions or by local people, since analyses always involve symbolic systems. Local analyses are constructed as descriptions rather than definitions; they transmit knowledge and interpret. Scholarly concerns in social activity have moved away from institutions to identities and concepts, where local observations of the causes of social experience point up differentiation and individual action within and against a social order. A concern with long continuities does not shed light on social evolution or other developmental concepts; it is not 'survivalist'. It does not deal in origins, and is not really interested in periods of time, but with what is presented as a continuing social and moral identity. Archaeological and historical theoretical concerns are constructed from strands that interact among and between themselves; local people sometimes use pieces of this information in negotiations with officials, as they use local histories in discussions with other local groups. If a society has a view of itself as existing through time, and is concerned with this, as it affects identity and social practice, then history, both local and official, has to be part of the analysis. The rest of this chapter deals with archaeological and historical information relevant to the presentation of long continuities in systems of land and water use and associated systems of production and distribution in the region, with comment on particular topics.

Archaeologists concerned with the development of society and social practice in the Bilâd ash-Shâm have suggested early dates for a tribal way of life. Betts (1992: 16) argues "that the late Chalcolithic/Early Bronze Age saw the establishment among steppic groups of a way of life similar to that of more recent bedouin, and that the socioeconomic systems that characterise modern groups had their foundations in those protohistoric periods. Such a suggestion can be argued plausibly on the basis of ethnographic parallels and evidence for pastoral, multiresource nomads in the Syrian steppe from the sixth millennium BC or the Late Neolithic .... it is also reinforced by evidence from Sinai and the Negev for similar periods" (quoting Rosen 1988). Early Bronze Age II and III political units in the Jordan Valley and in the eastern *harra* may have been tribally based (Betts and Helms 1991; 1992).

What of the long continuities in the physical environments assumed by local commentators? The landscape was domesticated during the Epipaleolithic, with the regional plant repertoire in place from 12,000 BC (Hillman, Colledge and Harris 1989), although plant resources were managed rather than farmed. Harlan (1995: 86–99) considers the evolution of agriculture to arise from the "suture between the Pleistocene and the Holocene", which in the Near East was at around 12,000–11,000 years ago. The post-Pleistocene adjustments in flora and fauna may have given a momentum towards the cultivation of plants and the domestication of animals. Harlan notes that the earliest plant domesticates, emmer, einkorn and barley, are all associated with oak woodland, although einkorn may be found at higher altitudes and barley in drier land. At the end of the Pleistocene, oak woodland was found only in narrow strips along the Mediterranean coasts of the Levant, and where oak woodland is now there was then *artemesia* steppe. Two final observations are made by Harlan (1995: 239); that agriculture came "as a result of long periods of intimate association between plants and man," and that "one could well ask which are the domesticated. Did people domesticate plants or plants, people?"

Rainfed farming of wheat, barley and legumes has existed since the early Neolithic (Zohary and Hoff 1988: 42, 60, 87). Van Zeist and Bakker-Heeres (1979: 168) suggest that surface water was being exploited for agricultural purposes at Tell Aswad in the 8th millennium BC. Olives, dates, figs, grapes, almonds and pomegranate were being cultivated in the Early Bronze Age at sites in the Jordan Valley (Zohary and Hoff 1988), while Lipschitz *et al.* (1991) discuss the difficulty of distinguishing wild and cultivated olive stones found in excavation, and that charred olive stones, presumed to be wild, have been found at c.8,000 BC in Israeli sites. Orchard trees, apart from apple, have not been found attested until classical times; their cultivation depends on grafting, and except for *pistachia*, are not native. Wild fruits of *pistachia, azarolus*, almond, acorn and *zizyphus spina-christa* are present in many Neolithic and Bronze Age sites. Colledge (1994: 243) gives evidence for cereal and legume cultivation in the eastern arid zone, and considers that many sites were used seasonally. Firm evidence for the management, if not herding, of goats is seen from late Prepottery Neolithic sites in the Bilâd ash-Shâm (Garrard, Colledge and Martin 1996: 208). The domestic sheep is present from around 6,500 BC, with the existence of mobile groups primarily dependent on pastoralism (Ducos 1993: 153). There is no firm evidence of milk production before the Chalcolithic and Early Bronze Age (Garrard, Colledge and Martin 1996: 210), although milk production is a more effective use of sheep and goats than meat production (Russell 1988: 152–7). Muzzolini (1989: 157–8) finds the fat-tail sheep in Egypt around 4,400 BC, and milk and wool production in Egypt and Nubia by 4,000 BC. Networks of exchange throughout the region are wellknown, with obsidian from Anatolia and shells from the Red Sea and Mediterranean being standard items. Colledge suggests (1994a: 255–6) agricultural crops and livestock may have been involved. Garrard and Gebel (1988: 426) record cattle and goat at later Neolithic Azraq, and consider goats, sheep, cattle and pigs the standard domestic animals of Jordanian sites. Woven cotton has been found at a site in the Jordanian *harra*, dating from 4450–3000BC (Betts *et al*. 1994). Braemer (1993: 422), working south and west of Damascus at Chalcolithic and Bronze Age sites also finds domestic goat, sheep and cattle, and notes diverse modes of livelihood and a variety of water conservation methods. Changes in climate in the past until the present are discussed by Shehadeh (1985) and Garrard, Colledge and Martin (1996) for Jordan, and Bruins (1986: 1994) for the Negev and the Southern Levant. Bruins (1994: 310) finds there to be no "simple relation between climate and human history .... Human determination may clearly override negative climate trends," although there are shifts in climate over time. Garrard, Colledge and Martin's work from the archaeological record shows an essentially steppic flora and fauna in eastern Jordan since the Epipaleolithic, with wetter and drier phases.

These archaeologists imply social practice and forms from livelihoods as revealed by excavation and ethnographic parallels. An alternative view of tribal organisation is that it was called into being as a series of local alternatives to states. Rowton's papers (1970s, 80s), using cuneiform records, focused on a model of enclosed nomadism, where sheep and goat herding groups were constrained in their actions by the demands of states on whose lands and markets they were dependent, unlike those of camel herding nomads (but see Lancasters 1997b). Postgate (1992: 83–4) largely accepts Rowton's thesis on the way steppe areas were managed in the second and third millennia. Postgate speculates (1994: 7) "about the mechanisms through which such a strong cultural continuity could be established and maintained" in the Ancient Near East where language, ethnicity, or political structures did not define territorial entities, and finds one in "the strong tribal links between Syria, northern Mesoptamia and southern Mesopotamia" relevant at various periods, such as Mari, the early first millennium, and the Umayyads. Postgate finds (1992: 85) that nomadic groups "lacking a permanent territorial base ..... are always identified by their tribes", and that there are "scattered references to clans or tribes within the long-established urban communities of the third millennium" (1992: 82–3). At this date, patrilineal descent is known in the cities of Mesopotamia. In the Old Babylonian period, the "normal residential unit was clearly patrilinear and patrilocal" (1992: 92), while the extended family could own land. Similarities between tribal social organisation and that of urban and village extended families are also noted. Like Rowton, Postgate works largely from cuneiform records and state iconography for his discussion of society; Briant (1982) establishes the stereotypic views of nomads contained in these sources.

An alternative, but complementary, perspective is that where particular modes of livelihood are assumed to require certain forms of social organisation. Ingold (1980) argues (to grossly over-simplify) that herding for meat leads to a situation where the dominant form of social interaction is competition between households, whereas herding for milk leads to co-operation, unilineal kinship systems and shared long-term reciprocal transactions. In the Bilâd ash-Shâm and wider region, herding can be compared to foraging, but with animals who turn inedible plants into milk for their herders. Herding is a core of multiresource economics, with services as another part. Acquiring service contracts and raiding[4] can be seen as hunting. Herding could be associated with both unilineal descent systems, co-operation and longterm reciprocity for its longterm structures, and with flexible and undefined bands, associated with foraging and hunting, in which actual herding and other livelihood strategies are carried out.

The association of camels with man, leading to camel pastoralism, is now put at the late third millennium in southeast Arabia (Potts 1990: 129, 256–7; Ripinsky 1975; Uerpmann 1989: 165; cf Kohler-Rollefson 1996: 287 who gives a date before the second to first millennium transition). Camel transport in long distance overland trade is put by Zahrins (1989) at c.1500BC, agreeing with Sauer and Blakely (1988) from work in Yemen. Late Bronze Age finds of camel bones in Jordan and an increase of Iron Age finds linking Jordan and Yemen in long-distance trade are reported by Sauer (1995: 42). Between herding sheep and goats and providing donkey transport to camel herding for milk and the supply of camel based services for trade, transport, the military, or water-lifting in oases, is a difference in scale.

Farmers who herd are often placed in a developmental framework of villages dominated by city polities, or a geographical framework of spatial hierarchies of city centres with dependent

---

[4] Raiding takes many forms, from the taking of a few unguarded animals by someone with none to start a herd for livelihood to expressions of political opposition and selfhelp in the face of broken contracts.

rural peripheries. The Bilâd ash-Shâm had large and small village settlements, some used seasonally and others permanently, from the Neolithic onwards but no city as such until Damascus from the second millennium BC (Pitard 1987: 191). In Mesopotamia city-states, based on intensive irrigated agriculture, formed at the end of the fourth millennium; Yoffee (1995: 284) considers "the countryside was created as a hinterland of city-states and as a fertile no man's land to be contested". In northeastern and northwestern Syria, cities based on extensive rainfed arable agriculture and sheep production developed in the third millennium, set up by countrysides to "accomplish what must be performed in central places" (Weiss 1985: 79). Pitard comments that Damascus "shared in the culture of and held the same kind of status as the other and insignificant city-states of the area of Canaan....... It was only towards the end of the tenth century BCE that Damascus, now the capital of a state called Aram, became a significant political and cultural entity," until it was incorporated in the Assyrian Empire. Assyrian domination incorporated Aramean local rulers, many of whose names "point back to old tribal forms of organisation" (Strommenger 1986: 322ff).

Arabs first appear in ninth century BC Assyrian records as camel-herding nomads. Ephal (1982: 4–5, 112–59) and Briant (1982: 153–61) show the Assyrian (and Babylonian) need of Arabs to manage the lucrative incense trade[5] and to supply armies on their western campaigns with camels, food and water. In the neo-Assyrian texts, Arabia was the lands from the western Gulf coast to the Gulf of Aqaba, and under minimal political control. By the Achaemenids, Arabia was an east-west arc from the Gulf to the Red Sea with commercial relations between Mesopotamia, East Arabian coastal sites, and Syria-Palestine maintained by Arab Nabataeans with their capital at Petra. On the northern edge of the Bilâd ash-Shâm, Tadmur/Palmyra, first mentioned in 19th century BC Assyrian trading archives, rose to prominence in the 4th century BC as an Arab trading principality (Bounni 1985: 381). In northwest Arabia, the 13th–11th centuries BC small settlements of Qurayya and Tayma on the north-south incense route resulted from

---

[5] Groom (1981) notes that many local plants in Syria and Mesopotamia were used as aromatics and ascribes the date of 500BC for a substantial trade in frankincense and myrrh from southern Arabia.

Egyptian commercial activity (Parr 1989: 42). During the Assyrian period, the only apparent urban settlements in north-west Arabia were a few small sites in the northern Wadi Sirhan, comparable to seventh century Iron Age II Jordanian sites of Amman, Dhibon and Busairah. Bienkowski (1995) considers Busairah the probable capital of an Edomite state, made possible from increased economic activity from mining at Fainan, the Arabian trade, and demands for tribute. Trade, especially in aromatic substances, was influential in developing commercial stations and settlements, and driven by regional powers such as Assyrians, Achaemenids, Seleucids, and Rome who did not always incorporate relatively distant regions. Members of local populations acted within states and were active in trade in a variety of occupations. The accounts of a four-camel Gaza caravan, carrying grain from Galilee, reed mats and pickled meat from Egypt and dates, are described in a third century BC papyrus from Egypt (Grant 1937: 126, n.1, quoting Westermann and Hasenoehrl 1934). The Roman Empire annexed Petra and Palmyra to obtain control of trade revenues (Eadie 1989; Sidebotham 1989).

Technological developments in the camel saddle around 500BC were regarded by Dostal (1959) as establishing a warrior bedouin society. Graf (1989), Lancaster (1988) and Macdonald (1991), have queried this thesis on ethnographic and epigraphic evidence. Caskel (1954) considered that the bedouinisation of Arabia resulted from the development and decline of the incense trade. Both Caskel and Dostal regarded 'being bedouin' to depend on the possession of riding camels, enabling the predation of oases, caravans and villages. Those currently describing themselves as bedouin consider being bedouin as living from the badia through acceptable social practice, to which camels provide a means. All bedouin are tribal and mobile, but not all tribespeople are bedouin, while mobile people need not be tribal. Parker (1986) reasons like Dostal and Caskel in his view that the Roman limes were barriers against predatory nomads. Banning (1986), Graf (1989) and Isaac (1990) see the limes as monitoring movement and ensuring internal order, while states needed a simple social category of 'nomads' to describe the presence of tribal groups both inside and outside the empires, and moving between opposing spheres of influence. Using the numerous Safaïtic and Thamoudic inscriptions, Graf (1989: 400) concludes that if the Saracens of the literary sources were 'the nomads', and if their lives were reflected in inscriptions in local scripts, they "portray a society deeply engrained in the life of the Hauran, resistant to Roman penetration and occupation of the land." Macdonald (1993: 335–346), while in general agreement with Graf's thesis that there is little evidence for a nomadic threat to the Roman Provinces of Arabia and Syria, finds his use of epigraphic material misleading, and sees the authors of these inscriptions as nomadic individuals who migrated seasonally into the Hauran, many of whom "must also have had familial, commercial and occupational relationships with the sedentary population on Jabal Hawran" (1993: 345). In his study of Palmyra as a trading centre, Galikowski (1994: 32) observes that "security in the desert was not, and could not, be maintained by the Romans. It was created and maintained by the nomad shaikhs themselves, when they realised the profits to be gathered from the existence of the great market of the Empire."

Villeneuve (1986; 1991) portrays the wider Hauran from the 4th century AD to the Islamic conquest as a countryside of villages, with Bostra as its urban centre. Agriculture was based on grain and vine cultivation, with cattle herding. The suppression of the Palmyrene rebellion and plague briefly affected village population adversely, but from the 4th century new villages were built on the eastern slopes of the Jabal al-Arab and the desert edges. Village institutions built and administered churches, community houses, and water cisterns and channels. House and village layouts suppose "an egalitarian social organisation" (1986: 113). There were long established networks between villagers and nomads; MacAdam (1984: 53, 62 n. 49) discusses references to renting or leasing of village land by clans, and to the renting of grazing land by the crown to local 'Arabs'. Nomads installed themselves in some villages, partly because of the spread of Christianity and partly from the increasing influence of first the Salih and then the Ghassanids (a family not a tribe) in the regional Byzantine administration during the late sixth and early seventh centuries (Shahid 1984; Graf 1989). Ghassanid notables established themselves in Jaulan, together with Monophysite monasteries; other Ghassanids were at Qastal in al-Balqa, Ma'an and al-Jibal (Sartre 1982: 183–7). Other tribal families had estates in al-Balqa and al-Hisma, and tribes that would be important in the political and economic life of the Bilâd ash-Shâm for many centuries lived there (Bisheh 1987; King 1992). Settlement expanded around Karak at this time, linked by Johns (1994: 4–8) to an increased demand for grain after the loss of

Sicily and North Africa by the Byzantines; he notes the boom occurred when the region was entrusted to Arab confederates from the steppe. At the same period, there are references to grain being traded to Palestine from east of the Dead Sea (Hirschfield 1992: 83–4). The inter-regional trade in aromatics from the south appears to have declined by the end of the sixth century AD (Kennedy 1986: 23), although intra-regional trade remained important. Historical sources and archaeological excavation show that after the Islamic Conquest and the establishment of the Umayyad caliphate in Damascus there was both more continuity and regional variation than were supposed.

Basic farming, pastoral, trade and service activities were in place. Social groups identified themselves in terms comparable to those of the more recent past. The Islamic Conquest may be seen as a religious re-focusing of existing jural and political ideas and practices rather than as a break. The countryside is affected to varying degrees over time and space by the range of extremes of climatic factors, natural disasters, and diseases of crops, animals and people, and by the actions, direct or indirect, of political and economic associations of city populations within and outside the region. The inhabitants of the rural Bilâd ash-Shâm assess themselves as primary producers of goods and services for their own use and for a series of shifting markets and constituents/customers/partners as parts of a widely distributed and flexible series of embedded networks. Relations between constituent parts of the market and local participants are negotiable, flexible, encompass alternatives, and resilient, generated as much from internal decisions as external factors.

The countryside, from its different areas, supplied urban areas with basic foods of grains, legumes, oils, vegetables, fruits, nuts and dates; live animals for meat; and dairy products. It was also the source of industrial supplies: wool, cotton and silk; skins and leather, used like rubber and plastics today; charcoal and fuel; salt; wood, stone and plaster; ashes of various plants for glass and soap; plants as gums, paints, varnishes, drugs and cosmetics. Animals from the countryside provided draught energy for working wells and mills, carried all freight, and were ridden. The countryside was the arena of transit of people and goods for administration, trade, military purposes or pilgrimage. The market links town, city and countryside. Agricultural and herding production is not expected from the same regions in predictable amounts each year from variations in rainfall, drying winds and unseasonable temperatures, and from crop and animal pests and diseases. Mobility of arable and pastoral production, modes of distribution, and alternative sources of labour are necessary, and achieved through share partnerships, short-term hired labour, and other local processes along and between embedded networks.

The aggregate of countrysides comprises the lands, villages and encampments owned, used and inhabited, and whose owners and users have rights of preferential access and control vested in its inhabitants, although these rights are abrogated by the state at some times. The inhabitants of towns make their living from income generated from outside the area; people may be servants of the state, traders, craftsmen or servants of the above. Many towns are rural service centres, as Cohen and Lewis (1978: 107,n. 3) describe sixteenth century Hebron, Doumani (1995) for Nablus, Burckhardt (1822) for Salt, Wallin (1854) for Ma'an, Doughty ([1888] 1936) for Khaibar and Taima, and Weuleresse (1946: 307) and Métral (1989) for Sukhne in Palmyrena. The oasis towns of northern Arabia were usually autonomous political entities, integrated into wider regions through the market and religion. Village inhabitants lived largely from their own resources, farming, herding, craft and labour; specialised goods were obtained from travelling merchants or at regional markets. Villagers provided their own administration, defence and justice, legitimised by their reputation as competent persons. Nomads provided for themselves through herding, services, and exchange, and administered themselves.

In the Bilâd ash-Shâm, Damascus and Jerusalem were the main cities. Many settled areas had a succession or alternation of centres, and the number and location of villages in an area shifted over time. Late Byzantine and the early Islamic periods are regarded as the time of the greatest extent of settlement in the region (Johns 1994: 4–5). The reasons for this density are attributed to the wealth of the area coming from agriculture both by villagers and large estates, industries based on agricultural products or supplying goods needed by processors and transporters, the early importance of Christian pilgrimages, and long-distance trade. The large-scale investment in landed estates by the Umayyads and their associates in Palestine is noted by Khalidi (1984), and in Jordan and Syria by Bisheh (1987; King 1992; Kennedy 1992).Within the period, archaeological excavations have revealed expansion and decline of physical settlements, linked to political and economic factors, and to natural disasters such as earthquakes and epidemic diseases. Agriculture, settlement, agricultural and industrial processing continued in the Bilâd ash-Shâm after the Abbasid Caliphate moved the capital to Iraq, although there was a gradual change towards a more rural status for formerly urban sites (e.g. Lenzen 1991; Walmsley 1991). Arab historians in the mediaeval period (Le Strange 1890) commented on the numerous villages in al-Balqa, Moab and the Shera. The importance of the lands east of the Jordan to both Crusaders and Ayyubids for trade route revenues and agricultural and pastoral products has been noted by many authors, with towns like Karak, Shobak, Ajlun and Salkhad flourishing as citadels and service centres. Ghawanmeh (1982) notes the abandonment and refounding of Karak villages in the Mamluk period. Ayalon (1993: 121, n. 39), discussing the economic decline of the Mamluk Sultanate in Syria and Egypt, concludes that to measure this decline from the available data on the number of villages is "absolutely unreliable for Egypt and utterly impossible for Syria". He quotes Sibt ibn al-Jawzi (1907: 397) who gave a total of 2,000 villages in the Bilâd ash-Shâmiyya in 616 AH/1219–20 AD. Under the Ottomans, patterns of settlement declined from the sixteenth century, although former permanent village settlements and towns remained foci of local economic and political activities. Tax collection by Government, when functioning, continued to be organised around former urban centres. Settlement expanded from the late nineteenth century to the present patterns, where the density of village settlement is said to be similar to the late Byzantines and early Islam.

    Chronicles of all periods attribute the abandonment of rural villages to bedouin incursions, taxation and conscription policies of central governments, epidemic disease and natural disasters. Travellers drew similar conclusions, and until recently, historians followed. Archaeological excavations and co-operation with historians has been important in evaluating land use history; Lenzen (1991; fc) at Bait Ras in northern Jordan, Johns and McQuitty (Johns 1994) at Khirbet Faris in central Jordan, and Dentzer and Dentzer-Feydy (1991) in the Jabal al-'Arab present important resumés. Some nineteenth century travellers comment on the switches between abandonment and settlement of Hauran villages (Burckhardt 1822: e.g. 221; Robinson and Smith 1841: iii, 176–80; Lewis 1987: 82ff). Burkhardt related such movement to access to and the availability of agricultural land and labour, while seeing the *malikana* taxation system, as in the Pashalik of Acre where villages were assessed for a certain yearly sum which each village was obliged to pay whatever the number of inhabitants, as "one of the chief causes of depopulation of many parts of Syria" (Burckhardt 1822: 341). Urban landlords could contract to pay the assessed taxes and used village lands to produce commercial crops (Burckhardt 1822: 341). Conder (1881: 367) saw the same process in the same area after the Ottoman wars with Russia in the 1870s. Most travellers note the seasonal movement by local populations, who both cultivated crops and herded, and who either only sometimes resided in villages or who never used stone houses but permanently lived in tents. Present villagers in many areas of the Bilâd ash-Shâm comment that earlier in this century they used their villages as bases for stores and summer water resources, and moved with their herds in winters. These people worked for themselves and/or as share-croppers or herders, and/or wage labourers for harvesting. In the present, similar patterns of movement for economic activity are undertaken, but as more people have houses in villages and towns, village settlement is more apparent and movement less so. At some seasons, many houses in villages along the Desert Highway in Jordan are deserted, while in the summer many people in the southern Ghor move up to the plateaux. Movement between economic sectors and associated residence patterns was and is linked to viable economic activity and fiscal behaviours by governments. 'Rural depopulation' was commonly said to occur by the movement of peasants to towns; but there are documented examples of villagers becoming nomadic herders, transporters or traders (Seikaly 1984: 406), and urban-based military-administrative personnel becoming transporters and traders with tribal affiliations (Doughty [1888] 1936: ii, 157–9). The chance recording of such movement implies its greater frequency in reality. Restrictions on herding and the loss of economic viability for camel-herding would appear to mean that most herders settle and move to agriculture and/or employment; but the flexibility of herding practices means that it may be combined with other income producing strategies with some network members living more or less permanently in villages while others move between villages and pastures, or between grazing areas. There are examples of families returning to being fully nomadic in the present (d'Hont 1992: 214).

The permanent or temporary abandonment of villages as units has been recorded. Doughty ([1888] 1936: i, 628–9) mentions the desertion of villages in Jabal Shammar from plague, while malaria caused seasonal movement in river valleys. Floods destroyed villages in al-Qasim (Doughty [1888] 1936: ii, 422). Unresolvable disputes cause villages, or parts of them, to be abandoned in Palestine (Finn 1878: ii, 215), Qalamoun (Porter 1855: 238–9) and in al-Qasim (Doughty [1888] 1936: ii, 381). In the Jabal al-Arab, villages during the nineteenth century were occupied in favourable years and abandoned when conditions were unfavourable for cultivation. Settlement and mobility, agriculture and herding, are and were decisions taken by their actors in relation to a variety of changing facts, assets and options, environmental, climatic, economic, political, and familial.

Changes in the number of recorded villages and their locations have been developed into the frontier of settlement model (Hutteroth and Abdulfattah 1977; Lewis 1987; Rogan 1991 among others). Musil (1928: 45), in a classic portrayal, associates expanding peasant settlement and agriculture with strong centralised government, and vice versa. An extension of agriculture is rather linked with production for expanding markets or tax demands rather than a strong government as such. In Musil's model there is no mention that herding tribesmen may become grain-growing peasants, or that herding and grain farming co-existed, although he records these at other places in his material. The sources mention camel-herders owning date groves, and sheep and goat herders growing cereals; some indicate camelherders also kept sheep and goats, and grew cereals themselves or with share-partners in suitable localities and rainyears. Tristram (1873: 303, 306) indicates this to be so for some Beni Sakhr in 1871. This could be construed as an effect of the re-establishment of Ottoman rule in as-Salt in 1867, but ignores longterm relations between the Beni Sakhr and the Balqa tribes, and information from the 1538 tax register, where Beni Sakhr farms are recorded as producing grain (Bakhit and Hmud 1989: 136). The same volume mentions farms or agricultural production undertaken by the tribes of Beni Karim, Beni Zaid, and Beni Mahdi in al-Balqa' and Beni 'Amr in al-Karak. Doughty ([1888] 1936: i, 487–8) describes the camel and goatherding Muwahib Anezes' terraced gardens growing barley and melons in the Khaibar *harra*. He also mentions, several times, "granges", isolated farmsteads occupied during the crop season by town or herding families. In addition, holding of agricultural land in oases or riverain lands by tribal families or sections and cultivated by share-partners was common as indicated by Doughty ([1888] 1936: ii, 132–3) and Jaussen and Savignac (1920: 8) at Khaibar, and Musil (1927: 360) near the Euphrates. The reverse pattern may be seen in villagers, like those of Qàlamoun, who had extensive sheep and goat flocks far from the villages at most seasons.

Strong central government is associated with general economic prosperity, and weak government with economic decline. 'Decline' and 'expansion' of settlement in local perceptions reflect changes in the amount of economic surplus coming into an area from outside through trade, subsidies or booty; whatever form surplus takes, it comes from external sources initially, although such surplus may well initiate increased economic activity locally. A more useful approach than 'growth and decay' or 'decline and expansion' is that of the users of land, whose flexible and resilient redefinitions take account of changing conditions. In local terms, the reality is what is possible, given available resources and what these are, how access to them is achieved, and in what terms these are defined and legitimised. If these are in local terms, then local considerations and practice dominate; 'ownership' comes through access, use, action, and is validated by defence and reputation. If in official terms, then resources are at the disposal of the centre, registration validates access, and the benefits of resource access are bestowed in return for services or payments. In parallel, jural and political ideas from interpretations of Islam influence attitudes by groups in different economic sectors to legitimisations of access and control to resources, both in the cities, towns and countryside of villagers and tribespeople. Religious movements such as the Khawarij, Qaramita, the interaction between Shi'a Ismai'ilism and Sunni orthodoxy of the mediaeval period, Wahhabism from the end of the eighteenth century, the dervish movements in Palestine in the late nineteenth century and current Islamic fundamentalism have all influenced political actions in town and country, and illustrate that the networks linking the two were stronger than once thought. (There are many sources for the above; e.g. Kennedy 1986; Havemann 1991; Bianquis 1986, 1989, 1991a; Gaudefroy Desmombynes 1923; Abbas 1979; Antoun 1989).

The repertoire of crops increased over the centuries. Fruit trees enter during the classical period. Watson (1983) discusses new commodity crops and the development of summer crops; these enabled a growth in population and a spread of settlement, with a greater trade in foodstuffs and industrial crops, and so increased wealth. Hard wheat, not mentioned until shortly before the Islamic Conquest, was well-established by the tenth century. Sorghum/ *dhurra* (also used for millets), which needs only spring rain, appears at the same time. Rice was grown in the Jordan Valley, probably between the 3rd and 8th centuries and certainly during the tenth. Sugarcane, grown in the Tigris and Euphrates valleys in the reign of the Caliph Omar (634–44 AD), is not mentioned further until the tenth century, when its cultivation was widespread. Old World cotton, a perennial variety, was probably cultivated in the Jordan Valley pre-Conquest, and used for thread and cloth, either alone or with linen, silk or wool, for stuffing quilts, cushions and mattresses, and for paper. All these crops, apart from hard wheat, were summer crops and needed irrigation. *Colocasia*, bananas, lemons and sour oranges also entered the repertoire. Watermelons are not mentioned until the twelfth or eleventh centuries, while Potts (1994: 260) quotes Stol (1987) who considers the Hebrew and Greek Old Testament term to refer to watermelon rather than melon generally.

Of the South American crops, green beans may have been the first, with *loubia* mentioned among the taxable vegetables in the Damascus market in 1548, along with onions, aubergines, cauliflowers, cucumbers, carrots and asparagus (Mantran and Sauvaget 1951: 17–18). The editors consider that *loubia* may have been used for another legume and then transferred to the New World green bean, while referring to Rauwolf's noting in Aleppo in 1573 of "Phaseola grands et petits tout blancs". Singer's (1994: 70–1) examination of fiscal registers and court cases in the seventeenth century Pashalik of Jerusalem refers to crops of green beans. Maize, as *zea mays* rather than as a European traveller's synonym for millets or sorghum, had reached the Euphrates by 1574 (Harrison *et al*. 1969: 197). Potatoes were seen in 1810, with a note that they had been cultivated "for some time past", in Kisrawan in Lebanon by Burckhardt (1822: 22). Tomatoes were common in as-Salt by the early 1870s (Merrill 1875;160) and grown in Tabuk about the same date (Doughty [1888] 1936: i,112); while Musil (1908: iii, 151) notes tomatoes and pumpkins in the vegetable repertoire for Karak. Wetzstein (1857: 476), describing the Damascus market, does not mention tomatoes but records saladings, roots and tubers, cucumbers, sweet and watermelons, *cousa* (a variety of courgette), two sorts of aubergine, *loubia* beans, beans, peas, *bamia* (okra) and so on. Merrill also notes prickly pear for hedges a little later (1875:

147). Sweet and chili peppers are not noted until this century, but vegetable references at all dates are few in the sources.

The repertoire of agricultural implements is both simple and effective, using a variety of hoes, mattocks, adzes and knives for irrigated and treecrop cultivation, and mattocks and animal drawn *ard* (the sliding plough without a wheel) on rainfed land for arable crops, with sickles for harvesting. Animal drawn sleds threshed the grain, which was then winnowed. Irrigated agriculture used ploughs or spades, spades and mattocks for cultivating and opening and closing flows of water, and levellers for making the beds for sowing (Thoumin 1936: 134–6; d'Hont 1994: 57–62), with date cultivation in some areas like al-Hasa using only spades, mattocks, hoes and a variety of knives and sickles (Vidal 1955: 154–5). These implements have a long history (Hopkins 1985; Schumacher 1889; Potts 1994), and are, with local variations, peninsula wide (Dalman 1932; Gingrich and Heiss 1986; Palmer and Russell 1993; Potts 1994). In the Rahba, east of the Jabal al-Arab, tribespeople grew grain crops on land flooded by runoff waters without made tools, sowing seed directly onto the soil after the first rain, and covering it using branches, and hand harvesting (Wetzstein 1860: 30–31).

Agricultural production, especially from alluvial irrigated lands (*sawad*), was a main source of wealth for governments and private individuals. Many urban dwellers lived on revenues – profits in shared enterprises, rents or taxes – generated in the countryside, apparent whenever there are appropriate written sources. The life of the Bilâd ash-Shâm centred on agricultural production for foods and industries, the distribution of surpluses by exchange, taxation and trade, and the consumption of production by different parts of society. In other words, the economic and political activities of individuals, families and social groupings centred around access to and control of resources of various types. Changes in legitimate or accepted access and control to resources through central government activities or the effects of external forces influenced the decisions over types of production, movement, and participation made by local populations. Instances of such changes may be seen to reflect the costs to the overall political economy of the transactions between its component parts in achieving production and distribution for consumption and maintenance. Societies where personal contacts between individuals is the preferred mode may have low transaction costs in administration, although longterm financial and political costs to the centre and particular local parts may be high. Low producing regions may be more economically administered by delegating administrative responsibilities either to local individuals and groups or to agents of the central government as payment for their services.

If local systems of the acquisition of benefits are addressed, the internal logic becomes clearer. Ownership of assets is better considered as access to and flexible degrees of control over resources of whatever nature. Ownership of a resource comes through adding to or developing it beyond its natural capacities. Thus irrigated land is owned outright by its developer; crops but not necessarily the land are owned on rainfed land; animals and water storages are owned but not the land in pastures. An industrial or processing plant is owned, as are the provision of places of exchange of goods, and the escorting of traders. These developed resources can be bought and sold, rented, inherited, used to obtain credit, and be held as shares, which themselves can be sold, leased, and so on. Although most developers are men, women do own resources. Evidence for the past comes from historical documents, often court records, such as those for seventeenth century Jerusalem and region analysed by Ze'evi (1995: 166–8). Here it is attested that women bought, rented, invested and defended their rights of inheritance in houses, fields, vineyards and orchards. As some of the properties were far from where the women lived, this implies some deals were for business rather than enlarging the family estate. In the present, many tribal and village women are known to own land, shops, businesses, houses, gardens and shares of flocks.

Providing the defence and protection of persons and their resources is the essential of ruling, whether by states or local groups. Participation in ruling, providing services of defence or protection so that people can live their lives, means the acceptance of responsibilities which in turn confers benefits. Benefits may be membership in a group, or financial or other recompense. These basic ideas are capable of much negotiation and refinement, and are also flexible, which allows for their resilience through history. The linkages between agricultural production, urban industries and processing, trade, taxation, and the payment of military and administrative service providers are clear.

The variable nature of agricultural production in the Bilâd ash-Shâm has been noted. Delegation to local individuals is found for rural tax collection in Roman Arabia (Isaac 1994). Shaban (1976: 16–18) considers the Umayyads inherited a fiscal policy based on the agricultural economy, where urban tradesmen and craftspeople did not pay taxes. and that this tax loophole was a cause of the rapid development of the textile industry, primarily a cottage industry, in every Islamic town. It also encouraged rapid urbanisation, since the new garrison towns both needed artisans and could pay them from their stipends. The *Qusûr* ('desert castles') of Jordan and Syria (Bisheh 1987; 1989; Helms 1990; Kennedy 1992; King 1992) are now seen as the development of large agricultural estates by leading families before and after the Conquest. *Qusûr* in desert fringe regions were located there for recreation and meetings with tribal leaders, and as local and regional trade depots. Kennedy (1992: 295) considers *qusûr* development may lie in the fiscal structure of the Umayyad Caliphate, where the tax on agricultural land was the base.

Iqta' were originally land grants by the ruler (who in Islamic law owned all land not developed and owned by others) to members of his family and their associates for services (Shaban 1976: 75; Khalidi 1984: 183). Later *iqta'* were granted to military leaders, often tribal, in areas where they exercised some administrative functions. *Iqta'* land concessions soon became "an equivalent of pay on the basis of [the land grant's] cadastral fiscal value" and "basically nothing but a wage collected at source, directly, without the intervention of the state treasury" (Cahen 1979: iii, 1088–91). *Iqta'* holders' lands were often scattered, and holders did not organise their lands. *Iqta'* extended to include inter-regional trade route revenues (Shaban 1976: 116). *Muqâta'a* were fiscally autonomous districts which paid the state a fixed and contracted sum, and applied particularly to tribal groups. Tax-farming was used in agriculturally productive areas closer to central governments; initially, the amount to be collected was set by the government and contracted to a local agent, but later the contract to collect taxes was allotted to agents of the military as payment for their troops, or put up for auction. These systems of taxation could mean the movement of land ownership away from their proprietors to tax-collectors. Local leaders, often members of important tribal families, were approached by central government to act as tax-collectors. Methods of raising taxes were never consistent throughout the Empire, and different systems often co-existed in the regions. At times when the central government was short of money, for example under the later Abbasids, almost every military leader demanded or seized control of fiscal affairs, while local dignitaries like nomad chiefs, wealthy merchants or large landowners, rose up to get the same privilege or to protect local interests (Shaban 1976: 121). Many of the 'tribal emirates' of the late Abbasid and Fatimid period start from the impetus of being agents of central government, receiving *iqta*'s, and working on the exchange of services for benefits or stipends for the taking on of responsibilities. Khalidi (1984: 184–5) describes how the Jarrahids, a family of Tayy', dominated Palestine in the late tenth and early eleventh centuries. From forts they held in the Shera mountains south-east of the Jordan, they acquired *iqta'* from the Fatimids of Egypt in Bait Jabrin, Nablus and Ramlah (with a flourishing olive oil industrial base) and administered their *iqta'* through administrative centres or *hilla*, as did the Kalb and Bani Kilab in southern and central Syria (Bianquis 1989: 459). At this date, Palestine and southern Syria were economically and politically linked with Egypt. Bianquis (1989: 664ff) sees the revolts and uprisings by the urban poor, villagers and tribespeople as based on a crisis in grain production.

The tradition of delegation to local notables for tax collection, supplies of animals for the armies, and administrative services, noted under the Fatimids (Bianquis 1991b: 91), continued under the Mamluks, who instituted the Amirate of the Arabs as a part of the bureaucracy; tribal groups had the responsibility of safe-guarding certain trade routes (Hiyari 1975). The question of control by city-based states over the countryside was rather one of networks of relations that supplied needed crops, live stock, and military and administrative services. Goitein (1967: 75) comments that in Fatimid Egypt, society and government administration were essentially urban, and peasants and bedouin were regarded as outside society. Under the Mamluks, although the emirs and shaikhs of the Banu Fadl Arabs "paid formal homage to the Sultans of Egypt, it was they and not the Sultans who were the real rulers of the badiya ash-Sham" (Irwin 1986: 49).

The early Ottoman tax registers (Hutteroth and Abdulfattah 1977; Cohen and Lewis 1978; Bakhit and Hmud 1989, 1991) set out the classes of revenue bearing lands in the Bilâd ash-Shâm as: the Imperial domain, where whole villages and many taxes in towns and villages belonged to the Imperial purse; that of Provincial Governors, paid from grants of a number of villages and certain taxes in the province; *timar* and ziamet, grants of rights to collect revenues; *mulk* or freehold real estate, land with buildings, orchards, vineyards and the gardens around towns; and *waqf*, pious or private foundations. Local leaders were incorporated in the system of *timar* and ziamet grants, and tribal leaders were given administrative and tax-farming posts (Bakhit 1982: 189–91, 200, 204ff).

Rafeq (1981, 1992) illustrates relations between Damascus and the villages of its countryside in the first half of the eighteenth century. The *'ulamâ*, governors and janissaries were economically active in the countryside, holding land as *iqta'*, renting property, acting as market supervisers, and hoarding foodstuffs (1981: 657). Village lands were held by landowners, farmers, holders of *iqta'*, taxfarmers, and beneficiaries of waqfs. Land could be bought and inherited, and villagers could buy land in other villages. Land ownership and the right to usufruct tended to move from villagers to urban dwellers, who had made money from exploitation, trade or money-lending (1992). The military were often also grain merchants. Land was worked by landowners themselves, with share partners, workmen who took fixed shares, day labourers, or '*ghallatiya*' (renters) employed as cultivators or herders. *Iqta'* land was rented, often by the Governor's agents or the Governor himself. State land not given out as *iqta'* was usually distributed to tax-farmers. *Waqfs* were rented or leased, sometimes by a number of villagers. The main products were wheat, barley, maize, and cotton; sheep; *kilw* ashes brought in for soap factories and export to Europe; and gunpowder from mountain villages. Villages "settled by small nomadic tribes" like Adhra and Shaikh Miskin (1981: 677) benefitted from supplying camels to the Pilgrimage. A similar picture is given for property in land and transfers in its ownership by Reilly (1989,1990) for Damascus and its surrounding countryside in the nineteenth century. The position of Western Palestine, for example, was fiscally different, as shown by Cohen (1973) and noted briefly above.

The Pilgrimage was of crucial importance to the economic life of Damascus and its Province, and to the political legitimacy of the rulers of whichever political state Damascus was a part. The eighteenth century Ottoman administrative reorganization made its Governor also the commander of the Pilgrimage (Barbir 1980: 45, 108ff). The revenues of the Province of Damascus were largely dedicated to the performance of the Pilgrimage to the Holy Cities (Barbir 1980: 110ff) until the decision to re-establish direct rule south of Damascus and in Arabia initiated under the Tanzimat (1839–76) and accelerated during the reign of Sultan Abdel Hamid II (1876–1909). The 1858 Land Law and the 1867 Vilayet Law were the main thrusts of the new bureaucratic rule and reinforced by a military presence. Until the governorship of Muhammad Rashid Pasha (1866–71), Ottoman administration in Jordan was essentially an annual visit by tax-collectors. Rogan (1994: 45–8) finds a strategic use of land title under the 1858 law in the extension of direct rule. In Jabal Ajlun, village settlement was the norm; Mundy (1994: 62) demonstrates that "the diversity written on the landscapes ...... echoed in the idioms of cultivators ... with a common structure .... reflecting traditions of membership in the village community, of social regulation of the cycle for ploughing, harvesting and grazing, and of collective responsibility for the payment of agricultural tax" was accommodated in the new framework of land registration. In al-Balqa', tribal groups had existing claims to agricultural lands (Abujabr 1989: 68–73; Rogan 1994: 48; Wahlin 1994a), and created villages by registration of lands around the already existing summer cisterns and threshing floors. The settlement of Circassian, Chechen and Turkmen refugees from the Causcasus between 1878–1906 on lands expropriated from tribes in al-Balqa' encouraged tribes to register and use all their potential agricultural land, and in some instances to settle share-croppers or hired Palestinian or Egyptian labourers in villages to keep their lands. The government's position, says Rogan (1994: 46–7) was made clear by Midhat Pasha in 1880 when challenged by Sattam al Fa'iz, the shaikh of the Beni Sakhr, over lands granted to Christians from Karak recently settled in Madaba; 'While recognising that Sattam had formerly given the lands of Madaba over to sharecroppers, Midhat claimed that he had created unstable living conditions for farmers giving them only one fifth or one sixth of harvest instead of the standard quarter .... he has no rights to lands not cultivated'.[6]

While grain was always a useful commodity, its importance increased after any extension of direct rule through its place in the tax structure, and the need to cultivate land to maintain possession. Rogan (1994: 51–2) discusses the three-point strategy of merchants moving into al-Balqa'; trade, moneylending and land acquisition. The advance purchase of grain and other forms of money lending, often resorted to by peasants to pay taxes, bound suppliers to merchants who thus obtained produce at favourable prices; when loans

---

[6] In customary law, the proportions of harvest granted depend on the inputs contributed by each side; one fifth-one sixth are customary for those who provide only labour, the landowner providing all else.

could not be serviced, merchants claimed the land as collateral. This is "clearly discernible in the inventories of personal property of leading merchants in al-Salt drawn up for the settlement of estates, which reveal extensive property holdings, large stores of grain and thousands of piastres in loans outstanding". Doumani (1995: 214ff) examines similar concerns of soap merchants, focused on the commodity of olive oil, in Nablus, where again merchants were also landowners, money-lenders and members of the new Ottoman Advisory Councils. Over three centuries, urban ruling families in Jabal Nablus shifted from being holders of *timar* grants to tax-farmers to acquiring wealth from trade and urban real estate (1995: 240). In her analysis of the Hauran uprisings, Schilcher (1991a) establishes that although agriculture was the most profitable form of investment in 1879, profit margins on grain were falling from those of the 1850s and 60s, until in 1892 world prices were below those paid to cultivators in the fields (1991a: 58). In 1879, the state introduced direct taxation, resulting in the first serious outbreak of rural violence for ten years, and followed by several others in the next twenty years. With the downturn in the grain market, entrepreneurs and middlemen competed for survival among themselves, and tried to squeeze the peasants, with the state now supporting one group, then another. Most grain came from small producing units, and as peasants kept their links with bedouin and mountain villagers, they "for the most part succeeded in retaining a land-tenure system and a mode of production (share-cropping) that conformed to their view of things" (Schilcher 1991b: 194).

In Jordan the British Mandate land programme of fiscal survey began in 1927, followed by the land settlements of 1935–52, aimed at increasing production and tax revenues (Fischbach 1994). The land settlement programme permanently partitioned all *musha'* (communally owned) land, mostly in Ajlun district, marking the "end of corporate social control over land ownership in Transjordan" (Fischbach 1994: 93ff). Droughts and crop failures of the '30s and '40s caused many villagers to go to merchants who were also moneylenders. These met the demand for loans from the profits of trading and smuggling during the Second World War (and see Amawi 1994), and by issuing mortgages on land, especially near Amman, al-Balqa', the Jordan Valley and the Beni Hassan area. Tribal leaders and tribesmen increasingly took up cultivation on state lands over which they already herded to replace lost sources of income and lost easy mobility and access to seasonal grazing and wells after the introduction of borders in the Mandated states and their successor national states.

Grain dominated the acquisition of rural surplus by state agents for administrative and military needs, but animals were important and featured in taxation at some dates although not at others. Will (1957), discussing the Palmyrene trade at its height; distinguishes between merchants using camel caravans, the leaders of the caravans, and the owners of the camels, who had great herding estates around Palmyra. Grazing reserves for animals, a kind of *hîma*, belonging to political and military leaders, are recorded before and after the Islamic Conquest (Donner 1981: 72, 298 n. 83; King 1994: 196–8). King, using Al-Rashid's (1986) analysis of excavations at al-Rabadha 200 kms. east of Medina, discusses the association of camel husbandry with fine glass production, soapstone manufacturing and copper smelting at the site. The importance of animals from the countryside in the urban economies of mediaeval Syria may be seen in Ziadeh (1970: 27–36), where revenues for the year 609 AH/1212 AD are presented. The dues on the trading of cattle, camels, horses and above all sheep outweigh all other transactions, and in the industrial sphere, tanneries paid more tax than any other industry. Lapidus (1967: 52) comments, in the Mamluk period, on the exchange of Syrian sheep for Egyptian grain by the amirs who were paid in kind "not only as a fiscal convenience, but a means of arranging complex market operations for which neither adequate capital nor organisation was otherwise available." Ayalon (1958: 259–60, 263–271) discusses the need of the Mamluks for supplies of meat to the armies, and of horses and camels. Horses came from the Arabian peninsula, Cyrenaica and Syria, and the best came from Syria through the shaikhly family of the Fadl tribe, who became wealthy from this supply; as camels were mostly baggage animals, there is little information about their supply in the sources. Irwin (1986: 185–6) notes the Mamluk need for horses, camels and sheep was a main reason for their interest in having good relationships with tribes, and shown by the institution of the Amirate of the Arabs where tribes were entrusted with the protection of regional routes (Hiyari 1975). Under the early Ottomans, this office was continued (Bakhit 1982: 200–1) as a *timar* grant, for which the holder had to supply over a thousand young camels and thirty young horses each year to the Ottomans, with the value of these animals going directly to the Sultan's revenues. The value of these animals was far greater than the value of the *timar.* The caravans between Syria and Iraq from the end of the sixteenth centuries to the early twentieth century are discussed by Grant (1937: 131ff) using the records of European traders and travellers. These caravans had camels carrying goods and camels carrying fodder, provisions and water, and there were also large caravans of three to five thousand camels for sale in Aleppo or Damascus. Camels for the Pilgrimage from Damascus during the eighteenth century and early part of the nineteenth were mostly rented from tribesmen (Barbir 1980: 187, 198; Rafeq 1981: 677; Rafeq 1987: 129–30; Burckhardt 1829: 247). In the nineteenth (Grant 1937: 229–30), camels were also bought as cheaply as possible from the bedouin by merchants who were also contractors of transport to the Pilgrimage. Musil (1928a: 278–281) describes the purchases of camels from the bedouin tribes by the Ageyl merchants of al-Qasim for sale to Egypt, Iraq and Syria; bedouin herders and guards were employed by Ageyl traders. The Ageyl (or Uqayl), according to al-Torki and Cole (1989: 75, quoting al-Misallam 1985: 27–44) had a long history, from before the sixteenth century.

Agro-industrial commodity production is affected by trade demand at home and abroad. External factors on trade include government monopolies on commodities; state embargoes on the import or export of goods; currency differentials; changes in demand from consumers, through altered purchasing power or fashion; technical processes and developments. Sugar was an important crop in Egypt and the Jordan valley in Mamluk times, when it became a state monopoly, but declined when sugar production developed in Sicily and southern Spain closer to its markets. Although cotton cultivation for local industrial production was important earlier (d'Arvieux 1735: iii, 98), the expanding market for cotton in western Europe was the basis of Dhahir ibn 'Umar's rule in north Galilee in the mid-18th century; earlier he exported grain and wool westwards to the Mediterranean. In this century, demand for cotton during the Korean War drove the mechanisation of agriculture in the Jazira of north Syria (Lewis 1987: 161). Nineteenth century French import policies affected commodity crops in the Bilâd ash-Shâm. When, in 1806, a high tax was placed on cotton imports, with that of cotton thread forbidden, Syria switched to silk production (Marsot 1984: 233); in 1828, France rescinded its ban on grain imports, and Syria began exporting grain to France. Muhammad Ali Pasha, the ruler of Egypt in the first decades of the 19th century, established a monopoly over foreign trade which he introduced to Syria and

Palestine when he and his son Ibrahim Pasha extended their control in 1830 (Marsot 1984: 234). Artisans and a small group of Christian traders benefitted, but the monopoly threatened traditional networks. Competition from Europe in textiles caused local industries to use European spun thread in their manufactures and to concentrate on the cheap and expensive ends of the market (Doumani 1995: 118–28). Changes in a particular market for agricultural commodities sometimes led to alternative outlets, although over-production could be a problem. Collected plant products, like *kilw* ashes for soap and glass manufacture, and gums for paints and varnishes, were undercut by products from the new chemical industries in Europe. Burckhardt (1822: 446, 601) comments that collecting gums was no longer profitable. Abujabr (1989: 135) suggests that the decline in the value of *kilw* was the reason for his family's decision to enter grain cultivation as a partner of Sattam al Fa'iz of the Beni Sakhr.

The transit of inter-regional trade across the Bilâd ash-Shâm has been a source of income at many periods. This trade has had both north-south and east-west axes. The desire to control the transit trade has led many regional and external powers to extend their authority into the region. The position of the Bilâd ash-Shâm is as one option in the crucial interchange between the trade from Europe or Byzantium and the north, Africa and Egypt, Central Asia, and that of India and the Far East. The discussions of Shaban (1978: 99ff) make clear how crucial income from trade and trade routes was in the early mediaeval period, and how much political actions were impelled by considerations of trade. He considers many of the major revolts, such as the Zanj, Saffarid and the Qaramita, to have been inspired by the desire for income from trade routes, while many local revolts were framed in contexts of rights to access to trade (and land) resources. Textiles from further east and those produced in the cities of Iraq and Syria were the mainstay of this trade in the Middle Ages, when Abu Lughod (1989) describes the world system of the 'long thirteenth century'. There were three routes between east and west, with the central one between the Gulf and Baghdad, and the Mediterranean; the central route had a short desert crossing between Baghdad and Damascus, while a longer crossing but with better water resources, followed the Euphrates north and then cut across to Aleppo. The Bilâd ash-Shâm prospered from Middle Route trade during the Abbasids and in the Crusader period, when the Sultan and Crusader rulers guaranteed the safety of each others' traders, ships, merchandise, money and freedom of movement (Abu Lughod 1989: 146). After the creation of the Mongol Il-Khanid and the conquest of Mesopotamia in 1258, European traders used the Il-Khan's favoured route of Tabriz and Hormuz rather than Basra and Baghdad for Indian goods. The subsequent decline of Baghdad allowed independent entrepreneurial enclaves to develop in the Gulf (Abu Lughod 1989: 208–9). This world economy collapsed with the withdrawal of China, with the closure of the central Asian trade route, and the departure of the Chinese fleet from the Indian Ocean after 1453. Abu Lughod (1989: 361) sees the takeover by the Portuguese of Indian Ocean trade as changing the rules; "perhaps the old world system had adapted so completely to the co-existence of multiple trading partners that it was unprepared for players interested in short term plunder rather than long term exchange". The entry of western traders with their new rules, new sources of wealth from New World mines, and readiness to develop new technologies of weapons and ships emphasised sea routes and was accompanied by a shift in the focus of wealth creation from the shores of the Indian Ocean and the Gulf to the Atlantic coasts. The development of the oil industry is the new source of wealth in the region, bringing the Arabian peninsula back into a world political economy.

Trade and the carrying of goods requires security of goods and persons, along with mechanisms for recompense and restitution; there need to be enforceable contracts between merchants, traders and carriers, and between traders, carriers and the providers of security of passage. State and/or local, often tribal, authorities can provide these. The provision of animals, water and fodder, loads and rate of travel have to be agreed. For the exchange of goods, security in the market-place is needed, whether this is a physical structure or an act of exchange; weights and measures, rates of exchange between local variations and currencies need agreement. Credit facilities, debt collection, and the settlement of disputes between parties were provided within the framework of tribal processes for local, intra- and inter-regional trade. Grant (1937: 176–9) gives examples of such agreements for inter-regional trade in the Bilâd ash-Shâm between the 16th and 18th centuries. Altorki and Cole (1989: 67ff) do so for the intra-regional 'Unayza caravan trade with Mesopotamia and al-Hasa.

Local and intra-regional trade had and has a greater relevance than some authorities are willing to concede. It has already been noted that local people consider that no one environment can produce a sure and secure livelihood of itself through the seasons and over the years. Movement of producers and products is necessary. Such movements reflect the variability within and between the region and its surrounding areas. At its simplest, landscapes produce different products; grains, fruits, dairy produce, live animals for meat and work, and so on. Within a year, given the variability in climatic conditions, some areas will produce good crops of certain products, other areas that may have had excellent crops last year will have no crop this year, whether the crops are from plants or animals. Harvests of grain, olives and other plants, vary in time because of variations in growing conditions and between varieties. Within a plant species, some varieties are good for one form of processing, while others are better treated otherwise. A processed product, such as *dibs* or fruit sugar syrup, can be made from more than one fruit, although one sort may be preferred; *dibs* is made from grapes, dates, pomegranates or figs, although grape *dibs* is the preferred product in the Bilâd ash-Shâm. Varieties of grains and fruits have different storage qualities. Waste products from some processes become wanted for other treatments; cereal straws and date stones are animal winter fodders, olive waste is winter fuel and animal feed, and dried pomegranate skins are a popular tanning agent, to give only a few examples.

It can be seen that with this inherent variability there are many opportunities for local and intra-regional trade. The 'selling coals to Newcastle' syndrome commented on by Crone (1987: 104) and others ignores local production and processing in the Bilâd al-Shâm. It is quite plausible to, for example, sell dates of one variety in order to buy dates of others with different properties. Southern al-Balqa may have a good wheat harvest one year, and export some; the following year the wheat crop there may be very poor, and people eat stored wheat, or barley or sorghum (depending on when there were rains), or import wheat in exchange for sheep, goats or their products. The low levels of profit and small-scale nature of trade are also commented on. In the region, the costs of trading are relatively small; hospitality while travelling to and from markets is free, and the trader had a donkey or camel or now a pick-up truck, or he walks. With low costs, low profits are acceptable, and low profits imply low degrees of risk. There are additional, unseen benefits like increased information and renewing old acquaintances on personal networks. If part of a man's capital is knowledge of possible markets, and access to networks, routes and transport, then trading fits. Constant small investments and the opportunity to act as an agent for family members increase the viability of small-scale local and intra-regional trade. Debts and credits among family network members in response to family obligations and opportunities for profit in trade and investment provide further motivation for trade.

Trade encompasses a varied range of scale, investment and long-term commitment. There are and were urban-based merchant families, who often have investments in the countryside for the supply of agro-industrial commodities; small-scale travelling traders, town-based, who work in the countryside and the steppe; and also numerous seasonal, rural and *badia* based men who trade, supplying people like themselves or urban traders. The boundaries between these are not necessarily tightly defined over the period of an individual's life, or over the generations. Money was not necessary for trade, as payment could be made in goods; dates, corn, clarified butter, wool, and live animals were the norm for payment in rural areas. The known and accepted social processes of guarantee, contracts, sponsorship and witnessing allowed the safe transport of persons and goods, credit, restitution of lost or stolen goods, and delayed payment allowed the exchange of goods and services. The system of shares, which could themselves be traded or sold, allowed easy access to investment and exchange. Commodity and futures dealing were the norm, whether on a large or small scale.

The Pilgrimage to the Holy Places integrated local and inter-regional trade with local grain and animal production, since its success depended on adequate supplies of riding and transport animals, grain and fodder, and security. Co-operation between government appointed officials and local leaders was necessary. Inter-regional trade normally accompanied the Pilgrimage routes, going and returning. Payments to tribal leaders to guarantee security and supplies of grain, fodder and water were known from Mamluk times (Faroqhi 1994: 54–73). For the Damascus Pilgrimage in the sixteenth century Bakhit (1982: 107–117, 204–26) provides administrative information and relations between the Ottomans and local tribal leaders, some of whom had high posts in its administration. Payments to certain tribes continued through the seventeenth century (Faroqhi 1994: 56–8). Rafeq (1970: 55ff) describes the administration of the Damascus Pilgrimage in the eighteenth century, when local tribes along the routes were important to its functioning in supplies and security (Barbir 1982: 169), as they were for the nineteenth century.[7] Under the Ottomans, the revenues of the Province of Damascus were largely devoted to the Pilgrimage. The Pilgrim caravan from Cairo joined the Damascus caravan south of Aqaba, and was an important factor in the rural economy of Sinai, southern Jordan and the Hijaz (e.g. Burckhardt 1822: 404–5, 436–7; Wallin 1854: 123–4; Doughty [1888] 1936: i, 246–7). A Pilgrim route from the east came through Palmyrena, where de Boucheman (1939: 86–8) comments that until 1913 the Pilgrim trade enriched the caravan town of Sukhne. Bedouin raids on the Caravan normally arose from contraventions in their terms of contract with the authorities (Bakhit 1982: 225; Faroqhi 1994: 65–71). Events in distant countries such as Persia or central Asia influenced pilgrim numbers, which affected the amount of goods from these areas entering Damascus and the Hijaz, since pilgrims often financed their journeys by selling goods brought with them.[8] Internal factors, such as the rise of the first Wahhabi state meant in some years there was no Pilgrimage (Burckhardt (1831; 200). Accounts of the trade accompanying the Pilgrimage may be found in various authors, among them Burckhardt (1822; 656–660), Wallin (1854; 123–4), and Musil (1908; i, 301). The importance of the Pilgrimage for local economies began to decline with the introduction of steamships, continued after the Suez Canal was opened, and finally came to an end after the Hijaz railway started. In the present, pilgrim buses pass along the main roads of Syria, Jordan and northern Saudi Arabia, but their only influence on local economies is at roadside restaurants and garages. The disappearance of the Pilgrimage as an integrative mechanism for production and distribution between urban centres and the countryside in the Bilâd ash-Shâm has been replaced by an expansion of central administrations as tax demanding and tax collecting bodies, with a parallel movement of urban-based merchants to the Levant and Palestine.

---

[7] Some of these payments and services appear to have been sub-contracted to other individuals and families from other tribes, as from a story we were told by some Ahl al-Jabal whose ancestors had, at some point, been expected to provide information on water sources in the *harra* fringes to the Sardiyya shaikh who was an officer of the Pilgrimage.
[8] This continues, with Chechen pilgrims bringing *tanaka* (20 litre tins) of mountain honey and Afghans carrying carpets.

The introduction of modern communications and transport affected the organisation of economic activity in the Bilâd ash-Shâm, with a consequent spill-over into political life. The advent of telegraphs and railways partially enabled the Ottomans to re-establish direct rule, although the decision to do so was based on the loss of grain growing (and therefore revenue producing) lands in the Crimea and the Balkans, the desire to block growing European control of Syrian foreign trade, and worries about British interests in Egypt and Arabia (Rogan 1991; Schilcher 1981). Schilcher links the Hauran conflicts of the 1880s and 1890s with the bedouin revolts in the Hijaz and Najd that eventually resulted in the emergence of the Hashemites and the ibn Sa'ud, and the collapse of Ottoman authority. Syrian (and Egyptian) grain was important to the Hijaz and Najd, and during the First World war "the British clearly used food as a weapon to win the tribes' support ...... by blocking the Red Sea ports and by inflating Hawran prices, distributing large amounts of gold there and among the tribes" (Schilcher 1991a: 84, n. 73). Steam ships, telegraphs, railways, and later motor vehicles all took traffic, control, and income away from the countryside and extended that of central governments. Rogan (1991; 1994: 49–50) describes the telegraph system linking Damascus and Medina, completed in 1901, and paid for by local subscriptions; "to cover the costs of laying the line, the state levied between 170,000 and 200,000 telegraph poles from the residents of the Vilayets of Beirut and Damascus, to be paid in kind, or where forests were lacking, in cash", and camels were requisitioned from the bedouin to carry the poles. The Hijaz railway (Ochsenwald 1980) built between 1900–08, was considered a strategic necessity for transporting administrators, soldiers and pilgrims between Damascus and Medina; the spur from Dara'a to Haifa carried Hauran grain to the Mediterranean. Rogan (1994: 50–1) considers that "in balance, the project was more harmful than beneficial to local interests", with the loss of a considerable amount of local income from the hiring of animals for carrying. The only clear beneficiaries were Nablus and Damascene merchants who extended their interests east of the Jordan, "falling over one another to get to al-Salt ..... in their delights of its lucrative resources which they have tasted through business dealings with the desert 'arab of the area" (Rogan 1994: 53, n. 54, quoting al-Qasimi).

The Vilayet Law of 1864 established a pyramidal system of local government, introduced in the Hauran in 1866 by Muhammad

Rashid Pasha. This created judicial districts in Ajlun and Salt. Muhammad Rashid Pasha's expedition to al-Balqa' in May 1967, aimed at the submission of the bedouin tribes, extracting tax arrears, and establishing direct rule. Salt submitted without a fight, with "massive quantities of grain and livestock commandeered in the name of tax arrears, and around three million piastres" (Rogan 1994: 38–9). Also in 1867, the judicial district of al-Karak was established on paper, and Muhammad Majali appointed as *qaimaqâm*. A second expedition against the Balqa' tribes in 1867 resulted in the defeat of the tribes, and the imposition of a 225,000 piastre fine and the expedition's costs. Direct rule was not extended to al-Karak until 1885, followed by Ma'an. Administrative changes in Jabal Nablus are analysed by Doumani (1995), the Hauran by Schilcher (1981; 1991), in al-Balqa' and al-Karak by Rogan (1991; 1994). Arab sources include Tarawnah (1992) and al-Jaludi and Bakhit (1992).

Many features of Jordanian, Syrian (and Saudi Arabian) society attributed to more recent political manifestations can be traced to the last seven decades of Ottoman rule (Rogan 1994: 32). Doumani (1995: 230ff) comments that many aspects of the centralising reforms undermined the basic pillars of the interdependence between locals and officials, and that as central government control increased, many "must have felt" a growing need to reconsider Ottoman rule and seek alternatives. Similarly, Schilcher, from her analyses of the Hauran, finds the rural disturbances of 1879–1900 triggered by external demands, affected urban notables and peasants differently. Urban notables "reacted negatively to the state's policies .... and turned to new political solutions such as decentralisation and separatist nationalism" while the peasants saw that "*vis-à-vis* the urban elites, the state was a potentially positive force." (1991a: 76).[9] In general, tribal and village groups had to accede to central government forms of taxation and registration of land to achieve access to land and resources. Various studies on the ways in which rural groups negotiate this access from Syria (Métral 1984; 1989; 1991; 1993) and Khalaf (1991), and Jordan (e.g. Bocco 1986;

---

[9] This is presaged by Singer's (1992) description of the three-sided relations between peasants who produced, government officials as tax collectors, and the Sultan as protector of the peasants and foresees Asad's aim (Seale 1991: 103) of producing a state supportive of peasants and destroying urban merchant domination of politics and rural production.

1989a; Lancaster 1981; Layne 1994) indicate that rural groups are active rather than passive in relations between themselves and central government. Wealthy urban merchants increased their importance in funding new rulers arising after the dissolution of the Ottoman Empire; Amawi (1994: 166) indicates this for Transjordan. Field (1984: 108ff) comments on the use of merchants as informal administrative agents by King Abdul-Aziz bin Sa'ud in Saudi Arabia.

The establishment of the Mandated states and the rise of ibn Sa'ud to the south affected the future of the Bilad al-Sham. Ibn Sa'ud wanted a corridor into Syria, while the fall of Hail to ibn Sa'ud in 1922 "convinced T.E. Lawrence and so the Colonial Office of the wisdom of prolonging Abdullah's rule in Amman as a buffer between Palestine and ibn Sa'ud" (Bocco and Tell 1994: 110). Conferences were held to establish borders between Iraq and Najd, and Najd and Transjordan, resulting in the Treaties of Hadda and Bahra of 1925. As Britain wanted a corridor linking Transjordan and Iraq, the Hadda agreement ceded Kaf to Ibn Sa'ud, so the Beni Sakhr, Sirhan and Sardiyya lost winter grazing grounds. The Bahra agreement gave a significant part of Mutair winter grazing to Iraq; these losses increased raiding into Iraq by Mutair Ikhwan. Ibn Sa'ud, committed by the agreements to stop raiding, fined the raiders and imposed punitive taxation on the trade routes to Kuwait. By this time, the Ikhwan saw themselves as the defenders of the tribal order, and so opposing ibn Sa'ud's project of a territorially defined state and his accommodation to the British order in the Middle East (Kostiner 1993: 11, 113–7). Tribesmen wanted "an autonomous and unbroken grazing space" (Kostiner 1994: 112) quoting the Ikhwan leader Faisal al Darwish. Grazing lands without borders and under tribal control remain as tribal concerns seventy years later. The necessity of crossing borders for grazing management and trade (free passage of which is guaranteed by the Hadda agreement) makes tribespeoples' lives more difficult and costly. What governments call smuggling is seen by tribespeople as legitimate trade made illegitimate by governmental restrictions and mistakes.

Politically punitive raids by the Ikhwan under ibn Sa'ud's authority into Transjordan left sections of the Huwaitat and Beni Sakhr losing thousands of animals and scores of men. Return raids, continual friction over winter grazing grounds, wells, date groves, the problems for tribes of paying both Jordanian and Saudi taxes, and the return of raided animals, eventually led to Glubb's appointment in 1930. He established the Desert Patrols and built forts for security and welfare considerations, and patrols acted as agents in collecting the animal tax. The IPC pipeline and roadworks also provided short-term relief work for tribesmen impoverished by raids, droughts, loss of grazing lands and the general depression in world trade. In the late '30s, Glubb provided relief work cleaning out cisterns and wells, and arranged for tribesmen to register land east of the railway. The outbreak of war in 1939 brought military construction works to Palestine and Aqaba, where 5–8,000 tribesmen worked in the winter of 1941 (Bocco and Tell 1994: 126). Harvests were good, and camels needed for transport, as spare parts for lorries were few. Transjordan's business community became richer from smuggling and the black market, increasing in confidence and expertise.

The interaction of climatic, economic and political factors on the lives of rural inhabitants is illustrated by the droughts that three groups see as causing changes (Lancasters 1993). The Beni Sakhr's drought came in the '30s; that of the Karak tribes in the late 40s; that of the Rwala in the late 50s and early 60s. Common to these droughts was that customary ways of managing drought – by increased movement, reducing herd numbers, raiding, wage labour or sharecropping in other areas – were not available because of recent centrally imposed political changes. In addition, market changes made people more vulnerable since their capital had either lost value (as with camels) or had increased (agricultural land) so that it was attractive to urban merchants supported by national laws rather than local custom.

Rural populations see regional economies, and so their politics, as no longer based on rural resources. Wealth now comes from access to the new resource of oil, developed by external interests for foreign markets. Regional governments acquire their income from taxes on transactions, custom duties, earnings from state assets, and aid from foreign governments resulting directly or indirectly from oil. Earnings as migrant workers in oil producing states benefit local rural economies directly while the development by governments or NGOs are regarded as benefitting only urban elites. Rural people have access to government financial resources by serving in security services and ministries, and from the provision of education, health, and physical infrastructure. The official agencies have less need to negotiate with local agencies to achieve resources for their maintenance or legitimacy. Much of the latter now comes from their acceptance by foreign governments and bodies. Local agencies in the countryside dispute their government's application of market forces, sources of political legitimacy, and the boundaries of political power. Disputes may be manifested in different arenas – mosques, university campuses, street demonstrations, newspapers and political groups; they are likely to be negotiated at a series of levels, but most effectively at face-to face local meetings using long-established customary processes and local premises of right behaviour.

CHAPTER 2

THE FRAMEWORK TO SOCIAL PRACTICE

Local practice is constructed through descriptive rather than determinative terms, definitions and frames of reference. People describe themselves and others, their behaviour, and that of others contextually, with each description founded on specific terms of reference. Common terms may thus have different though connected meanings, depending on whether a frame of reference is occupational, political, social, religious, economic, the speaker's own group or one distant to him, the relations between the speaker and his audience, and so on. Meanings shift over time and space because of other referents and concerns, themselves dependent on degrees of integration, complementarity or opposition with wider systems. Terms are not 'defined', but rather ideas used by participants as part of social practice, floating around shared and common referents. Key referents in the countrysides are concerned with prescriptive persons as autonomous and jural equals before God, and the contexts in which persons situate themselves for particular defining activities. The arena that these referents surround is not the sole arena in which people act. Other arenas of political, economic, legal and religious life cross over into that of tribal and familial definitions; this is not new. Within and between these arenas, situational referents of public and private, and official and local are among those that operate. Metonymy and metaphor are commonly employed, often resulting in public images of solidary bodies constructed in the idiom of genealogical descent, attacked through the spilling of blood and defended by vengeance taken for that blood. Actual reality is founded more on the reputations of autonomous, honourable individuals, who are members of families.

Most people of the Bilâd ash-Shâm describe themselves as Arab by descent, language and culture, and Muslim. Syria and Jordan have Christian minorities. Most Jordanian Muslims are Sunni; southern Syria has a Sunni majority and a Druze minority; in Saudi Arabia most are Sunni and Wahhabite. In southern Syria and Jordan there are Circassian Sunni Muslim and Armenian Christian minorities, and in Jordan a Chechen Shia Muslim minority.

Both countries have small groups of families tracing descent from Turcoman and Kurdish units. Gypsy groups[10] move between towns and rural areas. Migrant workers from Egypt, Syria and, recently, Iraq, and Sri Lanka and the Philippines, are relatively short-term inhabitants in Jordan; Saudi Arabia has numerous migrant workers from Pakistan, the Indian sub-continent, the Philippines and Indonesia, and Egypt, Syria, Jordan and Sudan. There is division between 'citizens' and 'migrants', although migrants of Arab descent may become citizens of neighbouring Arab states, and citizens migrate to other countries outside the Arab world. The people of this book live in the rural Bilâd ash-Shâm – southern Syria, Jordan and northern Saudi Arabia; or in urban areas but have active links to rural areas; or in urban areas they see, for one reason or another, as artificial creations of the oil-based economy, and that therefore continued participation in and maintenance of their long-standing rural networks are important to them.

People perceive themselves generally as part of a region-wide society, but particularly as members of more local groups. An individual has many ways of identifying him/herself; language, nationality, religion, locality, tribe or family, or occupation. States and their agents, observers, and scholars of various disciplines find it convenient to identify bounded units of identity to be counted, compared or differentiated. The number of identities available gave rise to the 'mosaic' metaphor which, as Eickelman (1989: 49) says, while "useful for conveying some of the bare geographic and ethnographic facts concerning the Middle East ...... is less than adequate in explaining the interrelations among these elements or their known historical transformations".

Internal and external referents of identity meet to create apparently clear-cut entities; tribe, city, village; nomad, settled; pastoralist, farmer, craftsmen, merchants. People 'should' be one or the other of such categories, but many shift between them over time, within families, or exercise action in more than one category simultaneously. There are relations between constructed entities, and entities may transform themselves. Categories drawn from

---

[10] Gypsies were recorded as '*zutt*' and '*nûr*' in the district of Ajlun in the 1538 tax register (Bakhit and Hmud 1989: 14); Burckhardt (1822: 240) met some gypsies in the Hauran, where they had come to sing at weddings; Tristram (1873: 164) encountered gypsies in al-Balqa tinkering, conjuring and fortune-telling. Gypsies in al-Qasr in 1992 said they were not Arab, and married "as far away as we can".

degrees of geographical movement, occupations, and types of political loyalty have often been assumed to be interdependent: thus, nomads are pastoralist and tribal; village dwellers are farmers and 'semi' tribal; townsmen are merchants, craftsmen or state appointees, and not tribal. These may coincide, but often they do not in the Arabian peninsula in general (Altorki and Cole 1989: 23–4, 58; Lancasters 1988), nor for the Bilâd ash-Shâm in particular (Layne 1995; Lancasters 1996). Just as 'tribe' and 'state' are ideas, constructions, so are 'groups' like '*bedu*', '*fellah*'/'peasants', 'traders', or '*hâdhr*'; the settled.

When people talk about these in conversation as understood, defined entities, or explain and define them for non-participants, the speakers know what they mean, and their explanations are clear. But when people start talking about 'a group' as an assemblage of known individuals, the clear definition clouds and dissolves into disparate threads of contradiction and justification. There is a small section of Rwala who settled in northern Jordan some four hundred years ago; everybody agrees they are 'the group name who come from the Rwala'. They consider themselves *bedu*, "because we behave like *bedu*" while other Rwala in the *badia* say this section "is *hâdhr* because they've been living in villages for a long time. They're Rwala, but they're *hâdhr*." Abu W. of this settled Rwala section considered that "everybody is *hâdhr* now, that is we all more or less live in towns and we all come under a state government. But we can still remain *bedu* (in the sense of behaving properly, treating everyone as an autonomous individual to whom one has moral responsibilities). We have our *madhâfir* (guest room), where we feed and are hospitable to everyone and anyone who comes, not just the people we invite – which is how the *hâdhr* behave. On the other hand, there's nothing to stop the *hâdhr* feeding all-comers, and there are those who do."[11]

---

[11] Much discussion on the oppositions of 'the civilised/urban settled/the *hâdhr*' and 'the outside civilisation/the tribes/the *bedu* (and the *fellah*)' by political scientists, historians and anthropologists is drawn from ibn Khaldun's Muqaddimah. In his analysis of classical Arabic historiography, Khalidi (1994: 222ff) comments that ibn Khaldun's most notable contrast with other historians of his time is his theory of power and its consequences, and his awareness of larger and more impersonal forces shaping historical processes. "Metaphors invite the reader to think of power as a 'commodity'......... For every kind of activity there is an appropriate sense of power and a level of power diffusion .... The

Participants often construct the frameworks of social practice of the Bilâd ash-Shâm and the Arabian peninsula around two kinds of descriptions. One emphasises identity constructed around and ratified by the 'community', the people who "live, work and marry together"; the second comes from the more formal oral tribal traditions or the literature of the learned urban families and the military, and portrays a stratified but interdependent polity. The oral traditions of the tribes speak of those who are *asîl*, are part of the Arab tribal genealogy, hold resources and have reputation; these are apart from those like Htaim and Sulaib who are tribal, have honour but none of the other qualities; slaves, now ex-slaves, also have honour; the non-tribal and urban world is of no concern in this portrayal. The description from the learned urban families and the military similarly ignores the tribal world, except sometimes as a distant origin. The abstractions appear to conform to one opposition of the countryside to the towns, and another of stratified polities to working communities. These become less sustainable when actual social practice by individuals and the groupings of which they are members is examined. The first could be described as informal and everyday identity; the second places identity in time and space, with validation from accepted narratives, whether written or oral, and from accepted political repertoires. The narratives of rural families and tribes describe relationships among certain sets of similar families and tribes, they do not include all possible groups. Is this stratification? The refusal of *bedu* families to permit marriages of members with slaves, *htaim*, *sulaib* or gypsy families has been taken to indicate that this is so. But given that marriage is seen to draw some groups closer and to push others away, and that the social world is seen to be composed of individuals, families, groups and tribes who are closer or more distant from the speaker, then stratification is less significant than a distancing, a disinterest. The principle of 'like with like' is well established in Arab life, operating within as well as between groups. Both descriptions have contextual validities. One is concerned with

---

([11] Contd.,) universe of ibn Khaldun is a structured whole, with gradations of reality and meaning. History is the record of this structure ...... The building blocks of the Muqaddimah in general and his History in particular were ready to hand in the historiography of his age." Some users of ibn Khaldun's work apply his discussions of these categories as if these were the only referents of persons and stable over time, which is inappropriate.

internal everyday life, where rural and urban communities construct identities in similar ways, and towns and countryside are linked through the intermeshings of embedded economic, political, and social networks. The other is more do with abstractions of identities in terms drawn from accepted historical traditions in the milieux of the educated, whether of the towns or the tribes.

Active 'communities' construct identity from the processes of getting a livelihood from local resources, from political loyalties focused on keeping the peace so that livelihoods may be assured, from the defence of the community itself and its resources, and from the social practice seen as fundamental to the achievement of its aims. Participants project social practice by individuals identified as members of groups constructed around descent and based on individual responsibilities within known moral premises. Members of active 'communities' see themselves as "equals before God", equal in the possibility of "living a good life". These communities may be 'tribal and bedouin', or 'tribal and agriculturalist' as in North Yemen (Dresch 1989), or 'villagers of known descent and agricultural' as in north Jordan (Antoun 1972). Other communities are perceived as close if like themselves in the terms of construction, or distant if not. Such active 'communities' are found within towns, rural villages and the *badia*, where the majority of communities are based around ideas of 'ruling' and 'ownership of resources.'

The urban, learned description focuses on status, with the groups thus constructed seen as vertically stratified discrete units. Lapidus (1967: 5), in his classic analysis of mediaeval Muslim cities, describes Abbasid society metaphorically as vertically stratified; its communities have no ties to each other but are bound directly to some higher centre of co-ordination. The communities of status of the mediaeval and Ottoman periods in the cities – like Damascus – were military and religious elites who mediated with state agents, notables who, with merchants (who might be members of these elites), directed the urban economy and were often rural landowners; and the mass of the people working in crafts and trades, who could also be members of religious bodies. This stratification accommodates the countryside through its generalised enmeshments with those of townsmen. Overall cohesion came from the incorporating state administration and Islam. Contemporary urban social groups continue to tap into residual legitimisations of former power groups, reformulated with shifts of personnel in current political ideology. Other towns are seen as closer or distant through linkages of learned and ruling families, from descent, affiliation, or shared or sponsored military action, or ruling.

The two descriptions are constructed from different concepts of what is to be described, and use different metaphors, so comparisons between them are like comparing oranges and apples. Both share the view that identity gains a participant entry to social practice. Identity and social practice are inseparable, although talked about in different ways. Both hold that family is central to social practice; and while individuals act as members of their families, an individual is responsible for his actions. Metaphors of descent derived from working communities or from learned status are used to legitimise power whether by groups exercising preferential dominance or by those who accept dominance by others.

These descriptions ignore differences in degrees of mobility, often seen as relevant. Altorki and Cole state (1989: 17) "throughout much of history, Najd was an area characterised by a high degree of autonomy from central control .... communities such as Ha'il, Buraydah and 'Unayzah were urban places with their own amirs under whose leadership the citizens provided for their own defence. These cities (and the villages attached to them) were like islands in that they were surrounded by vast stretches of desert that was controlled by autonomous Bedouin tribes. The basic bifurcation within the area was between the badiyah, "nomads", and the hadar, "sedentary folk", but symbiotic ties linked many of the hadar to the badiyah through ties of kinship and tribal origin." The sedentary population used three descent categories tied to political and economic life: *qabîli*/tribal, men and women of recognised tribal descent; *hâdhr* or non-tribal, born free but unable to claim tribal descent; and *'abd*, slaves who themselves, or their ancestors, had been bought. Major occupational categories were: *'ulamâ*, those learned in religion; *'umarâ*/rulers; *fellâhi*, tillers of the soil; *tujjâr*/ merchants; *jamâmîl*/cameleers; and *sunnâ'*/artisans. The rulers were always tribesmen. Only *hâdhr* worked leather, or gold, silver and iron, or as butchers; "other craft specialisations as well as the other main occupational categories were in practice open to all without regard to descent" (1989: 23).

For Oman, Wilkinson (1977: 189) uses the single distinction 'bedu' and 'hadar' (nomad and settled) for "convenience", adding there "are so many variables involved in the way of life of the people who inhabit the fringes of the desert that it is of limited value to classify their mode of life under such simple terms: the fact that people may not move around to earn a living does not, *ipso facto*, make them 'settled' anymore than livestock herding makes them 'nomads'." The 'settled' population of 'Unayzah were "for generations and long before the emergence of the contemporary economy ...... on the move both in and out of the city" (Altorki and Cole 1989: 81). Doughty ([1888] 1936: i, 339) notes similar movement by traders between the Hijaz and the Maidan quarter of Damascus. Musil (1928a: 125) comments on traders from the Iraqi towns of Kubayza and Rahba who made their living by visiting bedouin encampments. The town of Sukhne in central Syria was described by de Boucheman (1939) as a "*cité caravanière*"; today, Métral and Métral (1989: 156) write that "le commerce sous ses multiples formes y occupe la place centrale. Il s'appuie, tout comme autrefois, sur les rapports avec les tribus bedouines, les colonies établies en ville, l'élevage du mouton et une exploitation minière du steppe."

Musil (1928: 44–5) recorded from Rwala tribesmen their view of society at the beginning of the century. "The Rwala divide human beings into hazar (*hâdhr*), or those who dwell in permanent houses, and 'arab, or those who dwell in movable tents .... The dwellers in houses, hazar, are divided into karawne .... those who never leave their permanent dwelling, and the ra'w or ra'ijje ..... or those who change from their permanent dwellings during the rainy season to movable tents. After the sowing of the crops in the autumn, the ra'w or ra'ijje leave their villages and with their flocks of goats and sheep make their way into the steppe, where they dwell both in black goat's-hair tents and in gray tents of cotton fabric ...... when the harvest is near, they return from the steppe to their houses. The Arabs consist of Bedouins and swaja (*shwaya*). The swaja have two things black.... black tents of goat's hair and flocks of black goats and sheep. These flocks do not permit them to go into the interior of the desert .... The Bedouins are arabs who breed camels exclusively, or at least in the main, and for ten months of the year dwell in the interior of the desert." This view focuses on mobility and occupation, constructing identity around these and ignoring descent. Underlying it is a concern with degrees of political autonomy, provision of defence and assuring of resources associated with each category.

This description of bedouin and swaja/*shwaya* was discussed with Rwala tribesmen in 1994, with reference to changes in categories. The crucial factor "was the ability (or not) to provide one's own defence; what made *shwaya shwaya* was that they paid to be defended while herding sheep and goats in the inner desert during the winters and springs. When the same herders were not in the inner desert and provided their own security, they were not *shwaya* but bedouin." The dependence on context for description is clear. A comment made in the same conversations was that "we are all *shwaya* now; there are no *bedu*, because we all herd sheep (*sha'*, a sheep or ewe, root of *shwaya*) nowadays, not camels, and defending ourselves has been taken away from us by the state." On other occasions, the same men say "we are *bedu* because we provide for ourselves from our own efforts. We herd sheep and I have a pension because I served in the National Guard, but I arranged my service, I contracted with the government, the state did not command me to join. We manage our lives from what is available, and today that is sheep-herding for the meat market and employment in the services." In shanty towns outside Riyadh, the inhabitants say they are *bedu*, because they get their living through tribal and family connections and resource strategies, and manage their affairs in the same way; but on state agricultural settlements, people say they are *hâdhr* (hadar) because the government organises their lives for them. The opposition between providing one's own security and having it provided by others in the *bedu*/hadar swaja (*shwaya*) debate is echoed in Palmyrena where *shwaya* is currently used for those employed on government developed irrigated farms in the Euphrates valley. Marx (1996: 109–10) finds that "pastoral nomads offer a number of alternative theories about the constitution of their society .... every Bedouin in South Sinai is a member of even two such tribes, of which one which gives him rights to cultivate land and build a house anywhere in the tribal territory, while the other gives him access to grazing all over the southern part of the peninsula." Managing one's resources and defending them from within is '*bedu*', sheep-herding as such is '*shwaya*', being 'hadar' is to hand over one's autonomy to a government. These self-presentations indicate a dynamic resilience in rural populations in the face of rapid change. Tribe for tribespeople is not a solidary and corporate body that determines members' actions.

'Tribe' gives identity. Layne, examining concepts of tribal and national identity in Jordan, finds "identity as meaning constructed on an ongoing basis through the everyday practices of making a place in the world, that is, adapting a posture in the context of changing circumstances and uncertain contingencies" (1995: 29).

Concentrating on individuals as members of active social groups, the 'working communities', allows descriptions of social practice in the wider society of the rural *Bilâd ash-Shâm*. Jordanian *hâdhr* families say 'family' is 'like tribe', "we work the same way." A Muslim *hâdhr* family from as-Salt said "we became *hâdhr* when we became Muslim; before that we were *bedu* and tribal", introducing another dimension.[12] An urban Christian family "had always been a family, never a tribe; we work and marry within the wider family and with other Christian families like us. We've always been *hâdhr*, because we always live and work in towns." The construction of conceptually opposed groups, like *bedu* and *hâdhr* (e.g. Altorki and Cole 1989 in 'Unayzah; Lancasters in southeast Oman, 1992b; bedu and fallah in Seurat 1980: 109; Lancasters 1995), between land-holders and the landless in a Jordanian village (Antoun 1972: 35–6), or *qabîli* and *'arab* in North Yemen (Mundy 1995), is derived from the application of longterm referents to local contexts. Jabbur's discussion (1995) of *bedu* and *hâdhr* follows the learned literary tradition. These oppositions reflect differences in referents and provide a shorthand description of observable and public behaviour rather than producing useful models of the region, since diverse groups construct categories differently. Although such presentations of self lack validity as theoretical models, they report local and regional constructions of the social world by participants with long common traditions.

Political and economic transformations in the former Bilâd ash-Shâm, the present southern Syria, Jordan and northern Saudi Arabia, in the last hundred and fifty years have been rapid and far-reaching. It is tempting to see the new orders as fundamentally changing social, political and economic life. However, many people actively derive their current practices from those of their parents and grandparents, albeit influenced by changing technologies, economic conditions and political situations. The idioms of the past remain current in discussions of identity and political activity at the same time that new terms enter. The moral premises of social practice are perceived as outside or beyond time, although events in historical time are recognised as transforming political and economic manifestations. How people respond to changes is

---

[12] Muhammad Hamdan ar-Rwaili in Qara (1995) said his grandfather had been *bedu* and *jâhili* (pagan), who then settled and became Muslim, i.e. Wahabi.

effected through existing social practice in accordance with the moral premises. As a tribesman and senior local government official explained in a town on the Karak plateau, "The rules of the game have changed – that's because of the political changes, nation states, geopolitics, and because of the importance of oil. But the playing field hasn't; that's here, where we live and work. And nor has the game; that's getting a living from the resources of the region and keeping your family, and being yourself as a good man, and being part of your group (*ibn 'amm*), and your community (*jamâ'a*)." Marx (1984) and Kressel (1984) illustrate the persistence of traditional political and economic strategies practised by bedouin in Israel faced with massive cultural, legal and social change, as does Métral (1984:1989:1991) in Syria.

The apparent lack of institutions in Islamic and in Arab tribal society has often been remarked. Less commented on is the existence of structural processes, practised by face to face interaction without necessarily having constituted officials or formal arenas, although these may be designated through words and contexts rather than in physical plant or infrastructure. Jaussen (1948: 117–8) describes such processes in the contracts and treaties tribesmen made to get access to summer grazing and water on the Karak plateau, and in difficult years further west in southern Jordan and Palestine. Musil (1928a: 47) notes similar treaties among tribes and families in the inner desert of the Bilâd ash-Shâm, and Dickson (1949: 391ff) for eastern Arabia. Access, restitution and recompense were verbally agreed between representatives of each party and aurally witnessed, as in the present when 'Umûr groups use the *hamad* of eastern Jordan through Rwala agreement–even though tribal customary law is no longer recognised and grazing open to all. Other contracts, such as *rafîq*, *kafîla*, and *khuwa*, permitted the safe passage of goods and people, contracts being necessary for the very reason that face-to-face relations are effective only through processes of sponsorship and guarantee, and witnesses. Share partnerships have been and continue to be widespread in agriculture, business including raiding, commerce, and distribution (Firestone 1975: 185–209; Goitein 1967: 164ff.) creating enmeshed networks across occupation and family boundaries through personal relationships, mediated by sponsorship and guarantees. There are more informal long and short term relationships between rural inhabitants and merchants, service suppliers like mechanics and builders, government officials of various categories, and travelling groups of service providers like gypsies, *sunnâ'* or smiths, or Sulaib.

Maintaining social practice comes from individuals' active and continuing participation, while individuals have a social identity in proportion to their activity. This is a truism for any society, but is particularly apposite to Arab tribal society. A casual comment on the lack of attendance at a tent where a man was giving a feast drew the response that he had not attended those of others. Each participant operates on reputation, reputation as a 'good man' (Lancaster 1981: 44). To achieve this, he/she must act and be known to act as such (Antoun 1972: 110; Dresch 1989: 101). Actions and achieved reputation are discussed by his or her jurally autonomous peers in moral terms, or in terms of honour defended which comes to very much the same, as being in accordance (or not) with the premises. Social practice in this sense is unaffected by economic or political transformations.

The countryside of agricultural villages, the steppes and desert in Arab historiography (Goitein 1967: 75; Faroqhi 1994: 67–9), has been outside the realm of the civilised, although intermeshed with the urban centres its products help to support. The market and exchange bound the different economic sectors together. Altorki and Cole (1989: 81) saw the market as having been the integrative mechanism throughout the Arabian peninsula, but now transformed with the development of an internal market which distributes wealth from outside mediated by and focused through government. Similarly, Mundy (1995: 42), in northern Yemen, writes "by the 1970s it was apparent to men and women that the market (and marketable labour) was set to be the dominant institution in the new Yemen–as opposed to .... the market in which kin and domestic units form central organisational units within the emerging structures of political economy." Bocco (1989a and b) and Layne (1987; 1995) discuss change in Jordan in similar terms. Seale (1991) notes Asad's development of Syrian political institutions to integrate urban and rural arenas to free the countryside from urban merchants. Seurat (1980: 111) reports that the massive developments by the state in the Syrian *hamad* are to increase economic growth and to incorporate peasants and herders into the state.

Antoun (1972: 48ff) vividly describes his attempts to find definable units in a north Jordanian village. Looking for the consuming unit, he settled for "the holding of a common purse", although finding not even this wholly satisfactory. Seeking groups dealing with marriage, livelihood and defence, he had to recognise while the '*luzum*' was the group of participating members who arranged these matters, it also had absent or not fully participating additional members. Attempting to establish the domestic group responsible for production Fabietti (1990: 242), working with Shammar tribespeople of central Arabia, arrived at this definition; "the Bedouin domestic group (is)........a dynamic unit towards which converge resources originating from a variety of sectors, procured and organised by mobile individuals belonging to a parental group whose dimensions and composition are not definable a priori." This position fits such apparently diverse tribal and familial productive groups as farmers in Jordan (Antoun 1972; Lancasters 1995); tribally descended merchants and entrepreneurs of Sukhne in Syria (Métrals 1989); and fishermen in southeast Oman (Lancasters 1992b).

Yet rural society has named structural units, of which the most inclusive is tribe with various sub-divisions and further sub-sub-divisions. Such units are both relevant and irrelevant in the construction of working groups. The active groups concerned with livelihood, marriage and defence/vengeance can be identified by their smallest inclusive named unit. In Antoun's peasant village, this is the *luzum*, and support for co-members in livelihood, marriage and security is obligatory. For Rwala, it is the *ibn 'amm*, and the *khamsa* is the vengeance group; the *khamsa* shares a name the *ibn 'amm*, but the two groups' personnel is not identical as the *ibn 'amm* is larger than the more defined, but not unambiguous, *khamsa* (Lancaster 1981: 28–32). Like the *luzum*, the *ibn 'amm* is a named group. Like the *luzum*, it is ambiguous in its membership, although presented as a bounded group, and talked about as if it were localised. In fact, the active members of a *luzum*, like an *ibn 'amm*, are focused round particular cores of individuals who work together in a variety of enterprises although underpinned by ownership of land, water, herds or other resources. Members marry more inside this group, although there are always a proportion of marriages outside. Vengeance, defence, and support in disputes are key situations in which ambiguities in membership become visible.

To clarify, we digress to tribal and family public, formal structures. The peasant families of Kufr al Ma described themselves as '*ashâ'ir* Tibna' ('the tribes of Tibna') or '*ahali* Tibna' ('the people of Tibna') Tibna being the original village in the mountains. They never describe themselves as 'a tribe', either as an *ashîra* or a *qabîla*. They are an association of three descent groups together with some individual families; one of the latter is the ruling family, the Wazir family. For the different layers of inclusion, see Antoun (1972: 36ff). The Rwala see the Aneze confederation as the *qabîla*, the Rwala as the *ashîra*, one of the *ashâ'ir* of the Aneze; the Rwala then divide into *fukhûdh*, the *fukhûdh* into *ibn 'amm* groups. The *ibn 'amm* units are the smallest named units, and therefore the smallest identifiable groups; they are associated with vengeance. Those bound to take vengeance for one of their members and conversely liable to vengeance from the actions of a member are the descendants in the male line of an ancestor five generations back; this is the *khamsa*, which is not a named group. Within this, those responsible for taking blood vengeance and liable to it are males with five links between them, which limits this group to a three generations depth. Rwala know their fathers and grandfathers, but not often those more distant; they shift the criterion of descent from known to assumed, to the named ancestor of the *ibn 'amm*, reckoned to be of five generations depth. The five generation group is not however of five generations depth, as can be seen by comparing Musil's (1928a) lists of these groups with Lancaster's (1981: 169–171) despite some seventy years and so at least another two generations, few names change. The Rwala compensate for this lack of precision by a man "following his son into the sixth generation" of the group responsible for vengeance. But this group is not named, so although the responsibility for support in matters of blood realigns itself with the generations, the structure of the named groups does not.

The same sort of rule is used by the Beni Sakhr, and by the Beni 'Amr of Karak. Haj Khalaf said, "We, the Beni 'Amr, are a *qabîla*, we are in Jordan, Syria, Palestine and Saudi Arabia. The *qabîla* is made up of many *ashâ'ir*. We are of the *ashîra* of the Rumaithat, some of the Rumaithat are here in Karak, others in Saudi Arabia in Tabuk and in Qasim. *Ashâ'ir* are made up of *fukhûdh*, our *fakhdh* is the 'Aiyifat. *Fukhûdh* are made up of *'âyila*, our *'aila* is the Tarif. The *'aila* is more or less the group who support you, and mostly we marry inside it, and mostly the men we work with are in it. The *'aila* is from the third grandfather (i.e. five generations), so it is the group responsible for blood–like a *khamsa*, but we say *'aila*. But the *'aila* changes for a man when he follows his son into the sixth generation, so the group for blood is not exactly the same for everyone. My *'aila* was the Salama, but now I follow my sons (all adult) into the Tarif, who was Salama's son." Haj Khalaf's son was present. He had not known that his *'aila* ancestor,

Tarif's father, was called Salama; for him it was irrelevant. As real generations and chronological time lengthen, the groups of active men refocus the *'âyila*, but because the actual generations become elided the *fukhûdh* remain the same.

The Majali of Karak describe their social organisation somewhat differently from their Beni 'Amr neighbours. The Majali descend from brothers who came from Khalil (Hebron) to Karak as traders, stayed because of a blood feud, and eventually controlled the routes between Karak and Khalil. It was after this that members began farming, owning land and ruling. "We became an *ashîra*, 'a tribe', by always marrying our *bint 'amm*," though Majali do and did marry with other *ashîra* in the area. The Majali call themselves Âl al-Majali, in the same way that the Rwala call themselves Âl Rwala. The Âl includes the *fukhûdh*, and each *fakhdh* is associated with areas of agricultural land and villages. The group responsible for blood, within which most marriages take place, and within which land is transferred is the *fakhdh*; as the Majali prefer actual first cousin marriage, this group is of three generations depth in the main. The group responsible for vengeance is five generations deep, with those on the edges paying the 'camel of sleep' for less responsibility; the 'sixth generation' rule is followed.

An Ahl al Jabal family described their tribal structures as follows; "the *qabîla* (tribal confederation) is the Ahl al Jabal; then there are the *ashâ'ir* (tribes), and we are from Al-Sharafat. The Sharafat have two divisions, the Zraiqan and Khmaisa. The Zraiqan are the *jamâ'ât* (communities) of ibn 'Anaizan, and the Khamisa are the *jamâ'ât* of ibn Sfaiyan. Then there are the *hamâyil* (sing: *hamûla*); five in the Zraiqan and four in the Khmaisa(sic). The *hamâyil* of the Khamisa are the Mukhamis, the Hamil, the Shkhr and the Shafa'. The *hamâyil* divide into *fara'* (branches), like *fukhûdh* (thighs) only we say *fara'*, and we are from the Khlaif of the Mukhamis. Inside the *fara'* are families, and our family is the Hamil."

Fukhûdh names are used to differentiate inside tribes by tribespeople and by others who are closely connected or in close contact. The tribal name provides a widely recognised identity of known descent. Tribal identity allows participation in social practice outside the confines of known individuals by giving a means of identification for others within the wider society; it increases reach. An identified individual's reputation lets him sponsor and be sponsored, guarantee and be guaranteed, witness and so on.

All tribal or family structures are publicly concerned with men. They are discrete bodies, with membership from descent through men. Women are defined as daughters of their fathers, and thus members of their fathers' groups, but do not appear in the genealogy during their lifetimes, though occasionally they do so after their deaths. The preferred marriage is that a man marries his paternal first cousin, his *bint 'amm*. However, *bint 'amm* is used for all daughters of men of the *ibn 'amm* group, and groups prefer differing degrees of closeness for marriage. A wider domestic group binds itself together by marriages between its members. A family, by the marriages its children make, pulls some families closer and distances itself from others equally suitable in genealogical terms, and thus has an important role in generating the future direction of the group (for a discussion of Rwala marriages see Lancaster 1981: 43ff). Most marriages of a wider domestic group reaffirm existing linkages within and to other close groups; some take up fading links to more distant groups; and a few set in progress new links or re-establish old links to members of quite distant groups. With the preference for close or relatively close *bint 'amm* marriage, many individuals in a group have multiple ways of tracing relationships between them. Relationships through women take local and informal precedence over those through men; this was clear in southeast Oman where all those using the beaches at Ras al-Junayz explained their presence by links through women (Lancasters 1992b: 357).

From these links through women men build networks which reach beyond tribal identity on its own. Some networks are inside the tribe, others extend across tribe, and outside tribe. Our host in Karak demonstrated his reach; as well as close bonds within his *'aila*, he had close ties to one other *'aila* of his *fakhdh*, to another *fakhdh* of the *ashîra*, and to another *ashîra* of the *qabîla*; he also had links to another *qabîla* living in the same village, to three other *qabâ'il* (plur. of *qabîla*) in the neighbourhood, and to three more distant groups in the region; he could also use the links through women of other *ashâ'ir* to yet other *ashâ'ir* within the wider *qabîla*. As his wife said, "It's impossible to find a *dunum* (c. quarter of an acre) where you can't find a relation." Besides networks of action people use the private connections from relationships through women to bring people nearer to them, and to 'place' them in a more specific context. Other more informal associations, such as former neighbours, school, university or army colleagues, are also used in this way.

The small tribes in southeast Ja'alan in Oman construct a *jamâ'a*, a community, from "the tribes who live, work and marry here." As the *jamâ'a* is referred to by place rather than descent, and is ego-centred and fluid, it never competes with tribe (Lancasters 1992b: 359). Vengeance belongs to 'tribe', like marriage—although marriages help to construct the *jamâ'a*. The tribespeople of the north Karak plateau also talked about the *jamâ'a* but as a family's informal network rather than a coming together of people and families to manage a resource as in coastal Ja'alan. The people of the Karak plateau call themselves '*ahali* al Karak' or '*ashâ'ir* al Karak', the same terms as the 'people of Tibna', but while Tibna people say they are peasants who also herd, Karak people are more disparate in their self-descriptions. Some emphasise a tribal and bedouin descent, others tribal descent and shift between bedouin or *fellah*, yet others offer family descent and *fellah*. There are also the two political alliances of the recent past, described by Musil (1908: iii, 97–103) and Gubser (1973: 53), while other groups are independent of the alliances. In Ja'alan, all those participating in a *jamâ'a* had family and tribal members elsewhere, sometimes locally but also in the capital, in other parts of Oman or in the Emirates. This was also the position among Rwala working *ibn 'amm* groups, Ahl al-Jabal *fara'*, Shammar wider domestic units, Sukhne families, the *luzum* of the Tibna *ashâ'ir*, Beni 'Amr *'aila*, and Majali *fukhûdh*. Active working groups are not localised, bounded, or solidary, but individuals and their families.

Using and developing natural resources implies political and jural activity. In the Islamic tradition, which has roots in earlier Middle Eastern systems, owning is concerned with techniques of production—draft animals, ploughs, seeds, fruit trees, stores, domestic riding and herd animals. The medium of production comes from God, not from man, and therefore cannot be owned, although access to the medium may be restricted to customary users, identifiable through known tribal or family identities. At the same time, the medium of production is assumed not to be a scarce resource; a common declaration is that "everything comes from God and He is generous." Local scarcities through poor rains, disease or locust depredations are expected, and everyone has ways of compensating. Owning comes from developing a resource beyond its natural capacities which includes irrigation systems, wells, cisterns, perennial cultivation, mills etc. Arable cultivation dependent on rainfall does not of itself, in traditional usage, confer longterm ownership of the land but only of the crop; however, habitual use of land including customary fallowing builds up claims to control use. Not using land releases it to use by others and to their claims of control. Following this tradition, in pre-nineteenth century agrarian patterns in the grain economy of Greater Syria "land control was far more economically and politically important than outright land ownership" (Schilcher 1991: 176). Land control and claims on agrarian surplus exist alongside customary and jurally backed rights to usufruct for subsistence and the voluntary relinquishing of land (Seikaly 1984: 404 quoting the legal opinions of Khair ad-Din ar-Ramleh in the 1640s and held in customary law). Working of land was by its owners, through lawful share partnerships (Abdul Nour 1984: 80–1) or hired labourers. Combinations of all or any of these exist in rainfed arable and irrigated agriculture in all areas. Some owners of land also work other land through share partnerships, and work as hired labour for seasonal tasks such as harvesting, often in areas away from their own land. Share partners in cultivation may have claims on land in other areas, and hired labourers may have access to land through share partnerships, and/or have claims on family owned land in another place. Such mobility by agricultural producers occurred in the past as well as in the present.

The perception of owning land is split between land capable of agricultural development, and that usually suitable for herding, where development is of water collection and storage for herds and people. In Islamic state law, only land developed for agriculture, particularly irrigated farming, can be owned; all other land is considered as undeveloped, 'dead' land, and therefore at its disposal. Herding families regard their customary use of particular areas at certain seasons, their wells, and their provision of security through tribal processes, as a form of ownership. Developing physical structures and providing security so that people can live their lives are seen as tribal or family ownership in customary law. These areas of ownership or preferential access are often referred to as *dîra* or *watan*, both terms denoting customary use for the provision of livelihood and the necessary management of social relations. Different tribes use *dîra* and *watan* in a flexible and contextual manner. The Rwala use *dîra* variously (Lancaster 1986) as a local administrative centre, the area of influence of a man of good reputation while present, the total customary areas of use by known tribesmen, and to refer to areas where Rwala live and have owned or preferentially claimed assets. Currently, the *dîra* of the Rwala is

Sakaka and al-Jauf, the region where Rwala own resources and are influential, which is how Bocco (1985) describes Beni Sakhr *dîra*. A Sa'idiyyin in the Wadi Araba referred to the wells, gardens and grazing lands of his section of the tribe as his *watan*, comparable to the descriptions of tribally owned oases in the Arabian peninsula by the classic Arab geographers. Doughty ([1888] 1936: ii, 133–4) describes land use and ownership in the Khaibar oasis by different Aneze tribes and their village partners, noting that rights to receive *surra* payments and to be carriers for the Meccan Pilgrimage derived from "ground rights" ([1888] 1936: i, 117). Altorki and Cole comment on the development of Unayzah by members of the Beni Khalid and Subay tribes (1989: 15–6). Hail was developed by the ruling family of the southern Shammar as the capital of their tribal polity (al-Rashid 1991). In contrast, Sukhne is an open town resulting from an alliance between settlers and nomads (Métral and Métral 1989: 163). Although both Hail and Sukhne have springs and gardens, they relied more on servicing trade routes. Modern examples of tribally developed service posts in 'tribal territory' are Faydr and the ar-Rishas in eastern Jordan (Lancaster 1981). Bocco (1986a) describes the development of Muwaqqar in Jordan under the al-Khraysha division of the Beni Sakhr, and Lewis (1989) the earlier villages of al-Fa'iz Beni Sakhr sections on agricultural lands. Tribal 'territories' are scatters of preferential access to resources, together with shifting networks across the landscape of use and influence, rather than bounded and exclusive entities; Layne's (1994: 65) analysis of Abbad choices of where and how to live as showing that "spaces are defined by people and not by places" is comparable.

The Htaim tribes are said not to 'own' territory. Sometimes the reason given why people became Htaim is that they lost land they owned by defeats in battle, or the death of herds. Travellers give varying descriptions of them. Burckhardt (1831: ii, 20–22) comments on the dispersed encampments of Htaim, that they paid tribute in sheep everywhere for permission to graze and water, except in an area on the northeast coast of the Red Sea which they owned; here they herded sheep, fished, pearled and traded. On the Shararat, Burckhardt (1831: i, 29–30) says they paid tribute to several tribes, and had innumerable camels, which they took into the Hauran to sell for wheat. Wallin (1854: 126, 150) notes that the Shararat were said to have once owned Ma'an and al-Jauf, but took tribute in dates from al-Jauf with whom they also traded camels, wool, butter and rice from the Red Sea ports. Musil (1928a: 453), in many mentions of the Shararat as camel-breeders and raiders, once comments that there were paupers among them; they paid tribute to the Rwala among others. Guarmani (1938: 20, 74), travelling in 1864, and Carruthers (1935: 73) note both the material poverty of the Shararat and their great herds of renowned camels. Doughty mentions the high standard of abundant food among Htaim of the harra of Khaibar and the number of their flocks and herds ([1888] 1936: i, 83, 86, ii, 241) pointing out that they were rich because they did not have to provide for their own defence.[13] The smallness of Shararat tents was associated with the fact that their wealth was in camels, and that they moved so frequently in herding; while their apparent lack of material goods might be compensated for the fact they, like other long-distance herders, stored goods at preferred seasonal bases (Musil 1927: 177). There were poor bedouin, of whom Musil (1928a: 93 and 1927: 405–6) gives some examples among the Rwala and other tribes. Sulaib are also reckoned as Htaim, while Musil (1927: 216) says Sulaib families owned hill slopes in the inner desert as hunting territories. Families with no or little land in oasis towns often added crafts to agricultural work, as Doughty ([1888] 1936: i, 198–9) illustrates for Khaibar and for 'Unayzah (ii, 429; Altorki and Cole 1989: 48–9), while Sulaib were also smiths and wood-workers with bedouin tribes.

Ideas of 'owning' as claims of access to or control over resources and surplus link with those of 'ruling' by particular tribes and families. This appears to be 'power over', but 'ruling' is perceived by rural populations in two ways. Firstly, 'ruling' enables people to live their lives (i.e. power to) by providing security. This leads to the second, since providing security means practising internal means of solving disputes by mediation, agreements and restitution through the choices and actions of those involved; this is *hukm*, 'arbitration', and leads to 'rule' of mediation, consensus and self-help. Since tribespeople interact with individuals and families from other tribes, peasant groups and villagers, merchants and traders, and agents of central governments and of other 'ruling'

---

[13] The photograph of the Sharari tent in Carruthers (1935, opposite 74), captioned as "their tents, flimsy rags" is the kind of tent put up when moving nearly every day, which Shararat usually did. Rwala used similar tents in summer (Musil 1928a: 72, photograph 24, p. 93).

tribes, dispute settlement mechanisms are needed. 'Ruling' also means gaining access to external resources, such as transporting and guaranteeing trade, and supplying animals and labour to markets and trade caravans; this entailed agreements with traders and agents of merchants and governments. 'Ruling' is at a local context ability by individuals on their own behalf or that of their families or wider domestic groups to conduct their affairs within the purview of common and local social practice. At its most inclusive context, 'ruling' is practised by the shaikh of a tribe or confederation, acting as the representative of participating tribesmen, with the ruler of a centralised state or his agent in the agreement of particular action; at this level, tribal 'rulers' see themselves almost as tribal ambassadors (e.g. Lancaster 1981: 89–90). In both cases, rule is comparable to 'a civil association'.[14] The principle of *kafâ'a*, 'like to like' achieves activity between two apparently disparate groups needing representation; structurally equivalent persons brought structurally different groups of similar or different political identity close for a specific purpose.

Action by any individual behaving as "a good man" in "keeping the peace" or in maintaining individual autonomy or defending his honour, or enabling a second person to do so, is what enables the idea of 'ruling as *hukm*' to continue in the countryside, (e.g. Antoun 1972: 95ff) even if such practice is overlaid by the more visible processes and institutions of the state. Mediation may prove impossible between a state ideologically hostile to tribe and a tribe intent on preserving its security of livelihood and its honour.

Not all tribes and families in the countryside 'ruled', although every individual calling him or herself *bedu* or Arab acted in accordance with accepted moral premises in their management of affairs which entails social practice like 'ruling'. It is from this that Rwala say a Htaim can have just "as white a reputation as a Rwala" (Musil 1927; Lancaster 1996b). In the eastern *badia* of Jordan, Rwala,

---

[14] Oakeshott's model of the dichotomy of an enterprise association and a civil association as types of government is described by Mount (1992: 74–5) "the civil association is an association of persons who agree to subject themselves to a set of common rules and to a common government in order to pursue their own diverse purposes. Circumstances may, of course, intimate to them that they should join together in some common enterprise, to defend themselves against an external threat, such as war of famine; but these shared enterprises are not the prime purpose of the association."

Ahl al Jabal, Ghayyath and others say "the Rwala rule the *hamad*", while the others do not. On the Karak plateau, the Beni 'Amr ruled formerly, but were superseded by the Majali family heading of the western alliance, themselves challenged by the Tarawnah, leaders of the eastern alliance; most tribes and families belonged to an alliance while others were outside and neutral. The Wazir family of Tibna ruled as the head of an association of peasant villages in northern mountain Jordan (Antoun 1972: 36).

'Ruling' implied a reach across and between tribes, families and agents of central government. Ruling was exercised by a particular family or individuals within a family, whose actions should be based on a knowledge of what was acceptable to tribal members. Support for tribal and family rulers was manifested as participation in their actions *vis-à-vis* other similar units, and/or by financial support; this is comparable to the support or participation demanded by former agents of central government and regional polities, and by present nation states. This kind of 'ruling' takes time, skills and wealth; the amount of reach involved in the kind of 'rule' exercised by tribesmen is concomitant with his own and his family's resources, abilities, and inclinations. Rwala say that reach is enabled through inclusion of the peninsula wide tribal genealogy, which gives *asîl* tribal descent. This genealogy is rather a type of history described in a genealogical idiom used as metaphor. Others 'rule' by ruling and ignore genealogical participation, (as one of the Zeben Beni Sakhr remarked), or the shaikh of Khanzira village quoted by Jaussen (1948: 127), who saw intelligence and the possession of weapons to be the key. *Asîl* means 'original', 'authentic'; Layne (1995: 53) says, quoting Owaid (1982: xiii), that the Abbad of the Jordan valley see *usûl* (plural of *asîl*) as "rooted customs, decent characteristics; principles and.....right behaviour."

For tribal or family leaders, 'ruling' is a full time occupation, and may demand the employment of staff, technical and capital equipment, and the need for wealth from outside to provide for these as well as for hospitality and generosity to all comers. Wealth for tribal administration often comes from outside. Rwala shaikhs, like others, traditionally derived income from guaranteeing markets, *khuwa*, agricultural property in oases, raiding, dues from trade and at some dates, tax collection for the state as a condition of agricultural land holdings (Musil 1927: 109, 216–7, 239–40, 326–7, 424; 1928a: 60, 270, 278–81, 510–1): for the seventies, see Lancaster (1981: 84,112–116,121–131). Income, as stipends or honoraria, from the state usually demanded dues to the state; dues often exceeded stipends in early Ottoman (Bakhit 1982: 201) and late Ottoman times (Musil 1927: 390, 431). Unlike some other Aneze tribes, Rwala did not have moneys from Pilgrim routes (Burckhardt 1831: i, 4–8). Tribal shaikhs were and are considerably less wealthy than other social leaders. Merchants are the traditional holders of wealth; Coon (1965: 203) says that many of the *uqaylât* (sing.of *aqeyl*: merchants between the towns and the tribes) "could buy out several shaikhs (of bedouin tribes) several times over", a view echoed by Altorki and Cole (1989: 82). The Hashid shaikhs in North Yemen had incomes from commercial agriculture outside the tribal lands (Dresch 1989: 209-12), as well as subsidies. 'Rule' by tribal shaikhs or family leaders also demanded an education in tribal genealogies and histories, customary law, local politics and trade practices, and an awareness of regional economic patterns and political activity. Being a 'ruler' is and was a demanding career, and only a few members of ruling families become rulers. Rulers can lose support and reputation; both have to be earned and actively maintained through generosity, mediation, and sponsorship of tribespeoples' enterprises. Most tribal leaders see their current role as spokesmen on the tribal behalf. All tribal sections and families have their respected men, whose opinions are given due weight.

Herders and peasants, tribespeople and townspeople were and are part of society, through economic, legal/jural, religious, social and political intermeshings. The maintenance and continued functioning of these intermeshings is said by members of tribes active in this field to be one of the main shaikhly functions as tribal political leaders. Shaikhs are not leaders at the apex of solidary, corporate groups, nor were they. Men called shaikhs are different persons in different tribes and groups. For the Rwala and many others, shaikh is used as a term of courtesy to any elderly and well-respected elderly man. More specifically, experts in different fields of tribal customary law are shaikhs. 'The shaikhs of the Rwala' are the leading men of the ruling family who act in the tribe's interests with other tribes and agents of nation states. While Rwala and other tribesmen seek shaikhs' advice with problems, they do not expect the shaikhs to act on their behalf, except as ambassadors to national governmental bodies. Many tribespeople talk very little about their shaikhs. Some Sa'idiyin said "We have no shaikhs – or everyone of us is a shaikh." Some say they do not have shaikhs, although others from the same group say they do but never use them;

others say shaikhs are useless or motivated from self-interest. Tribespeople point out that any individual is capable of sorting out disputes in which he becomes involved through his own efforts; he can find a mediator, a temporary protector, a sponsor, guarantor, witness, by himself or by going through a third party. That this third party may be an officer of a local state security force indicates the change in the rules of the game. Some Ahl al Jabal largely ignore the shaikh of the Ahl al Jabal, who is well known and has held ministerial office, and instead go to '*fuqarâ*' (sing. *faqîr*), healers. These *fuqarâ* are Ahl al Jabal, of whom some are healers of physical conditions, while others are healers of the body politic. Other Ahl al Jabal ignore *fuqarâ* and shaikh, resolving internal difficulties through representations by reputable persons close to themselves. For representation to persons or bodies outside the tribe, people use the sponsorship of a reputable tribesman close to them. Layne (1994: 59–60) briefly illustrates the work of Ghawarnah shaikhs in the Jordan valley.[15]

An association of 'ruling' with providing one's own security, livelihood, and 'being *bedu*' is common throughout the Bilâd ash-Shâm. Tribespeople often say "being *bedu* is hard, it is a lot of work to behave honourably", and that individuals and families drop out both now and in the past. Doughty ([1888] 1936: i, 80, 241) comments "Many poor families both of Aneze and Hteym (*sic*) join themselves to that humbler but more thriving nomad lot", and "the Heteyman (Htaim) mostly pay Khuwa to all the powerful around them; thus being released from their hostility, are commonly more thriving than the Bedu of the same dira."

Khuwa (brotherhood) was taxation in return for protection, like paying insurance, and could contribute a significant part of

---

[15] In the Bilâd ash-Shâm, few tribes have *sayyid* or *sada* families living with them, (for Sukhne; Métrals 1989) with such families having mediatory and other functions, while Peake (1958: e.g. 153, 162–3) says that families descended from sayyids have been important in some towns, such as Ramtha, and small tribal groups in the north of Jordan. For examples outside the region, see Bujra (1968) for the Hadramaut, Gilsenan and Vinogradov (1974) for north Lebanon and northern Iraq, and Dresch (1989) for northern Yemen. The idea of inviting a member of a family renowned for mediatory skills, or for the ability to bind disparate groups closer is common; the ability may be come from descent from a holy family as in the *sayyid* and *sada* families, or from descent from a noble and reputed family, as in accounts of why particular tribes have particular shaikhly families, or from ability as in the story of the Wazir family of Tibna in Antoun (1972: 16).

the income of those leading tribal families who provided protection. Although *khuwa* is often reported as between one tribe and another tribe or village, it was agreed by a representative or 'brother' of a tribal family with that of the contracting tribal or village family (Lancaster 1981: 120–1; Musil 1908: iii, 66–70, 117; Musil 1928a: 60; Thoumin 1936: 153). Some groups had families that both took *khuwa* from one tribal or village group but themselves paid *khuwa* to another. *Khuwa* agreements were contracts, with a lack of protection leading to no or reduced payments. These arrangements allowed, for example, village sheep herds to use the *hamad* in the winter while the incomers were relieved of the need to provide their own protection, a real burden on production. *Khuwa* also functioned as a means of exchange of goods against services (Musil 1908: iii, 67–70, 117) in the Karak region. Here too some groups were both *khuwa* takers and givers, so *khuwa* provided an additional network of distribution to trade. Wallin (1854: 122–3) describes *khuwa* for Ma'an and its surrounding tribes as a dynamic system where claims for goods in return for protection were continually tested, and where resistance by villagers was admired by tribesmen, and "generally goes a great way in making the relations with them more intimate...... This greatly contributes to facilitate the intercourse between the two parties, and a livelier trade". Trade might take products out of the area or circulate them around in exchange for money or credit for goods, *khuwa* redistributed goods and services around the area between different productive groups. *Khuwa* was compared by Rwala and Ahl al Jabal tribesmen to present day customs duties and other payments to the state; "*khuwa* was fairer, because you agreed to pay first, and you got protection in return. If they didn't protect you, then you didn't pay or you got your money back. If you didn't like the contract, you left. But with state taxes, there's no option, and if you don't pay, they cut off your water and electricity."

Taxation was often a condition of membership of settled communities, one side of a contract that provided protection as noted by Antoun (1972: 107–8) for Kafr al Ma in northern Jordan. Altorki and Cole (1989: 49) mention that the citizens of 'Unayzah paid taxes on produce to the amirs when the city was autonomous, and later to ibn Sa'ud. Through most of the Ottoman period, taxation from the Pashalik of Damascus financed the Hajj to the Holy Cities, through which much economic activity in the province was channelled. The tribes were taxed under the later Ottomans for access to urban markets; tribal leaders collected the taxes and forwarded them to the Ottoman government with a proportion kept back as payment for collection (Musil 1928a: 58–9). Seasonal markets held under the protection of tribal leaders in summer grazing grounds also levied taxes (Musil 1928a: 269–70, 280).

The existence of taxation, *khuwa*, share partnerships, markets, traders and transport assumes a production of surplus in the countryside over the centuries, however its distribution was managed. The rural economy has always supplied markets outside itself, and for commercial reasons. The rural economy was not a subsistence economy, although individuals may have operated in a subsistence economy for themselves in occasional years. An opposition between a subsistence and a commercial economy is false, nor is it useful to think of a continuum between them. Participants make choices in their production, in their consumption, and in how they distribute exchange goods or surplus. There are obligations to be observed, but there are also opportunities for profit in some ways that are lacking in other years or in other locations. Flexibility and resilience are crucial to long term survival.

An individual's concern is to provide for his family in an honourable way; while there are preferred methods of doing this, it is more dishonourable not to support your family than to do so by following an unpopular occupation. Keeping the family is not only a matter of basic subsistence, there are social costs to be made; a man has to be generous and hospitable, and contribute to costs incurred by members of his wider family group, as well as being able to draw on them for expenses he may incur. While people have access to the means of livelihood from the inherited capital of the wider family group and their individual shares in it, whether these are agricultural lands and/or herds of animals, every individual also pursues other economic opportunities. Service and craft opportunities have existed at all times, being part of local and regional enmeshments. As a man pursues opportunities arising within local and regional economies, reaching them through processes embedded in social practice, and is more or less able, lucky, or hardworking than his brothers and cousins, it follows there are, as always, degrees of economic differentiation within and between families and wider domestic groups. Field (1984: 298ff) illustrates this in his discussion of the establishment of a merchant business and its subsequent distribution of profits; the pattern appears totally random but is in fact based on a combination of Islamic laws of inheritance and informal family practice. If non-investing family members fall on hard times, they would be supported, and in any case may well be given presents from the profits; but these are not of right, but of affection. Since the wealth of the region has so greatly increased with the oil economy, there are more opportunities for wealth to be made and for an economically diversified population.

Limitation on individual or family economic activity within the arena of the wider family or the tribal community is achieved by the need for an individual to personally manage his enterprises. He need not carry out all the work, he can have share partners, employees, or hire an agent,[16] but it is his responsibility to supervise, to make his share of decisions, and take his responsibilities. He must be an active participant, and this limits the economic activities of any one individual. In addition, his reputation demands this active and reputable participation in the tribal or family arena. Outside that arena, the constraints on economic behaviour are less. Most rural people work within their own arenas.

Differentiation because of economic diversity between members of a family or a wider domestic group has always occurred, as interests diverge and interaction lessens. Often family ties within a family or wider domestic unit remain strong and active even when there appear to be, superficially, gross economic differences between members. Political activities by particular families are said to give rise to social differentiation; the most often quoted examples are tribal leaders who have become rich through the acquisition of newly developed agricultural lands, or through political skills in developing an area, or in merchant families, where it is common to have the members who invested in and set up the business and became rich to have quite poor brothers and close cousins. Although there are and have been undoubted transformations and development in the past, there are also undoubted continuities in social practice and social structures. Capitalism, in the sense of using capital to develop an enterprise for profit, has been around for a long time. Many entrepreneurs observe and take account of their social duties, since people do not blindly respond to

---

[16] *Wakîl* / agent; a man cannot rely on his brothers or cousins to look after his enterprises unless these are shared businesses or there are agreements, since they themselves are occupied with their own work. So he employs a paid agent.

external factors or internal mechanisms but choose paths of action from their reading of the moral premises underlying social practice at any time. Behavioural choices are made that lead to changes in social practice incompatible or less compatible with those of their familial social milieu.

Such internal tensions and conflicts, particularly in relation to questions of access to resources within accepted social practice, can cause splits within tribes or families, with movement of particular sections or realignments of groups following movement. Conflicts over the right use of tribal resources and the right use of ruling may have been at the root of the change of rulers of Karak from the Beni 'Amr to the Majali at some undetermined point in the late seventeenth or eighteenth centuries from the material presented by Burckhardt (1822: 381–2), Dissard (1905), Musil (1908: iii, 70–84) and current Beni 'Amr commentators.

That alternative courses of action are always available to participants within accepted social practice is a major force in maintaining the cohesion of rural society. It is possible to leave and to stay if a particular course of action can be successfully argued as compatible with underlying moral premises and to be effective. Action that goes against the good of the family or community can result in informal refusal to co-operate leading to a *de facto* exile (Lancaster 1981: 76–7), formal exclusion (Ginat 1983), or in certain situations, death (Burckhardt 1831: 325–6). In the seeking of vengeance, the close kin are allowed time to flee (Abli 1996; Burckhardt 1831: 1, 151; Musil 1928a: 491). The emphasis on face to face relationships is relevant to the continuation of social practice, since it demands known identity and reputation, verifiable between persons not known to each other by the mediatory offices of a sponsor or guarantor known to both parties. Maintaining identity in known terms is important to the workings of society, whether in the countryside and traditional arenas, or in the functioning of modern nation states. Iraq forbade tribal and family affiliation but then found it had no way of checking on individuals in the army and security services (Baram 1997), so tribal identity was re-introduced.

Participants regard the moral aspects of social practice to continue effective access to local and state resources, and to be the basis of individual, family and regional identity. Existing social practice is thought to provide for the future more surely than any observed alternative, as it is the route for the accommodation of political and economic transformations within its premises. It is accepted that specific behaviours can change; examples are obvious to all, but the aims and values of wider domestic groups are assumed to be unchanging, since it is from these that the logic of social practice is derived, and from these that identity is derived. The identity and status of 'others' is consistently predicated on the assumption of different and opposed moral premises.

Plate 1 Household vegetable garden – Sakaka

82

Plate 2 Date market – Sakaka

Plate 3 Hard grazed for years, plants re-appear with suitable rains – ar-Risha, eastern hamad, Jordan

Plate 4 Villa in Sakaka: note water-tower decorated as an incense burner

Plate 5 Mechanised large scale farming – near al-Jauf

Plate 6 Water filled *mahafir* in the *hamad* – east Jordan

Plate 7 Camels going to water in Wadi Ghwair: old gardens on right bank – southern Ghor, Jordan

88

Plate 8 Bronze age water storage still in use – the harra in north Jordan

Plate 9 A ghadir in the eastern *hamad* – Jordan

Plate 10 Gathering and milking sheep – eastern *hamad* of Jordan

Plate 11 Threshing machine – Karak plateau, Jordan

Plate 12 Vet injecting sheep, helped by the owner's daughter – Southern Jordan

Plate 13 *Jabban* making cheese – eastern *badia*, Jordan

94

Plate 14 Yarded dairy herd, central Jordan: bales of straw from Saudi Arabia

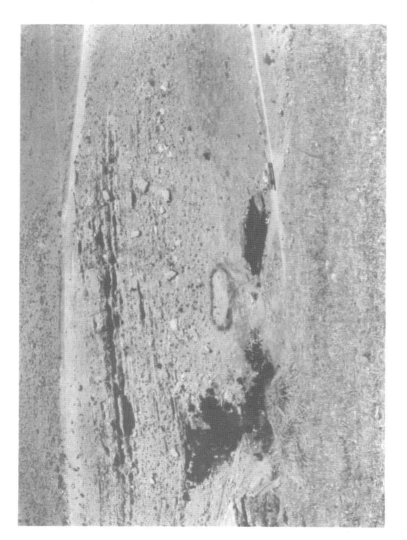

Plate 15 Herder's tent, *haush* and pick-up: water seep nearby – Karak plateau, Jordan

Plate 16 Spring-fed gardens at Dana – Southern Jordan

CHAPTER 3

PHYSICAL ENVIRONMENTS, LANDSCAPES AND NATURE

Users of the physical environment turn natural facts into culturally and socially constructed landscapes through visual perception, use for livelihood, and from selected past events of the area; the resulting landscapes have these components of topography, livelihood and historical associations. One form of historical referent ties a landscape into a distant common past, while another ratifies current use and political realities. A few general landscapes are common throughout the wider area, generated over time from the three factors of topography, use and historical association; well known examples are Najd, Qasim, Hijaz, and the Bilâd ash-Shâm. Each conveys in the minds of speakers and their audience a characteristic landscape of varied topographical and geographical features, effective modes of livelihood, and certain historical and political events and groupings. These 'landscapes' have become 'regions' through the workings of historical geography together with an indigenous appreciation of their local biogeography.

Inside and alongside these 'regions' are more local landscapes, elaborated by their users, existing alongside one another and often overlapping in parts, just as the activities of their users do. From such multiple usages topographic features may have different names for users of different groups, focusing on a particular characteristic of the feature and described either straightforwardly or in metaphor, or derived from an event occurring in the area as a reference for it. For example, in the *harra*, an isolated black peak with two smaller ones is known as Jabal al-Abd (mountain of the slave); near Turaif is a peak with three smaller peaks round it is called either Umm wa Ayyalha (mother and children) or Umm Wu'al (mother of ibex, i.e. a good ibex habitat); Wadi at-Taxi or Wadi al-Cadillac is a wadi bed in the *hamad* so called by the Rwala because al-Aurens ash-Sha'alan crashed a Cadillac saloon there one night; the area around a rainpool used to be called al-Munawikh, the little halting place, but its name was changed to al-Manajih, the place of a successful ending, after a possible fight was averted between the Sha'alan family of the Rwala and the Milhim of the Hessene on one side, and ibn Hadhdhal of the Amarat, ibn Mhaid of the Feda'an and ibn Mirshid of the Sba'a on the other. Using topographic features for place names means replication; e.g. Buraika/Buraiga (cisterns), Hdaib (a symmetrical hillock on a spur, Guraigira/Kuraikir (a wide hollow with an entrance), Umm al-Guttein (mother of dried figs), ar-Rumman (pomegranate trees), and so on. Naming villages after qualities of the landscape, such as al-Al (the high or excellent place), Musharifa (the noble place), Saliha (the good or healthy place), has the same result. Virtually every place name has or has had a meaning.

Topographic, geographic and use terms often become conflated, so that each implies the others. Physical environments do not determine use, but the environmental facts of geology, soil types, temperatures, rainfall, dew and humidities are taken into account in using a landscape. The wider Bilâd ash-Shâm has its mountain spine of differing geological configurations and associated drainage systems, plains and valleys; gravel and basalt steppes; and the northern part of a sand-dune desert. In use terms, each has areas capable of development above its natural capacities; those useful for grazing and/or rainfed agriculture in most years; and those used for grazing only seasonally or more rarely. Restrictions on development above natural capacity come through limitations of soil type and available water, whether from rainfall or from storage over time.

Users speak of the parts of the Bilâd ash-Shâm in a mixture of topographical and historical terms: Damascus and the Ghuta; Qalamoun; the Shamiyya *hamad*; Hauran; Jabal al-Arab; the *harra*; the *aghwâr* (sing. *ghor* – a depression); Jabal Ajlun; al-Balqa and Jabal Nablus; al-Karak (and Khalil); ash-Shara; Wadi Araba; *as-suwwan* (flint-strewn steppe); al-Hisma; Wadi Sirhan; al-Basaita; al-Jûba; al Nafud. Qalamoun the southern Jabal al-Arab and the *harra* the *hamad* a section of al-Jûba and its environs al-Karak and a section of ash-Shera and Wadi Araba are known from personal research with the exception of Qalamoun, of which we know from members of the Rwala; it is included because the area has strong links with the *badia* which it borders.

Qalamoun is the name given to a series of mountain ridges, valleys and villages running more or less parallel to, but lower in altitude, and east of the Anti-Lebanon, north of Damascus. Winter rain and snow brought by the prevailing southwest winds on the summits and slopes of the higher Anti-Lebanon mountains and infiltrate into the subsoils. The geological structure, aided by the topographical unity of the gradients and directions of slopes, enables movement of water through the calcareous marls and conglomerate rocks with periodic sandy formations. The storage of this underground water has been helped by karstic erosion. The inclination of storage layers towards the bottom of the syncline, and parallel to the topographical slope, leads the water in the downstream direction, where it rises to the surface either naturally as in the gorges at Yabrud or artificially by the means of *foggara* or *qanat* (underground channels) known locally as '*sarab*' (Haj Ibrahim 1990: 296–7). The amounts of rain falling in the valleys of the Qalamoun are low; Haj Ibrahim quotes figures of around an average of 115mm per year from a ten year period, with Deir Attiya having the highest average of 145mm, so that irrigation for cultivation is essential. Precipitation on the summits averages 600mm, mostly as snow, while the higher plateau to the southwest receives above 200mm.

The water available for irrigation is augmented from an east-west structural fault linking the hydrographic basins of Wadi al-Majarr in Qalamoun to those of the Anti-Lebanon slopes to the west. The fault allows the numerous seasonal torrents of rainfall or snowmelt from the west and southwest to flow towards Deir Attiyah and Majarr al-Qalamoun and enter the underground water table. The western Majarr al-Assal occupies the higher plateau between two parallel ridges of Qalamoun, and separated from the middle plateau of the Majarr al-Qalamoun by the town of Yabrud with its springs. East of the Majarr al-Qalamoun, with its towns and gardens of Nebk and Deir Attiya, is another mountain ridge. Southeastwards is another series of small springs and *foggara* with gardens at Kutaifeh, Ruhaibah and Jerud. The areas of cultivated land irrigated by the spring-fed *foggara* are called '*ghouta*'. The typical Qalamoun steppe vegetation resembles that of the *harra* and *hamad* to the south, with *artemesia herba-alba* dominant and its associated perennials and annuals; scattered trees of *pistachia atlantica*, *amygdalus* and *cratageus azaroleus* are present.

The Majarr al-Assal and al-Qalamoun have been called the gifts of the Anti-Lebanon. The Jabal al-Arab (or al-Druze) is described as a veritable water tower standing high to the east of the Hauran plain. West of the Hauran is Jabal ash-Shaikh which waters al-Jaulan, the rocky western part of the Hauran. The Hauran is a fertile plain of red volcanic soils and basalt outcrops with rainfall usually sufficient for cereal production. The Jabal al-Arab at 1800m in altitude has a higher rainfall, and above 1200m its springs are perennial. Oak woodland with *pistachia atlantica*, *amygdalus*, and *cratageus* cover the western heights north of Suwaida, and isolated groups of trees grow on the mountain's eastern flanks. Variability between rainyears is high, and various forms of irrigation are used to ameliorate this. Braemar (1990: 453ff) gives detailed information on climate and irrigation in the area. Irrigation systems channeled rain and snowmelt runoff to fields or cisterns. The eastern and southeastern foothills fall away through ravines and wadis into the *harra* or black basalt steppe.

The *harra* lies east and south of the Jabal al-Arab; basalt deposits also cover the land north and north east of the Jabal but these areas are referred to as al-Ladja and as-Safa. The *harra* is punctuated by the remains of volcanic cones, of which the highest is over 1,000m and, with a series of associated cones, form a watershed, the Tulul al-Ashghaf. The drainage systems/wadis west of the watershed flow to the Shabaikha depression, Azraq or to the Qatafi, where they link with flows carried by the Rajil and its associated drainage systems from the east and south of the Jabal al-Arab. Those on the east flow to the Mingat system that continues north into Syria, ending in the Rahba depression; drainage systems from the *hamad* to the east also end here. Along the eastern face of the *harra* and so between the two drainage systems of the Mingat of the *hamad*, and the short but numerous and steep wadis from the eastern *harra*, are a series of wells and the semi-permanent water pool of Burqu', with a smaller series of pools at Wisad to the south. The western edge of the *harra* parallels a line from east of Bosra, Hallabat, Azraq and the eastern side of the Wadi Sirhan to north of Jauf; on the east, from east of as-Safa to Burqu'; then west of the Wadi Mingat and Khabra Abu Hussein to west of Turaif. North of the Saudi-Jordanian border, the basalt fields and volcanic cones become more scattered. More *harra* lies east of the southern half of the Wadi Sirhan.

Rainfall in the *harra* is below that needed for arable farming, except in the far northwest of the area. Even here rainfall is too low in one out of every three years. The wells and semi-permanent pools depend on rainfall and runoff water for their recharge, while rainpools (*khabra*) and partly aided natural reservoirs (*ghudrân*, sing. *ghadîr*) are directly dependent on rainfall in a relatively close area. Between the old volcanic cones and dense scatters of basalt boulders, areas of soil extend along drainage systems. Some are like meadows (*shi'bân*, sing. *sha'îb*) flourishing when watered by rain or runoff from upstream. Others have soil particles too fine and densely packed for plant growth; these act as rainpools (*khabra*) or as dry flats (*gi'ân*, sing. *ga'*) Excavated artificial pools, *mahfûr* (pl. *mahâfîr*), may lie between *shi'bân* and *gi'ân*. In some years, a few depressions accumulate enough soil-moisture from rainfall and runoff at the right times for dry farming. Runoff need not be a one-way flow of water but two-way, depending on rainfall amounts and intensities, the absorption capacity of soils, and surface gradients. Two-ways flows occur around Shabaikha, and explain the accumulation of soil moisture levels in some years; Shabaikha means inter-meshing, network, possibly reflecting the network of drainage systems criss-crossing the depression.

The Jordanian *harra* is associated with the Wassamat al-Bahel section of the Ahl al-Jabal. Tombs built for some of their dead look down from the high places to campsites along the *shi'bân*. The Ahl al Jabal use the *harra* for grazing goats, sheep, donkeys and camels. Goats, donkeys and camels thrive, liking the *harra*'s varied vegetation and winter shelter. Sheep find the *harra* hot in summer, and are often moved west to Ahl al- Jabal villages or the uplands; in winter sheep, not needing protection from cold or wet, are herded east to the *hamad* for grazing on early grasses and annuals. Ahl al-Jabal live in the *harra*, and others use the area; individuals marrying into the Ahl al-Jabal gain access. Others, like the Zbaid, have families who have long used areas of the *harra* as part of preferred seasonal movements; Ghayyath families use the northeastern parts; Sardiyya herding families move east through the *harra* to the *hamad* for winter grazing, returning west in early summer; some Rwala know the *harra* well. All these use the grazing and water in *khabra* and *ghudrân*, but never water at the wells, owned by the Ahl al-Jabal families who dug and maintain them.

The literature and non-users portray the *harra* as a difficult environment, hot in summer, limited in resources, and restrictive of movement. Its Ahl al-Jabal inhabitants describe it as a fruitful haven, sheltered in winter from the winds that blast the *hamad*, with water resources, and good and varied grazing through the seasons in most years along the *shi'bân* and on the black bouldered slopes. Winters can be cold with snow and fog, but when the sun shines the black boulders create micro-climates. Summers are hot, as the black boulders reflect the heat, and the change between day and night time temperatures is less than in the *hamad*. The fertile soil of many slopes, the micro-climates, and an increased water penetration into the soil from the stones, produces early and abundant annual growth – given rain. There is easy access to the eastern slopes of the Jabal al-Arab for summer water, and to the Rahba depression where some Ahl al-Jabal cultivated barley and wheat in most years. The depressions of Shabaikha, Mahdath and along the Wadi Jawa were also cultivated sometimes. Cultivable soil is called *zawîyat as-sauda*. In the villages of the Jabal and the Hauran Ahl al-Jabal sold salt collected at Azraq and dairy produce, and bought clothes, flour, coffee and tea, sugar, and household goods. Dates were collected from the oasis towns along the Euphrates, or from the oasis of Qurayat at the northern end of the Wadi Sirhan. In bad years, Ahl al Jabal took their herds to Jaulan or to the northeast part of the Wadi Sirhan; now, to the northern Ghor or the mountain slopes west of Ajlun and Jarash.

The *hamad* is the limestone plateau east of the *harra* and south of Qalamoun and the mountain chains between Damascus and Palmyra, stretching south to the al-Jûba depression and the Nefud and Dahna sand dunes. The *hamad* is a series of gently rolling hills and wide plains, gradually losing height from west to east. Many areas have a flint pavement, and limestone outcrops are frequent. Heavy silts and clay soils in low-lying areas are common, and stretches of windblown Nefud sand overlie *hamad* gravels and limestone and *harra* basalt to the south. The *hamad* drains generally west or east from its wide and hardly discernible north-south central spine of a series of isolated flat-topped hills rising from the plain; these narrow but long heights rise steeply on one side but shallowly on the other. The spine is drained by a series of shallow valleys, called *rijlât* where they flow into a main channel, or *rishât* when they lie in an undulating gravel plain.

Users typify the vegetation of the *harra* and *hamad* as similar, dominated by *shîh* (*artemesia herba-alba*). Some western locations have notable trees or shrubs, like Wadi Butm with its *butm* (*pistachia atlantica*); a few grow in Wadi Uwainid and on the edge of a *ga'* along TAPline; Wadi Rattam has flourishing *rattam* (*raetam raetam*); a few *za'rûr* (*cratageus azaroleus*) grow in Wadi Jawa. There is a large local corpus of knowledge of plant species, their morphologies, habitats and locations, and seasons for grazing, food for people, medicines for people and animals, and craft use. Plant names often refer to characteristics of the plant itself, its smell, forms of root, stem or seed, colour, feel, or taste, and to its other qualities. Plant names are not always translatable into Latin botanical names, since species botanically separate are not so distinguished by *badia* users who have different names for plants at various stages of growth, when dry, or the fruit. Tribes have different names for the same plant, or use the same name for botanically separate but visually similar plants. Apparent variations in name are sometimes due to consonant shifts between dialects.

Only those plants in the *harra* and *hamad* important for seasonal grazing are mentioned; spring flowers were seen (see appendix) in the exceptional spring of 1995, a *rabî'a at-tafha*. Most species have preferences for particular soil conditions. Some slopes are yellow with *khafsh* (*brassica tournefortii*), while others are green with the grass *sam'a* (*stipa capensis*), or a *sha'îb* pink with *shiggâra* (*matthiola*). Locally dominant species such as *brassica tournefortii* have their associated species; others, like erodiums, grow in most habitats. Where sand overlays the floor, as near the Saudi-Jordanian border, *artemesia herba-alba* is replaced locally by *rimth* (*haloxyletum salicorni*), *ajram* (*anabasis lachnantha*), or *ruta* (*salsola vermiculata*), depending on soil structure and drainage. On deeper sand in the Wadi Qattafi there is a stand of *ghada* (*haloxyletum persici*). Around al-Azraq and al-Umari to the south, as well as at other scattered locations with a higher water table and salt soil, grow tamarisk, *msha* (*nitraria*) bushes, *ghudraf* (*salsola volkensii*), reeds and tall grasses, and *roghol* and *roghaila* (*atriplex* sp). Most slopes have little perennial growth, but are covered with annual grasses and flowering plants if rain falls at the right times.

Badia users insist there is no over-grazing, since to do so would be self-defeating. Instead they emphasise the resilience of the vegetation. The physical structures and reproductive processes of arid zone plants are adapted in a variety of ways to their harsh and unpredictable environment (Mandaville 1900: 23ff). Local users comment on variations in root systems, speed of flowering and setting seed in some species, long periods of dormancy for others, plasticity of plants within a species according to conditions, speed of response to rain, and length of dormancy of seeds. They also point out that many plants are unpalatable when flowering or setting seed. Their patterns of grazing management utilise those areas that last the shortest time before those that remain for longer. Not every area of plant growth is grazed. Plants in ravines or in isolated patches of earth in dry mud-flats are ignored, and these serve as plant reservoirs, as Musil (1927: 182) also observed. *Badia* users insist that when there are sufficient rains at the right times, the vegetation will reappear, both annuals and perennials. The 1995 spring in the eastern *badia* justified their claims. *Shi'bân* known to have been ploughed for barley five and seven years earlier were thickly covered with a wide variety of valued annuals and regenerating and reseeded perennials. Ploughing for barley cultivation was minimal in the abundant spring of 1995, but undertaken in rainyears when the rain favours barley cultivation but not sufficient or welltimed for maximum growth of wild annual and perennial plants.

Plants are divided into cultivated and wild – although some of the latter are aided by people's efforts in distributing seed in suitable areas, burning, coppicing and grazing. 'Wild plants' are divided into 'trees, bushes and perennials' and 'annuals or spring grazing plants.' The emphasis is on grazing, wild foods for people, shelter, medicines for people and animals, cosmetics, water purifiers, water and soil indicators, crafts, soil improvers and fuel; plants actively harmful to people and animals are also noted, while there is a further category of 'plants of little interest'.

Badia users have a large vocabulary for rainfalls, amounts, timings, intensities, areas covered, and the saturation of surfaces, recorded by Musil (1928a: 7–10), and in current use. A star calendar indicates the timing of rain. The first star, Canopus (*shail*), appears in the first days of October, and rules for forty nights; Canopus is followed by the Pleiades (*trayya*) for twenty five nights, which are succeeded by Gemini (*al-jawza*) for another twenty five nights. This season of *as-sferi* lasts until the end of December. The season of *ash-shitta'* is ruled by Sirius (*ash-shâ'ira*) for forty nights, and then by Arcturus (*as-smâk*) for fifty nights; the rule of stars ends in mid-April. *As-saif*, early summer, lasts until early June, after which *al-gaith*, high summer, continues until early October. The hoped-for rains take their names from the stars or seasons in which they fall. Rains are divided into autumnal (*al-wasm*), winter (*ash-shitwi*), and early summer (*as-saifi*). *Al-wasm* rains are *as-shailawi*, *at-truwi* and *al-jawazi*. *Ash-shailawi* rain produces annuals; with *at-truwi* rain as well, grasses reach their full growth. *At-truwi* are the most important rains and the decisive factor for future grazing. *Jawazi*, *shitwi* and *smâk* rains are useful but grazing cannot be assured unless there are good *wasm at-truwi* rains. *Saifi* rain destroys the annuals, but ensures good perennial growth and fills the *khabra*, *ghudrân* and wells.

Types of rainfall include: *matar*, downpour lasting several hours; *ghayth*, rain of at least four days over a wide area; and *wabel*, a downpour for several days, flooding whole plains. The spring of 1995 in the *hamad* of eastern Jordan and northern Saudi Arabia illustrated the preferred rainfall pattern; rain fell at the end of October, November, January and late February, each time for several hours or days, and gave "the best spring for twenty years." Years with patchy or insufficient rain give patchy and short-lived springs, or no spring but "a year of want," *sanat al-jufâf* or *sanat al-mahal*.

There was only one permanent well in the Syrian *hamad* itself at Sab'a Biyar, with poor water; the wells at Mlosi some one hundred and twenty kilometres in western Iraq were also used, and those at 'Umari, south of Azraq. The semi-permanent pool of Burqu' at the west of the *hamad* provides water, while seasonal sources of water are the natural *khabra*, the enhanced *ghudrân*, and made *mahâfîr*. The *shi'bân* have, with sufficient rain, good grazing for camels, goats, donkeys and sheep; on the slopes, between the flint cover, grasses and annuals flourish. The rains give winter and spring annuals and perennial growth, while moisture held in the soils, aided by summer dews, brings an autumnal flush to perennials. The clayey soils have good water retention qualities, and while most *shi'bân* depend on runoff entering a shallow depression from which there is no exit, a few – for barley crops – have had earth dams constructed to hold back water.

Its people see the *hamad* as a good place to live; movement is easy, where it rains there is good grazing for sheep and camels, and water in *khabra* and *ghudrân*. People move with the seasons, indicated by conditions for use rather than calendrical progression. If there is no rain, there is no spring. Winter is cold, often below freezing at night, with northwesterly or easterly winds, and can be wet. Fog, snow and hail are common. Sheep cope well while camels are taken south to warmer areas. Summer is hot, dry and often windy, although the drop in temperature provides relief at night. Animals must be provided with water and feed, and shade for sheep. In the past, people usually moved out of the *hamad* south to the oases of al-Jûba and Taima, east to Karbala and Najaf, northwest to Jaulan or Hauran, or to the Marj and Qalamoun. Some, like Sulaib and some Rwala and Amarat families, said they remained in the *badia* using water in *mahâfîr* or rain fed wells in some years. Movement across national borders has become more difficult. People and their animals stay in the *hamad* longer as deeper wells have been dug by governments or individuals and dry feed is now available for sheep.

The *hamad* is used by groups at various seasons for sheep, goats and camels. In the late autumn, sheep herders move south, southeast or east into the *hamad* for the winter and spring; camel herders use the *hamad* in late spring, early summer and autumn. The Jordanian *hamad* is essentially a route between preferred winter and summer bases for many Rwala, as the *hamad* to the east was for the Sba'a and Feda'an. It is a preferred winter base for sheep herders from the Ahl al-Jabal, Sardiyya, Umur, Fuware, Jumlan, Ghayyath, Zbaid and Beni Khalid, while Beni Sakhr and Beni Hassan use its western part.

So why is the *hamad* associated with the Rwala (and the Sba'a and Feda'an) rather than with its winter users? Is it the assumed difference in political organisation between sheepherders and camelherders? Did camelherders have more military capability and impose their will on sheepherding groups? Or is it that the greater aridity of the *hamad* necessitates camel herding with its increased range so that areas between preferred winter bases (where camels calve) and summer bases (where there is access to water and markets) are essential? Rwala tribesmen use all three points in accounting for their 'rule' of the *hamad* (when they were there)[17], adding that for them the *hamad* is a crucial resource rather than a preferred asset. The *hamad* is seen as a nexus of communications and a focus of preferred behaviour by Rwala, with the two perspectives complementing each other. The *hamad* is 'where we can be free'.

In the *harra* and *hamad*, knowledge of water movements and the associated soil structures is important, so herders may predict the location and duration of stored water. In March 1993, a group of Rwala and a Muwaili, with considerable argument over points of fine detail, discussed water flows in the *hamad* of eastern Jordan. The consensus was that all the Ruwaishids eventually flow into the Mingat, which ends in the Rahba. Wadi ash-Shaikh, starting from Anqa, flows into the Hifna, and then joins the Jarra', itself flowing into the Mingat. A large number of small wadis from east and west flow into Ga' Abu Hussein, which flows north until it is full, when it reverses direction to flow south, and joined there by

---

[17] 'Rule' parallels presence plus effecting a purpose, seen in the metaphor of the 'rule' of the stars in rainfall.

the Dumaithat al Qsair and Dumaithat al Mahfûr; these continue southwards, are joined by another wadi system, and all of these flow into Khabra Athaman; Dumaithat at-Tais also flows into Khabra Athaman. The Athaina flows west, then north along the Iraqi border, and then into Iraq in a north-easterly direction. All these wadi channels have *ghudrân*, deepened beds that hold a store of water. When Anqa overflows, most of the water moves east and slides around the north side of Jabal Anaiza, but the water from the immediate western edge of Anqa flows into Wadi ash-Shaikh. Mahruta is a dead end, and most of the water enters from the south. Anqa and Mahruta both have collections of *mahâfîr*, and there are other *mahfûr* in Abu Hussein, in Anwar's Ruwaishdat, and in the Dumaithat. Burqu' is filled partly by underground water movement from rainfall to the west in the *harra*, and partly by flows from south and north, of which the northern flow is the stronger; when Burqu' is full, this flow reverses itself to move north. There are many two-way flows, mostly within wadi systems between sections based on gradients. The branch of the ar-Risha at the Rwala encampment of ar-Risha an-Nuri comes to a dead end on a *ga'* and *sha'îb* a kilometre or so to the south; for the water to flow on into the Ruwaishid to the south, the water would have to be about ten feet deep; since the ground is wide, the flow cannot become constricted, so water never reaches that depth.

The *hamad* continues south of the Jordanian-Saudi Arabian border. Just north of the border in both *hamad* and *harra* are areas of red-gold Nefud sand overlying rocky hills (*barqa*). The watershed, formed by a series of flat-topped ridges (*hazm*) running north to south, lies some seventy kms. east of Turaif, separating the *hamad* from al-Wudiyan. East of this, all water flows into Iraq. Northeast of Turaif is Khabra Bardawil, some fifteen kms. long, at the northern end of the al-Hor depression; there were *mahâfîr* at Sfai in the centre of al-Hor, but TAPline drilled a borehole and also used the site as an airstrip, destroying the *mahâfîr*. Ar-Rdai'aniya is a collection of five *mahâfîr*, which fill with water flowing from Umm Wu'al some fifteen to twenty kms northwest. The *mahâfîr* stand on the *ga'* and *faydhr* of ar-Rdai'aniya; a *faydhr* is a place where the water has no outlet, has soil, and therefore becomes a fertile depression. Water enters the *ga'*, goes westwards, then sweeps back eastwards and so enters the *mahâfîr* with the excess remaining on the *ga'*. Water may remain there for up to six months, and people spent the summers there. There are more *mahâfîr* at Katayfa and at

Aqrim, southwest of Turaif, in the eastern edges of the southern part of the *harra*; some kilometres north are the two of ad-Dawqara and a single *mahfûr* at al-Jirami. *Mahâfîr* are associated with areas of good grazing but no *ghudrân* for water; the *mahâfîr* can be seen as enhanced *khabra* or totally artificial *ghudrân*.

The *hamad* is used seasonally. Traditionally, sheep-herders left before summer. Camelherders, like Rwala, used the *hamad* in spring, summer if there was water, and autumn. The *hamad* was part of seasonal movements which varied for different herding groups, themselves flexible in personnel. Where people herded depended on the location of rainfall. If there was insufficient rain for grazing and water in the *hamad*, Rwala herding families used al-Labbah to the east, the Nefud sand dunes to the south, al-Wudiyan to the northeast, or Busaita and the Wadi Sirhan to the west. Rarely would none of these areas had sufficient rainfall. Al-Labbah has few wells, and its perennial grazing is predominantly *rimth* (*haloxyletum salicorni*) and *arfaj* (*rhantherium* sp). Al-Wudiyan's dominant perennials are *rimth* (*haloxyletum*) and *shîh* (*artemesia herba-alba*), like the *hamad*.

The *hamad* falls away into the al-Jûba depression; al-Jûba means a depression or pit. Al-Jûba is triangular in shape, its southern edge lying against the Nefud, and the eastern and western sides being the escarpments (*jal*) where the plateau falls away almost meeting at Shuwaitiyyah on the triangle's northern point. There are remains of volcanic cones, lava flows, and sandstone hills eroded by windblown sand. Al-Jauf and Sakaka are the largest oases and towns; some of the smaller oasis villages, like Twair, Qara and Swair, are old and others are new, like Nathiyim and Rifa'a. New oases are usually at the sites of traditional wells, but using deeper water. Soils, between rock outcrops and sands, vary but are mostly suitable for agriculture. Al-Jûba is warmer in winter than the *hamad*, with its lower altitude and sheltering escarpments to north and east, although there can be occasional cold spells with frost. Rainfall is very variable, and falls as heavy showers with occasional longer downpours, or light drizzle. Years of really good rainfall for an abundant spring are rare. It is very hot in summer, and can be windy; south winds cause real distress to animals, crops and people.

The Nefud to the south of al-Jûba is a sand desert with large dunes. People regard the Nefud, like the *hamad*, as a healthy and satisfying region. The crests of the dunes are often crowned with *haloxyletum persici* or *ghada* trees, while the hollows between the dunes are filled with plants, looking like miniature meadows, if there has been rain. The dominant vegetation between the dunes is *abal* (*calligonum comosi*) and *adhir* (*artemesia monospermae*). Locally, *rimth* (*haloxyletum salicorni*), *adam* (*ephedra elata*), *'alqa* (*scrophularia* sp), and the grasses *kasba'* (*centropodia forskalii*), *nasi* and *sobot* (*stipagrostis* sp), and *najil* (*cynodon dactylon*) are important. The long, almost semi-circular dunes, called *flûj*, sing. *falj*, are sometimes three to four hundred metres in length and forty to eighty metres high. Between the dunes are deep pits like funnels (*farsha*), with the deepest parts of the elliptical hollows (*ka'âra*) between the dunes. These hollows are usually orientated west to east, and at the east they widen into sandy flats (*nawâzi*). As Musil's guide told him (1927: 152): "In the Nefud there are roads everywhere, and yet in the Nefud there are no roads. Whosoever does not know the Nefud must not venture thither, and who loses his way in the Nefud loses his life too." There are occasional deep wells in the Nefud; various Rwala sections have wells along its northern, north-eastern, north western and western edges. Since al- Jûba is traditionally considered unhealthy, while the Nefud is seen as healthy, herding families using al-Jûba for water or markets preferred to graze their herds in the Nefud.

Al-Jûba connects with the Wadi Sirhan to the west by a sandy plateau bordering the southernmost hills of the *harra*. The Wadi Sirhan is a wide enclosed drainage basin, running southeast to northwest from about thirty kms. west of al-Jauf to al-Azraq in Jordan; the easy communications of the Wadi Sirhan continue northwest to Zerqa and north to Bosra and the Hauran. The lowest section of the Wadi Sirhan is at the former oases of Kaf and Ithra. There are many old wells recharged by drainage from both sides of the Wadi, although the western side with longer drainage systems provided more than the shorter eastern systems from the *harra*. New wells for agricultural development have been drilled to use deeper water at around 560m (Sudairi 1995: 7). There are now several farming villages along the Wadi Sirhan, and two towns, Tabarjal at the south end and Qurayyat which has replaced Kaf and Ithra, in Saudi Arabia, while Azraq is the Jordanian town at the northern end. Tabarjal is inhabited by Shararat, Qurayyat is largely Rwala, and al-Azraq has a northern centre that is mostly Druze, and the southern centre was established by Chechen, although there are now Palestinian, Ahl al-Jabal, Beni Hassan and a few Rwala inhabitants in both centres.

The Wadi Sirhan, with its water, oases, winter shelter, and grazing, was an important area for many tribes. The natural vegetation is like that of the Azraq and Umari depressions; south of al-Issawiya there are sand dunes with *ghada*. The western side was used by Beni Sakhr and Huwaitat, while the eastern was used by Sirhan, Sardiyya, and Ahl al Jabal, with Rwala and Shararat in the centre. In the summer, the area was used less because of the heat, mosquitoes and flies, and lack of grazing. Like other depressions with water not far below the surface, such as Azraq, al-Jûba and the Ghuta, the Wadi Sirhan was considered an unhealthy area.

As well as natural features and vegetation, the landscape has animals. Some are feared, some admired, others largely ignored; some are hunted, most are eaten in emergencies, others avoided. Many are used as personal names, incorporated into stories and used as images in poems, or carved on rocks, or became part of the repertoire of tattooing patterns. Oryx (*mahâ*) lived in the sand-dunes of the Nefud, although none now remain; gazelle and ibex are rare, but survive in isolated areas. Hunting by car with machine guns, often by people from outside the area, is blamed for their demise. Ostrich no longer exist outside reserves. Wild boar, lions and onagers disappeared earlier. Panthers and cheetahs are also rare, and prefer mountainous areas; a leopard is said to visit the Fainan area from Israel and one or two have been observed in the *harra*. Militarised boundaries are quite effective as animal reserves. Wolves, hyæna, foxes, hares, and wildcats live in small numbers in various parts of the *badia*, and move in response to the supply of food. Hedgehogs, porcupines, jerboa', jird and species of mice are common in some years.

Many bird migrants pass through the *harra* and *hamad* in winter if there are rainpools and feeding potential. Cranes, storks, godwits, avocet, lapwings, duck, geese, coot and terns are regular visitors, and some are commemorated in recent rock carvings. Ravens and small finches are habitual residents, while swallows and other insectivores pass through. Eagles and falcons appear in the late summer, especially steppe eagles. Little owls are common, while other owl species are visitors.

Local people regard the animals, plants and features of the natural world in a somewhat analogous fashion to the way they consider other human societies. Animals and plants are not people, the categories are quite separate and distinct, but there is little arrangement of categories in an hierarchical order; each item in the observed repertoire has identity, reputation and associates, and is seen as closer or more distant to the speaker and his audience. Closeness and distance depend on actual frequencies and intensities of interaction, and on the value of the interaction. The most distant items need not be physically far away but are without interest. A distinction is made between 'wild' and 'domesticated' for animals, plants, soils and water resources, but 'wild' and 'domestic' are construed flexibly and contextually. The terms *barri* (wild) and *baladi* (local, domestic) are used about groups of people as well as animals and plants. Townspeople see *bedu* as *barri* while a Rwala will say Shararat are *barri*, by which he means he is genealogically far from the Shararat, and as a group are of little interest to him, although he knows well several as individuals. A farmer will describe a plant as being *barri*, a weed, in a discussion on cereal cultivation, but as *aishb* (annual grazing plants), when talking about grazing for sheep on the same fields.

From ideas of closeness and distance between people, animals, and plants it follows that there are correspondences between how people treat each other and how they treat animals and plants and, indeed, how they see animals and plants treating people. There are commonly held ideas of reciprocity between humans and the plants and animals of the natural world. A Huwaitat herder east of Qadisiyya said "If people are good to the land (region), the land is good to them" i.e. if you treat the land with respect it will look after you. Azazme and Rashaiyida herders emphasised constantly the reciprocal relationships they saw between trees and perennial plants and their livelihood as herders; "the trees are the foundation of life; without trees there would be nothing", echoed by farmers around Ajlun and by bedouin in Upper Egypt (Hobbs 1992: 103ff). The Huwaitat remark quoted above carries additional meanings of man's capacity for destruction of natural resources from greed, which is a common theme among herders and farmers alike in the Bilâd ash-Shâm, among fishermen and herders in southeast Oman, and among the Ma'aza herders of Upper Egypt (Hobbs 1992: 102). Greed is regarded partly as unrighteous behaviour by man from time immemorial but exacerbated and in some ways demanded by living within centralised states. Reciprocity and treating many animals and plants as quasi-equal partners becomes overtaken by the need to intensify land use and production so as to participate in centralised political and economic activity. To an extent, reciprocal relations between partners in production is an ideal, but one that people constantly refer to and discuss. Production between persons known and close to each other are usually joint enterprises; those between distant persons are hierarchical in form and often exploitative in purpose.

The same sort of shift from closeness and reciprocity to distance and unequal relations is paralleled in herd animals or plant resource management. Even within a herd, some animals are closer than others, with the lead sheep, the ewes of the core matrilines, their daughters and the rams being closer than poor performing animals, or those bought in to increase numbers in good years – unless an animal in this category somehow endears itself to the herder. Between herds, milk flocks might be assumed to be closer to their owners and herders than meat flocks, since meat flocks are considered to be more commercially orientated. However, meat flocks still depend on matriline cores and produce dairy products for the household, while milk flocks produce for the market. Flying flocks, bought to graze off unharvestable barley crops, are initially not close and regarded as commodities, unless or until part of such a flock becomes a nucleus of a longer term household flock. Camel herders also regard some animals to be closer than others; personal riding camels and favourite milking camels are closer than other milk animals and baggage animals. In the *badia*, camels, sheep and goats would not survive without the aid of people in getting water from wells for them, while people drink milk rather than the often somewhat saline well water. This reciprocity is often remarked upon, and herders frequently comment that their animals "are part of the family". The ruthlessness shown to family members who fail to behave "properly" extends to its animals who persistently fail to perform satisfactorily – unless such an animal redeems itself in some way. Herd animals are *haywanât*, 'good', 'clean' animals who graze or browse, and chew the cud.

This category of *haywanât* (yet another word which can have different meanings in different contexts) extends outside domesticated animals to gazelle, ibex and oryx, wild animals living in social groups that are hunted, eaten, sometimes taken as pets, and who figure favourably in poetic imagery. Gazelle hunted en masse and driven into traps for commercial purposes have different images to gazelle hunted by saluki and falcon, or by a solitary hunter on foot where there is a face-to-face relationship between hunter and prey. Ostrich were hunted for the sale of their feathers, but by single hunters, and carry favourable connotations of speed, endurance and beauty in poetry. Dogs, wolves and hyænas go together as *kilab*, unclean animals, but with exceptions and ambiguities. Salukis, hunting dogs, are frequently not regarded as dogs, while guard dogs are unclean but valuable and valued. Wolves, like hyænas, are shot when they threaten flocks but can cross into human society, whereas hyænas only cross into human society in association with jinns who, by definition, are not human. This is the position in Rwala stories, other tribes may have other conventions. A well-known story about a Rwala and a wolf has variations, but turns on a young man raiding for camels accepting help from a wolf as though they were brothers; both contribute to the success of the enterprise and return to camp, where the wolf's part in the raid and his becoming a brother is announced to all; unfortunately, someone absent during the announcement shoots the wolf, who is avenged by his human brother.

Wild animals become artistic images. Many carvings portray animals with young, or groups of animals in motion, made from deep appreciation of the animals. Animals are the characters in stories and fables, such as those retold in Musil (1928a: 20–31) and now. Animal images denote bravery, speed, and endurance, and are used in love poems for the attributes of the loved one. There are also animal images used as wry comments, or as topical jokes in narrative poems. Animal imagery is frequently used in analogies, such as that recorded by Musil (1927: 132) of a Sharari comparing the Shararat to ravens and the Rwala to eagles. Such images have layers of meaning. Eagles convey ruling, for example, but ravens are better survivors; and both co-exist. Many sayings depend on animal images, such as "Shall a wolf tremble at the lamb's fart?" and "As the ostrich said to the bedu; when you come hunting birds, behold! I am a camel, but when you come searching for your camels, see! I am a bird" (quoted by Doughty ([1888] 1936: ii, 155) from a man explaining his relations with the Turkish authorities). People also have their own stories about animals, drawn from personal experiences.

Plants, too, are liked not only for their useful qualities, but for sweet scents or shade provision, or for their endurance in harsh conditions.

Domestic animals are held in affection. Many name their camels and sheep, even if only by a physical or behavioural characteristic; riding camels, horses, some guard dogs, donkeys and many lead sheep have names, often of desirable qualities. Men sing to their camels, with clear bonds of affection between them. Rams are decorated, and lambs, kids and baby camels adorned and played with. Goats, cows, mules, horses, donkeys, dogs, cats, hens, geese and turkeys all have their place in their owners' regard. There is a reciprocity between people and their domestic animals, in which animals are recognised as individuals of their species who are both part of the family in production and consumption, and with personal characteristics and reputation. Domination over animals is rarely expressed; more common are comments of balanced or generalised partnership. Such attitudes to animals and plants, wild and domesticated, are widespread in the countryside of the Bilâd ash-Shâm and are like those of the Ma'aza of Upper Egypt, whose detailed schema of the animal kingdom are laid out by Hobbs (1992: 87–90).

Like al-Jûba, and Qalamoun, but unlike the *harra*, *hamad* and Nefud, the Karak plateau and its environs is a place where people live all the year round and cultivate, although they move within the region over the seasons with their herds. The plateau is part of the general mountain spine of the whole western region, but here the uplifted sedimentary limestones have been covered with fertile red soils of volcanic origin to different depths. Limestone outcrops are common and important for the constructions of cisterns for water storage. The plateau, which usually has winter rainfall sufficient for cultivation, is cut by wadis or shallow drainage systems which flow into the main drainage systems that delineate the region, with the Wadi Mujib on the north, the Wadi Nukhaila to the east, Wadi Hasa to the south, and on the west, from south to north, Sail Khanzaira, Wadi Karak, Wadi ibn Hammad and Wadi Jarra. The slopes of the plateau are in some places precipitous, but in others open out into wide bowls, a series of shelves, or flat-topped spurs with soils of varying fertility. Aquifers in the limestone break out as springs and seeps, which vary in flow; local people account for this from fluctuations in rainfall both locally and in the wider region, and by the over-use or new development of other springs along the aquifers.

Al-Karak is divided from al-Balqa to the north by the spectacular Wadi Mujib which unites two other drainage systems, one from the east, and the Nukhaila from the southeast. The Nukhaila marks, approximately, the change from the soils and rainfall of al-Karak to those of the *badia* to the east. The Wadi al-Hasa divides al-Karak from Tafila and Shobak, while the western slopes of the plateau fall down to the Dead Sea and the *aghwâr* of Haditha,

Mezra'a and as-Safi. Karak is the chief town, and in the early nineteenth century there were only three permanently inhabited villages in the plateau, all in the hilly region southwest of Karak with strong commercial ties with Hebron (Khalil) and Gaza. Many small towns and villages have since developed around the summer cisterns and lands of the various tribal groups of the plateau.

The plateau is one of a series of environments, whose variety is enhanced by the changes in geology and in altitude over a relatively short distance. In the west, the land is 300m below sea level; 25 km. east, the plateau is at 900m, and falls away to 300m another 35 km. further east. Access to different resource areas was necessary for livelihood before there was employment in the services of a centrally administered state. Immediately east of the Dead Sea are the irrigated lands, the *Aghwâr*, warm in winter and extremely hot in summer. The wadis Mujib and Hasa flow all year, though reduced in summer. Along the gravel slopes bordering the Dead Sea shore and further south, beyond Safi, are areas of *tlah* (acacia), *tarfa* (tamarisk), and *sidr* (*zizyphus spina-christi*) forest. *Ushr* (*Calotropis procera*) trees line the road by Bab adh-Dhra'. *Sidr* (*zizyphus spina-christi*) trees provide shade in the cultivated fields, while *difla* (oleander), *tarfa* (tamarisk), *nakhla* (palm) and *ghusub* (canes) grow at every seep or spring. Musil (1908: 3) describes the area at the turn of the century.

Traditionally, the *aghwâr* were cultivated in the winter, harvested in April and the inhabitants then moved up the slopes with their herds to the plateau for the summer, returning in the autumn. At present, this pattern is followed to an extent, but arable crops have been largely replaced by commercial crops of citrus, bananas and tender vegetables like tomatoes, peppers, beans and aubergines. The Ghawarnah, the people of the *aghwâr*, are cultivators who also have herds of cows, goats and sheep. People from the plateau, some of whom move to the valley for the winters, own or rent farms there. The smaller wadi systems, the Mujib, Jarra, ibn Hammad and the Nukhaila, were cultivated in similar ways; the land was 'owned' and used by people from tribes now based on the plateau. In winter the valley bottoms were used for grazing and for arable crops, in summer they were empty.

Wadi slopes were used in winter for grazing, since they were sheltered from the cold on the plateau, and for arable crops in the spring, when the animals were taken east. Around the springs grow *hamat* (wild figs), *nakhla* (feral dates), *sidr* (*zizyphus spina-christi*), *tarfa* (tamarisk), *haisabân* (*moringa peregrina*) (only in Wadi ibn Hammad on a sandstone rock face), *kûli* (willow) at Ain Al-Jûbaiha, and *rattam* (*raetam*) bushes, with *difla* (oleander), *halfa* (*imperata cylindrica*), *ghusub* (canes), *shrayt* (reeds) and *habaq* (a mint-like plant) by the waters' edge. Higher up the western or northern slopes are occasional *butm* (*pistachia atlantica*), *lwaiza* (*amygdalus*), and *sirr* (*noaea mucronata*) trees. Beni Hamida in Sirfa said that the "slopes used to be black with trees, blacker than Jarash or Ajlun", and that the Turks destroyed the oaks, *butm*, *seyal*, and *sidr*, supported by Musil's account (1908: i, 88) of oak trees below Sirfa being cut down for building government offices in Karak. More trees were lost to townspeople needing firewood, building wood and charcoal, and after the fifties by clearing land for the rapid expansion of agriculture. The plateau has some areas of grazing where the main perennials are *bilân* (*poterium spinosum*) and *shîh* (*artemesia herba alba*).

Summer grazing plants in the Wadi ibn Hammad from the permanent water up to the plateau included *sidr*, *rattam*, *butm*, *hasaibân*, *hamat* and *nakhl* trees; *shibriq* (spiny restharrow); *kutaila* (an unknown plant); *alayg* (similar to *capparis spinosa*); *chitâda* (*astragalus spinosa*); *najil* (grasses generically); *qattâf* (*atriplex leucolada*); *mrâr* (*centaurea sinaica*); *hamadh* (? *salsola schweinfurthii);* *agûl* (both *prosopis farcta* and *al-haji maurorum*); *ajram* (*anabasis articulata*); and *shîh* (*artemesia herba-alba*). People moved up to the plateau away from the mosquitoes at permanent water sources, and the heat. On the plateau they used water from their cisterns, and perhaps harvested a second crop. Also on the plateau in the summer were Beni Sakhr, Beni Attiya and Huwaitat camelherders who wintered in the Wadi Sirhan.

The differentiation made between environments continues, but with a shift away from a concentration on livestock and arable towards the demands of urban markets and employment within the nation state. Gubser (1973: 50–1) finds a correspondence between the best (because of highest rainfall) red soils and politically dominant families, with less important tribal groups having poorer (because drier and yellower) 'white' land, or, as in the *aghwâr*, with the politically dominant families owning the land and Ghawarnah doing the work. The relationship is not so clear-cut. Observers see an environment with crops, villages, tents, orchards, sheep, irrigation pipes and ponds, red earth and white earth, sun-dried land (*ardh shamsîyya*) and irrigable land (*ardh rayyân*).

The landscape is drawn and read by its users in political or moral terms, usually the reverse of each other, often largely rhetorical, in historical terms, or in a mix of all three. Many people read their landscapes in terms of their predecessors' preferences for particular forms of livelihood, the choices made by these and themselves, in relation to those of others of their family and tribal networks, together with the machinations of those outside their networks, all in the context of what is possible at any one time.

While a named entity, Karak and its people are part of the wider region, with families and tribal groups having kinship links up and down the mountain spine from the northern Hijaz to southern Syria, west across the Jordan to Hebron and Beersheba, and east into the *badia*. Trading links connected the people of the plateau, with their produce of sheep, horses, dairy products, grain and raisins, with urban centres in Palestine, Syria and the Hijaz, and with *badia* tribes. Contemporary networks often focus on service in particular army units, or employment in ministries, as well as on shared school and university links; if these are underpinned by ties of kinship or connections through women, such networks may be channels of influence, trade or entrepreneurial opportunities.

The northern section, al-Hishe, of the Shera mountains is another area of varied environments read by its users as a continuous landscape. In the west, below sea-level is the low-lying and sandy Wadi Araba; the precipitous and tree-covered sandstone western escarpment of the Shera rises abruptly to 1,500m at Jabal Qadisiya and at Shobak, but is 200m lower on the narrow limestone plateau which then gradually descends to the *badia* in the east. The western mountain face is deeply cut by Wadi Dana and Wadi Ghuwair which join at Fainan to form Wadi Fid'an. Although rainfall amounts are approximately the same as those of Karak, rain falls as storms rather than steady downpours. Springs and seeps come to the surface where the pressure of stored water, or gravity, works on the interface between permeable and impermeable layers. Major permanent springs are at Dana, Ain Lahdha, and below Shobak, in the mountains, and at Ain Fid'an on the low-lying plain; there are lesser permanent springs in the foothills at Ghuwaibah, Ain Sulamani, and at places along the Wadi Dana, Wadi Dahel (a doline system), and Wadi Ghuwair systems. The wadi systems flow during and after rain, greater at the higher elevations, and store water along their courses. The Fjaij plateau south of Jabal Qadisiya has good soil for arable crops, but the area is better known for its fruit and goats than grain and sheep. People say five rains for the plateau and slopes, or five flowings for the low-lying land above the Wadi Araba, are needed for good crops and natural grazing; these occur on average one year in five. People expect a good year, two or three ordinary years, and one or two poor years.

The changes in altitude produce zones of vegetation, also affected by rainfall and soil types. The low-lying gravels in Wadi Araba, around small seeps, carry *tarfa* (tamarisk), *difla* (oleander); the Hamra al-Fid'an sands and gravels have *ghada* (*haloxyletum persici*), *tlah* and *samra* (*acacia* sp), and the important grazing grass *thumâm* (*panicum turgidum*). In Wadi Ghuwair and Wadi Guwaibah *rashrâsh* (black poplar) grows at the water's edge, with *safsâf* (willow), *sidr* and *samra* (*zizyphus spina-christi* and sp), *tarfa* (tamarisk), *difla* (oleander) with *ghusub* (canes). By the seeps and springs in the drainage systems grow *zizyphus*, tamarisk, *hamat* (wild fig) and *nakhla* (feral date palms), with *yâsir* (*moringa peregrina*) in deep cracks on rock faces. Feral date palms also grow at the salty seeps at rock outcrops in the sands. Further up on the slopes the important grazing plants are *rattam* (*raetam*) and *ajram* (*anabasis lachnantha*); higher again, the trees are *butm* (*pistachia atlantica*), *ballût* (*quercus* sp), *lwaiza* (*amygdalus*), *samr* (*cupressus* sp), *'ar-'ar* (*juniperus*) with *artemesia herba-alba* an important grazing constituent, and *kharûb* (wild carob) at springs. Eastwards on the plateau are single oak or *pistachia atlantica* trees. With the great variety of grazing, medicinal and other plants at all altitudes and soils, people characterise vegetation by the most visible and useful species. Incidental queries as to the identity of unknown plants always resulted in a name and information on its preferred habitat. *Samh* (*mesembryanthemum forsskalei*) needs good rains and grows only on coarse sand overlaid by fine gravel; *qathîm* (*helianthemum lippii*), *sulaimiyya* an unidentified plant with medicinal uses and *jarrâd* (probably *gymnocarpus*) were all associated with *ardh baidha* or white, limestone soils; *rimth* (*haloxyletum salicorni*) grows on shallow sand; *arta* (*calligonum* sp) and *thumâm* (*pennisetum divisum* or *panicum turgidum*) need deeper sands.

The inhabitants of al-Hishe identify with their landscapes, which are drawn and read from past associations and present acts. Cores of each tribal section, rarely the sum of all members, have prior claims on specific water and irrigated agricultural land resources developed by members, and seasonal use of locations in various ecological areas. Tribal groups currently using al-Hishe were present when Burckhardt in 1812 and Irby and Mangles in 1818 traveled through the region, although each group says families both accreted and left over time. Sa'idiyin camel and goat herders, often politically part of the Huwaitat, had their wells along the Wadi Araba and used the eastern side of the Negev. After 1948, Sa'idiyin on the Jordanian side of the border were unable to use the Negev, and concentrated on herding and military service in Jordan. Since the seventies, members have developed irrigated farms around their wells at Tlah, Guraigira and Bir Madhkur among other places, and between Safi and Gharandal; goat herding remains important, using the Wadi Araba in the winter, the slopes of the mountains in spring and autumn, and summering around Shobak and Wadi Musa. Amarin and Menaja'a, from Ibn Gad Huwaitat, were also camel and goat herders providing services to traders and caravans between Palestine and the Hijaz. Their western base was Rakhama, southeast of Hebron, and the eastern was Bir Dabagha in the mountains south of Shobak, while they had easy access to the northern Hijaz where other sections of the tribe were. The very small groups of families of Usayfat come from Beni 'Attiya who, with the Huwaitat, are the major tribes of southern Jordan. The Usayfat lived on the western slopes of the Shera and around Shobak. The Mal'ab section of the Rashaiyida live in the Wadi Ghuwair system, while other sections own wells (and now gardens) further south in the Wadi Araba. In the last ten years, in their tribal lands the Mal'ab have developed gardens at Fainan using water from the Ghuwair system at least in part to forestall people from Shobak developing gardens there and so being able to claim the land and water; the point is that the south bank of the Fainan and the Ghuwair are Rashaiyida tribal lands, and the Shawabke (inhabitants of Shobak) could purchase land as individuals but not as a named group, just as the Shawabke own the land around Shobak where Rashaiyida can own land only as individuals.

These groups say that herding was or is their main livelihood; agriculture was secondary and for household use. 'Ata'ata of Dana and Sa'udiyin of Busaira say they were always farmers who also herded. Both are mentioned by the early travellers in their present locations, and at that date were said to be Beni Hamida. Musil (1908: iii, 62) reports that while Sa'udiyin maintained a Beni Hamida connection, 'Ata'ata had not. Accretion and dispersal of families was common to both. 'Ata'ata own Dana village with its springs and terraced gardens; their arable land lies on the plateau to the east, where they built cisterns and storehouses. Around the main group of cisterns by the main road, the King's Highway, the new 'Ata'ata village of Qadisiyya has developed. They moved west down the Wadi Dana in the winters for herding. The three stranded resource strategy of herding, arable and garden/orchard cultivation was successful because each part uses different natural resources, demands peak labour at different times of the year, and because each family was flexible in providing labour to the different enterprises.

Serahin Azazme use Fainan and the Wadis Ghuwaibah and Dana, moving between the Araba in winter and the plateau east of Qadisiyya in summer. They came from the Negev after 1948, and although Azazme are recorded as having used the Shera and the Araba at least since the early part of the nineteenth century, they lived in the Negev where they had their wells. They do not 'live' in the Hishe, nor do they 'own' land or water, except as individuals. A very few own small pieces of arable land around Shobak, bought by fathers or grandfathers in the forties. Serahin herd goats and camels almost exclusively, while a few sharecrop arable land or take part-time employment. Herding needs access to water and browse, both freely available; nothing has to be 'owned', access is what is required, and this is available because 'there is room for us'.

Al-Hishe is a good place to live. An Azazme, sharecropping for a Sa'udiyyin south west of Busaira, said "Life is fine; we have good, clean air; good water in the cistern, grazing for our goats. We have everything we want." At Dana, an 'Ata'ati commented while looking out over the gardens, "This is a good place. Clean air, good water, our gardens and the animals. What more could we want?"

People describe their landscapes in terms of possible livelihood or politics. The Agaydat of the central Euphrates characterise their various riverain landscapes in terms of locations and harvests that are certain or chancy, and which need active treatments or not to achieve these harvests (d'Hont 1994: 41, 44). In the landscapes described above, such categories are implicit rather than explicit. Political idioms of the past were raids and battles, victories and defeats; current idioms are land registrations, fraudulent claims to land, purchases, sales and gifts of land. Every inch of land, whether it be prime agricultural land, the smallest seep, or a stretch of dry grazing in the *badia*, is enmeshed in narratives which can be used to explain present use or to develop a possible claim.

There are also stories which have, apparently, no such application. One of these is about the mountains in al-Hishe. "There was a king, a long, long time ago, called Fainan and he sent a present of watermelons to the king of Kula (a plateau south west of Busaira). And unfortunately the watermelons were not at all good inside, so the king of Kula took offence and was angry with king Fainan for sending a bad present. So the king of Kula sent a present to the king Fainan, in boxes loaded on camels, and the camels each carried two boxes, one on each side. In each box was an armed man. As there were so few men to be seen with the camels, the people of Fainan let the camels and their guides in. Whereupon the armed men leapt out of the boxes and slaughtered all the people of Fainan. And Jabal Khalid is so called because the king's son, Khalid, was killed at the bottom and he was buried at the top. Jabal Fatma gets its name because the king's daughter Fatma was killed on the slopes. And the names of Jabal Safra and Jabal Hasaya commemorate the places of the killing of his other two daughters." Is this a story to amuse the children, does it have memories of a long ago quarrel, or is it a metaphor to show that the social relationships binding ecological regions can go wrong? Who knows?

The stories of the past associating ownership of a particular area with a successful armed attack or defence recall the achievement of recognised ownership by the actions of its people in terms of the accepted code of tribal society, where honour means autonomy defended, and autonomy implies the right to livelihood and its means – land, animals, women – are to be defended by force of arms if need be. Such narratives have a basis in events that actually happened and with actors who existed, but the historical facts of chronology, exact causes, consequences and personnel become simultaneously inflated, deflated and conflated depending on who is telling the story, to whom and for what reason. There are many versions of each story; every narrator is interrupted by cries of "Lies, all lies" and alternatives for every detail. All agree that "there can be no correct version, as each man tells what he thinks is right, and only God knows the truth." The only exception is when a narrative is told at a formal gathering when all statements have a political content, and solidary consensus is the norm.

Examples can be found in all the landscapes, but Karak provides a good illustration. Recent – since the sixteenth century – history records the land of al-Karak to have been lived in by the ahl al-Karak, the tribes of Karak. These divide into the Karakiyyin, the politically dominant leading tribal families and their tribespeople; by the nineteenth century these were divided into the eastern and western alliances, and neutrals. The Majali led the western alliance, and Beni 'Amr and Beni Hamida were neutrals. The Majali are from a Hebron family who had always traded with Karak, and these particular individuals fled to Karak to avoid vengeance after a killing. At that date, the Beni 'Amr ruled Karak. Due to internal dissensions over the meaning of ruling and its honourable expression, the Beni 'Amr had split into several parts, most of which left the area. The Majali and their allies among the Christian tribes fomented trouble between the Beni 'Amr and the Beni Hamida, resulting in the Beni Hamida and Majali fighting and defeating the Beni 'Amr. Then the Majali and Christians turned on the Beni Hamida and defeated them. Present distribution of agricultural land gives the best to Majali, the next to the Christian tribes, while Beni 'Amr and Beni Hamida have lands on the northern ends of the plateau on poorer, drier soil, and one group of Beni 'Amr in the southwest corner of the plateau. Majali account for this distribution by success in battle, battles triggered by Beni 'Amr tyranny, and Majali gifts of land to the Christian tribes as a reward for support. Christian tribes agree the land distribution reflects victories in war, but that their lands came by right, from their participation as equal allies, not as gifts. Beni Hamida comment they have always held their current lands, although they had more lands to the south, and their lands are not inferior to those of the Majali since they herd and farm, while the Majali and Christians are interested only in market orientated agriculture. Musil (1908: iii, 86) says that earlier Beni Hamida and Beni 'Amr plateau lands (mentioned in the Ottoman registers for 1538 or 1596, Bakhit and Hmud 1989, 1991) were cultivated by Karakiyyin, who acquired much Beni 'Amr land through debts. Beni 'Amr agree, adding that much of their tribal arable land on the plateau was shared with Beni Hamida and lost to the trickery of those who were agents in the government organised land registration (the Majali). Individual Beni 'Amr own small pieces of land in areas traditionally associated with tribal lands.

The idiom changes from acquisitions in battle to acquisitions of registration, from the accepted public language of tribal society to the official terms of a bureaucratic state where land is centrally controlled through a Land Registry, and all undeveloped land is state property. The two systems overlap, or rather the tribal idiom returns to life if the state system is considered to ignore just claims. A long-running dispute between Mal'ab Rashaiyida and Shawabke revolves around the nature of ownership and access to irrigable land on the south bank of the Fainan. The Rashaiyida position is that this is their tribal land, just as the arable land around Shobak is Shawabke tribal land, and each has a right to land in the other's domain only as an individual through purchase, marriage gift or inheritance, or access by sharecropping. They say they made an agreement with the Shawabke over the disputed land; the Rashaiyida would farm half themselves, while the Shawabke had the right to sharecrop with the Rashaiyida on the other half. That is, the Rashaiyida owned all the land, but gave the Shawabke access to half of it. The Shawabke claim that half was given to them outright so they owned it. In 1992–3, the Rashaiyida were cultivating two areas, using gravity fed water piped from the Wadi Hammam (the downstream section of Wadi Ghuwair). In the autumn of 1993, the pipes were extended west to a new area where water catchment basins were built and land ploughed and planted. This aroused the Shawabke who had done nothing about their option. They took their complaint to government offices at Aqaba, who contacted the army to send a bulldozer from the Jordan Valley Authority to clear the just ripening tomato crops. The Rashaiyida were furious at the prospect of their investment in pipes (nine kilometres at c.£1,000 per kilometre) and their crop being destroyed. They stopped the bulldozer from ripping up the crops and pipes by standing in front of it, trying to set the bulldozer on fire, and puncturing its tyres, while the shaikh of the Rashaiyida and his cousin went to Aqaba and Amman to see officials. More Rashaiyida arrived from Karak by pickup, all bearing rifles. In spring 1994, the tomato crop had been harvested and sweet melons planted, while the bulldozer had returned with more threats and counter-threats. The affair has now been settled through Crown Prince Hassan, with the Rashaiyida position upheld. The Rashaiyida feel that not only were they in the right, but that they acted more honourably by being ready to defend their land by arms, and seeking the mediation of the King, whereas the Shawabke acted at secondhand, pushing paper around the bureaucracy and getting others to act for them.

The public codes of state and tribe are the public face of social unity. Everyone accepts the conclusions, even if grudgingly, and understands the reasons for the conclusions; they are made in the public idiom which is one more way of confirming a social landscape. Underlying these are less visible but equally valid individual or group methods of achieving recognition of resource ownership, but which are not marked by the exercise of public armed power.

Giving resources, sharing or reaching a compromise over access to their use are in fact common and are honourable behaviour, but not the stuff of public poetry or narrative. Marking landscapes by narratives and poems recalling past events and actors is common, but for such works to reach a larger audience than their author and his immediate listeners requires particular associations of events and actors. The actors must be widely known, and from more than one tribal group; the events must resolve, either temporarily or more permanently, a situation of disorder – usually caused by improper behaviour such as unjustified aggression, greed, a lust for personal power without regard for the autonomy of others, moral or physical cowardice. Such narratives and poems that commemorate the re-establishment of the moral premises have to come from a situation where these premises were in abeyance. Continuing proper social practice does not generate poems or narratives that become subsumed under the rubric of tribal history, but instead confirms the reputation of the practitioners as 'good people'.

An example is the giving of land by the Beni 'Amr to a small group of Beni Ogba when the latter returned to Karak from Israel. The Beni Ogba were the important Karak tribe of the Mamluk period, from whom come the Beni 'Amr. At some point some Beni Ogba went to the Gaza region, where over time the original Beni Ogba families accreted other families around them; these families are known as the Beni Ogbiyya. The group who came back after 1948 consist of one family of Beni Ogba and the rest are Ogbiyya families. The Beni 'Amr agreed to give the returnees a share of Beni 'Amr tribal land, on which there is now a village. There are many stories like this, which tell of the sharing of natural resources with incomers, although there is a frequent twist of the incomers taking over particular resources such as the right to protect travellers or the leadership of the group.

Reasons for movement of groups vary, often because of irresolvable disputes with their former associates, from difficult economic conditions from climatic causes or shifts in trade patterns, or tribal or state wars. There are recognised procedures for making agreements between contracting parties for long or short term stays using the resources of the host. Sometimes payment is made, sometimes the arrangement is reciprocal. Individual families can also make similar arrangements. The facts of the landscapes in which people get their livelihood demand mobility between ecological zones, and sometimes outside the habitual range. Landscapes are social facts, constructed out of their users' interpretations of the facts of the physical environment and historical geography through social practice. While there is an overall acceptance of general trends of the social facts of use and ownership, there are diverse applications of these by social groups to the environments within which they operate. The resilience of the social practice concerning landscape relates at least in part to variations between ecological zones and variabilities seasonally and over the years.

The marking of landscapes by users through their activities has been indicated in the brief descriptions of the areas above. Al-Karak and al-Hishe inhabitants have been shown to use past and present political actions to explain specific ownership and use of some resources, in all cases of agricultural land and its waters. The association of an area, perhaps at one season or another, with a group is also accounted for by narratives and poems; the stories of the Beni 'Amr retold in Musil (1908: iii, 70–84) frequently mention that something happened at a place "and from that time people call that place the place of x". This is also the way that Rwala, for example, mark their landscape. Visible signs are wells with tribal and sectional marks carved on stones, former campsites, cains, topographical features named by Rwala for events and personalities, castles and markets; in addition, there are now villages, dams and gardens, quarters of towns, and radio masts. A group of fifty wells at Mughaira belonged to various sections of the Mur'ath, and each well was named; one was called Shabbibiya and belonged to the Nayyif, other names were Qatta'iyya and al-Khalaf, and a well with a stone drinking trough dated 1390 AH /1975 AD belonged to the Ma'ashi al-Jabr. A *ghadîr* in Wadi ash-Shaikh, north of Jabal Anaiza, is known as Ghadir ash-Shaikh since a Sha'alan shaikh cleaned out the debris and enhanced it as a water catchment. In the region of Turaif are the Qusur Sattam rock-cut pools where Sattam, a former Emir of the Rwala, liked to spend high summer. An old area of Taima is called Suq (market) al-Rwala, and in Zarqa in Jordan, the area by the bus station was/is known as suq ash-Sha'alan, as Emir Nuri ash-Sha'alan spent summers there early in the present century and guaranteed the market. The castle at Kaf in the Wadi Sirhan is referred to among Rwala as Qasr Nawwaf (the son of Emir Nuri). The service bases established by various Sha'alan in the eastern *badia* in the late sixties were differentiated by the names of their founders, so that there was ar-Risha Anwar and ar-Risha Nuri. New villages sometimes name themselves after the section of the majority of their inhabitants; an example is one called Nathiyim (an enclosed valley with good soil) or Faydhr al-Mu'abhil (the fertile valley of the Mu'abhil section). Two high radio masts on the road to Iraq are known as Sha'alan One and Two.

The way in which they talk about their landscape adds cumulatively to the marking. Some of this verbal marking is recorded in poems, and poems are used to confirm use of areas by people; examples can be found in the poem by Yusuf ibn Mjaid and transliterated and translated by Musil (1928: 580–6) or in the poem by Fahad ibn Sbaih in Musil (1928: 579–580); both poems are well-known. Any story being told demands the exact location of every event and each participant, while the teller is constantly interrupted and corrected by his listeners. Everyday conversation is marked, to an extraordinarily high degree, by the recounting and commenting on locations, routes and personalities. Yet to view, unaccompanied, the *badia* landscape with which the Rwala associate themselves is to see little except the natural features, perhaps the remains of a fireplace or campsite, a tribal mark on a stone, or tyre marks on the gravel. Musil (1927: 402) during his journeys in the *hamad* remarks that "It takes a long time to exhaust all the talk which such an abandoned camp furnishes."

The Ahl al-Jabal mark their *harra* landscape with structures. The eye is struck by the cairned tombs (Lancasters 1993b) on hilltops above the *shi'bân*, proliferations of black walled corrals, modern carvings of prayers at modern graves, and graffiti, animals, lorries and jet fighters on rocks above the grazing lands. There is indeed a superfluity of stones and boulders, and much of the basalt is easy to carve on; but the *hamad* has stones of flint and limestone, used by Ahl al-Jabal when present. As with Rwala, conversation is marked by establishing the present whereabouts of relations and former neighbours and their herds, and the preferred seasonal camping places of particular families are well known. Directions are given in the same general way, lining up landmarks, and noting changes in soils or vegetation, and marks on stones showing turns to campsites of former seasons. Stories are told in which exact locations of events are important, as they are for the tribes of Karak and the Hishe. A landscape is created by past deeds and maintained by present action, and although there are common themes as to how this is expressed, there are also variations between tribal groups. At its simplest, Ahl al Jabal often use visual images, Rwala use more verbal images.

The construction of landscapes by their users and developers, and conscious association between land and people has been made in spoken (and remembered) and in written (and recorded) words. The re-use of these associations is accompanied by metonymy and metaphor, where the group can stand for the place and vice versa, and the whole for a part, and the reverse. Reading recorded speech and hearing remembered words ignores the seasonal uses of land by different groups. The spaces of an environmental landscape are created into localities by social action; hearing narratives and seeing graffiti portray this. As Layne (1994: 149) tellingly argues in a somewhat different context, spaces are not defined by places but by people. Bianquis (1986: 1–14) discusses the construction of space in the maps and geographies of mediaeval Islam. Maps showed a schema of halting stages on journeys, rather than portraying space and natural features. The geographers conceived rather of many concentric spaces, defining solidary groups. A rather similar argument is followed by Ghazzal (1993: 13), when describing the seventeenth century topography by ibn Tulun of his birthplace, the Damascene suburb of Salihiyya. Ibn Tulun focuses on the 'lieux communs' of the town, with their contiguous spaces that no official power dominated, so that each appeared as an ensemble of autonomous professional and confessional groupings, usually unrelated. Each grouping had its own genealogical line of saints or *ashrâf*, etc. which went back to the Prophet. The constructions and transmissions of these lines were the main work of historians and biographers of the period, and allowed the legitimisation of power of each grouping and between them, through the events of the origin, maintenance and prestige. The mediaeval Damascene historian Ibn al-Qalanisi (ed. and trans. 1959) presents his description of Damascus in a similar manner. These parallel constructions of historical-geographical works on tribes, where the construction is in tribal genealogical idioms, but the spaces are associated with resolutions of dispute and the establishment of tribally endorsed moral order; a major example is al-Bakri's preface 'On the living-places and migrations of the Arab tribes' to his Geographical Dictionary (Wustenfeld [1868] 1993).[18]

---

[18] Recent examples are Peake's (1958) "History and the Tribes of Jordan", translated into Arabic with corrections by Toukan (nd) and Wahlin's two volumes (1993a and b) on the historical geography of the villages and tribes of the northern al-Balqa.

The *badia* was seen as important for communications by the nation states established after the demise of the Ottoman Empire. Jordan has its long eastward extension as a link between its Hashemite rulers and those of Iraq, while the right of free passage across the *badia* was established by the Treaty of Hadda. The *hamad* and *harra* is cut by borders between Syria and Jordan, Jordan and Saudi Arabia, and Jordan and Iraq. The oil pipeline from Kirkuk to Haifa, now defunct, paralleled the road as far as Zarqa, and two pumping stations became the nuclei of ar-Ruwaishid and as-Safawi. The road is now accompanied by power lines between Jordan and Iraq, with the ar-Risha gas field near the border. TAPline used to carry oil from the oilfields of eastern Saudi Arabia to the Mediterranean at Sidon; Turaif and Ar-Ar were originally pumping stations. Border and police posts, roads, electricity pylons and water pumping stations, civil defence posts, hospitals, schools and town councils further mark the activities of central governments of nation states. While some of these markings bring benefits to citizens, their exact positioning in a location can cause many problems (e.g. Antoun 1972 at Kufr al-Ma'). Concerning state land, *miri* land, people often complain about the use of a local resource for 'the general good'; road building, which needs gravel workings, are a frequent source of discontent, when *ghudrân* in wadi beds are destroyed for access to gravel. The damming of side wadis in the mountains for water for Amman causes unfavourable comment from those whose gardens are left without sufficient water for irrigation. National borders are perhaps the greatest causes of resentment among tribal and herding populations, who find their access to their former seasonal areas, markets or agricultural land made difficult or impossible. All groups in this chapter have found their lives changed by the establishment of national borders. Losses of agricultural lands and markets can be ameliorated but restrictions on herders' movement make herding both more expensive since water and dry feed must be purchased, and put stress on the maintenance of the physical environment. Similar views are expressed by mountain and valley tribespeople about the establishment of government grazing reserves or wildlife conservation areas. They point out "Why should we be destroying the trees and the grazing? We live from the trees. If there are no trees, there will be no plants and no grazing. It is people who don't get their livelihood from the area who destroy it".

CHAPTER 4

WATER

The preceding chapter establishes that landscape is constructed from social terms. A similar perspective is present in the classical Arab geographers; Miquel (1980: 102–3, 135) points out that, for example, the term 'wadi' is not so much defined by geomorphological characteristics but rather by the role it plays in a man-ordered landscape. The *Book of Making the Hidden Waters Well Up* written by al-Karaji at the beginning of the fifth century AH/ninth century AD, briefly described by Landry (1990), deals with scientific, judicial and technical aspects. Its information concerning water flow, the nature of sub-soils, origins of underground water, and surface indications of underground water, is comparable to those given by people in the countryside now. Views on the nature of water are derived from observation and practice, and consistent with a long tradition. In Islamic and customary law, a division is made between water that comes from God and therefore free to all, and water made more productive by man, and which therefore can be owned. The enhancement of water for ownership, so that it becomes the object of commerce, is its containment in a manmade receptacle, a jar, cistern, pipe, well, reservoir, made channel and so on. Rights remain over the use of owned water to satisfy the thirst of people and animals; there is a lawful right to water for subsistence but not for profit.

The legal position is set out by Métral (1987). Following the Quran and the Ottoman Code, water can be classified as common, private or as belonging to the public domain. Common waters are: those flowing underground; wells not dug by a known person; seas, lakes and marshes; and rivers of the public domain. Underground basins of still waters, rainwater, or water that gushes up belong to the owners of the land in which the water is. Continually running waters are classified as either of the public domain or private waters. The determining factors for public domain or private are tied to whether the bed in which the water continually flows is owned, and to where the water eventually flows – into an unowned channel like the bed of a public domain, river or the sea, or onto earth which is owned. Public domain waters differ from common waters, since public waters are by definition unalienable whereas waters originating as common can be alienated.

The establishment of rights over waters is made by occupation and work, and by intention. Ownership of waters confers exclusive use and the right to sell such use. To be a commercial object water must be contained, and while the water can be sold, what is being sold is the service rendered and the work undertaken to capture the water. The same logic is followed in establishing rights over naturally flowing waters which eventually drain into inland drainage basins, like the River Barada at Damascus; these are under the same juridical rules as a canal or spring, and the co-owners of the lands through which the waters are channeled form one or more joint societies. Ownership of water is acquired by inheritance, sale or gift of the property that receives the water.

There are also rights of use: the right of thirst may be exercised freely and without payment over all waters naturally renewing themselves, in both public and private domains; this right is extended to the watering of domestic animals and land from running water, and from wells and cisterns, provided no damage is caused. People have the right to enter a property to reach such water if the owner has not provided it at the boundary of his property. The right to irrigation is affected by whether the water is common or public water, or private. For common and public waters, the right of irrigation may be freely exercised by all, while newcomers must respect the rights of earlier occupants, and not destroy existing installations. 'Public' means open to all rather than the state asserting control over these waters. Traditional usage, confirmed by Islamic law, recognises that upstream water has priority over downstream, that riverain lands have priority over more distant land, and that previously cultivated land has priority over land about to be developed. For privately owned water, the right to use for irrigation belongs only to the owners, and others have only the right to quench thirst. The indissoluble relation between land and private water, where one could not be sold without the other, is illustrated by the problems encountered by the public utility companies during the French Mandate in Syria. Métral quotes Tresse (1929) who cites the litigation between the Damascus Electricity Company and the co-proprietors of Yazid, and between the

Hijaz Railway Company and the co-owners of Kanawat, Yazid and Kanawat being two channels of the River Barada.

Ottoman law, like the Islamic tradition, encouraged the development of the category of 'dead' land into productive land. The right of *harîm* or 'defended' land, a defined area around the wells, springs or trees, became the property of the developer for the protection and maintenance of his hydraulic installations. The developer also had the right of usufruct on registration of land in fiscal registers.

There are irrigated oasis lands in the *ghutas* of Qalamoun, the Ghor, and Jauf and Sakaka, but Islamic and customary law applies equally to water storage and use in rainfed agricultural land and in the *badia*. The rural populations of the Bilâd ash-Shâm hold that all water comes from God, whether this water is primordial or fossil, underground aquifers, or rain and snowfall. It is only the development of means of access to underground water by wells or *foggara/falaj/qanat*, and the means for storing rainfall or snowmelt water in cisterns or wells that can be owned. The right of satisfying thirst of people and animals from wells cannot be denied, although if present in the area, the owners of a well have a prior claim on its water.

Water use necessitates water storage. How this is achieved and to what purpose the water is put depends on the geology, soil characteristics and types of rainfall. Water collection and storage are divided by techniques into three main categories: natural collection; natural collection with enhanced storage; and constructed collection and storage. Traditional water use is concerned with surface water as flow or stored, or with water from rainfall or snowmelt that has percolated underground, entered aquifers and so travelled from the area of rainfall along clines or gravity. Deep fossil water has been exploitable only with modern drilling equipment, although there was an earlier assumption of primordial water (Landry 1990: 274).

Natural collection of water is exemplified by the *sayl* (flood water) flowing in wadi systems, and by the *khabra* (rain pools) where water collects in low lying places. *Sayl* waters often flow into depressions where the raised levels of soil moisture enable the growth of annual and perennial plants or, in cleared land, allows cultivation. A part of these waters may be diverted into channels to flood small areas for agriculture or to fill cisterns and wells. *Khabra* hold water because their soils have such fine particles that an almost impermeable base is formed. These soils are a component of *badia* surfaces, and water is often held in these pools for two or three months. Herders use such water for the household and for the flocks. *Khabra* are usually large in area but shallow in depth, so access can be difficult. Herders sometimes build stepping stones so that cleaner water can be collected for the household, while animals drink from the edges; the water is now also pumped by diesel pumps into tankers to be driven to tents and sheep and goat flocks.

The category of 'natural' collection with enhanced or constructed storage includes rock basins, *ghudrân*, and *mahâfîr*. All occur in the *badia*. Of the rock basins in the Jordanian *harra*, the largest and best known is Burqu', but there are also a series of four pools at Wisad, the pool at Luthaima, and pools at Jawa. Jawa has been dated to at least the third millennium BC (Helms 1981), and is now seen to share a water harvesting technology (although not implying a direct or causal relationship) with many third millennium sites across northern Syria (McClellan and Porter 1997). Burqu' (Betts *et al.* 1991) and Wisad are thought to have been modified for storage in the Neolithic. There are in the *hamad* and the *harra* many surface remains of structures for channeling and storing runoff water, but their dating is open to question since there are rarely many remains of datable artifacts (Betts and Helms 1989). Water storage structures at the Umayyad sites adapting natural features in the Syrian and Jordanian *badia*s have been discussed by, among others, Grabar (1978) for Qasr al-Hayr al-Sharqi, Gaube (1979) for Amman, Kharana and Qastal (also Carlier and Morin 1986), and Bisheh (1985) for Qasr Hallabat and (1989) for Mashash. Further east in the *hamad*, Betts (1993) dates two *mahfûr* systems to the early Islamic period. Kennedy (1995) discusses water collection and storage works in the Jordanian southeastern Hauran during the same periods; these used dams at the sides of wadis to hold floodwater, rather like the techniques used recently in the *harra* by local families. In the present, Point 4 and the Hamad Basin Development Plans also modified water storage at Jawa and Burqu'.

While runoff water is important for filling these pools, Burqu' also gets water by the seepage by gravity of rainfall to the west through rock formations. It is an almost, but not invariably, reliable source of water; Rwala and Ahl al-Jabal tell of going to fight a battle there, but as there was no water, the fight was abandoned and everyone returned to their tents elsewhere. The four pools at Wisad flow into one another; the lowest pool, the most long-lasting, is reckoned to provide water for three or four months to those using Wisad as part of their preferred grazing and watering places. (The number of tents using Wisad for water has never been seen to exceed ten). Luthaima is much smaller, although there are additional *ghudrân* nearby, and has never been seen being used for more than casual watering; the remains of tent sites with milk processing structures show that it can be used as a spring and early summer site. All these pools are referred to as *birka*, and compared to artificial *ghudrân*, while the Azraq depression as a whole was described as being like a huge *birka*. *Barka/bartza/barbak* refer to artificial rainwater catchments made by nearly encircling a depression with low stone walls; an example is at as-Sib.

Ghudrân (sing. *ghadîr*) are places in the wadi systems where the base rock floor is exposed, or where there are gravel beds slightly below the surrounding levels of the drainage system, so that flowing water collects. They occur in areas of shallow gradients where the drainage flows through basins of gravelly or loamy soils; in both *hamad* and *harra* there are drainage systems with kilometre after kilometre of *ghudrân*. Many occur in potential grazing areas. People improve *ghudrân* by removing debris, silt and stones; raising the height of the downstream level so that the depth of water retained in the *ghadîr* is enhanced (figure 1); and digging a particular type of shallow well in the gravel beds of *ghudrân*. At a large *ghadîr* in the *sha'îb* Mahdath, shallow stone walls had been built to direct runoff into the *ghadîr*, and an additional wall across the line of flow to slow water down. Wadi ash-Shaikh is so called because the large *ghadîr* in the wadi was cleaned out by one of the shaikhs of the Rwala. The presence of *ghudrân* enables the use of much of the *harra* and *hamad* during winter and spring by herders. As well as taking animals to the *ghudrân*, people now pump water into tankers and drive to flocks some distance away. Formerly, camels and donkeys were used to carry water to sheep and goats using distant grazing (Lancasters 1991).

In the *hamad*, Sab'a *Biyâr* and Mlosi were the most reliable wells between the Shamiyya and al-Juba before the introduction of modern well drilling machinery enabled the tapping of deep water. The first deep wells in the Jordanian *hamad* were dug by the Iraq Petroleum Company at H4 (now Ruwaishid) and H5 (now Safawi) in the 1930s, and IPC also pumped water from Azraq to

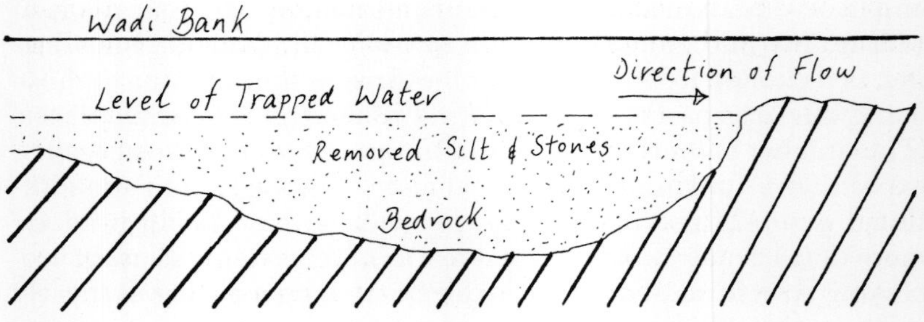

Figure 1 Diagram of a Ghadîr

Mafraq (Longrigg 1954: 76–7, 86–7). The Jordanian government, following 1948 after which IPC were unable to continue using the pipeline, eventually took over these wells and pumping stations, like Azraq, Mafraq, Bishriyya, and Ruwaishid. Later, the government drilled deep wells for seasonal use by herders in the *hamad* at Ga' Abu Hussein and Hifna, and in the *harra* at Qattafi. At Ruwaishid, Mingat was closed at the end of the seventies, and a new well at Jisr ar-Ruwaishid was opened. The villages in the western edge of the *harra* and along the road between Mafraq and Safawi, and Safawi itself have piped water from Azraq; Ruwaishid gets its water from the Ruwaishid pumping station using a local aquifer. Herders collect water from government wells and pumping stations when natural supplies dry up, although the seasonal wells are often not open. The *Hamad* Basin Project has drilled wells between Burqu' and the ar-Rishas in the *hamad*, and the *Badia* Research Project has a well at Shabaika, but these wells are capped. Local people do not know why these wells were dug nor why they have been capped. Gas and oil prospecting companies allow herders to collect water from wells they have drilled. North of Ruwaishid are three private wells drilled by Rwala shaikhs in the late sixties; these are used by Rwala and tribesmen close to them and by others by arrangement. Water is around three hundred metres deep, but in pockets of variable quality; Faydhr and ar-Risha Anwar have sweet water, while at ar-Risha an-Nuri, water is bitter. At 20 Jordanian dinars (c.£20) a metre, a well in this area costs at least six thousand dinars. There is also a deep Ahl al-Jabal well in Wadi Ghussein. There is considerable discussion by local *badia* users as to how much water there is in aquifers under the *badia*, and at what rate recharge occurs. Salameh and Bannayan (1993) discuss both surface and groundwater resources for the *hamad*, as well as the rest of Jordan.

Another traditional form of water storage in the *badia* are *mahâfîr*, described as a form of artificial *ghudrân*; they provide water in areas of good grazing lacking *ghudrân* or *khabra*. *Mahâfîr* means excavations, diggings. The pools are made by digging out earth at the lower ends of wide *shi'bân* or *gi'ân*, and using the excavated earth to build retaining walls to increase the storage capacity of the pool (figure 2). Some walls are stone-lined in places where the pressure of incoming water is greatest. Often there are shallow lines of stones to direct flow to the *mahfûr*. *Mahâfîr* built at the ends of *shi'bân* are assumed to have held water in the *sha'îb* to increase moisture levels for plant growth. *Mahâfîr* vary in shape from oval to horseshoe, and in size from pools dug in a drying *khabra* to use the last of the water for a camel or donkey, to pools of a few metres in diameter for sheep or goats, to some of fifty or more metres in diameter with walls four metres high. Betts (1993) and Agnew, Anderson *et al.* (1995: 71) have diagrams of *mahâfîr* systems. Some *gi'ân* or *shi'bân* have only one, others as at Anqa (figure 3), Mahruta, ar-Rghaban, as-Sib, as-Sbaihi and Ga' al-Ghazzi in eastern Jordan, and at ar-Rdai'aniyya, Kutaifa, and Aqrim, among others, in northern Saudia Arabia have many. Early and Mediaeval Arabic inscriptions have been found associated with some *mahâfîr*. Mur'ath (Rwala) families and a Dahamsha assert that in the recent past they spent the summers at these *mahâfîr* with their animals, although regarded with scepticism by others who say they always went out of the inner desert in the summers. Calculations by hydrologists (Agnew, Anderson *et al.* 1995) indicate that the *mahâfîr* at Mahruta could store 111,500 cubic metres of water, and at Anqa, some 55,000 cubic metres; after six months, with evaporation causing the only significant losses, about 30,000 cubic metres would be left at Mahruta and 14,000 at Anqa. Although water needs of camels, goats and sheep vary seasonally, local users' claims of using these and other *mahâfîr* for months are upheld. Two new small *mahâfîr* were built using a bulldozer at Jadda al-Ginn in the *harra* in the early nineties, for watering sheep and goats, while at Shabaika a Sharafat family had a *mahâfîr* dug by bulldozer in 1985 for watering camels. Other modern *mahâfîr* were made by the Iraqi army in Ga' Abu Hussein in 1970, but these were badly sited and did not work. Tribesmen using the *hamad* have considered cleaning out

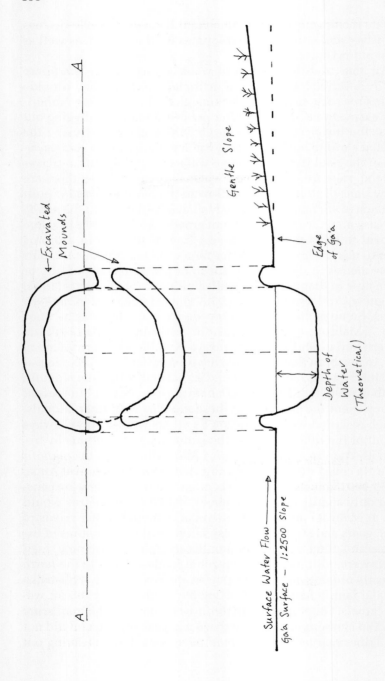

Figure 2 Diagram of a Mahfūr

Water 137

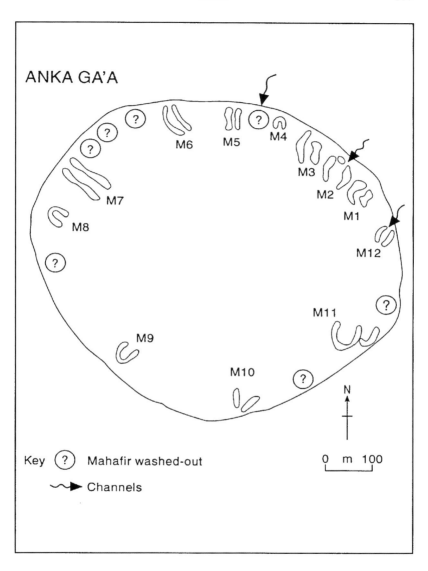

Figure 3 Anqa Mahafîr

mahâfîr such as the set at Anqa, but have abandoned the idea because of problems of preferential access to such water under the laws of nation states. A solution is considered possible "if enough *mahâfîr* from Palmyra to Nejd were put back into full working order", since on a big enough scale, ownership becomes redundant.

The words used for 'a well' are confusing, since people often use *bir* (plur. *biyâr*) for well in general, almost as any man-made container of water dug in the ground. Usually *biyar* (pl. of *bir*) fill from underground seepage, rather than from runoff which is how cisterns or *birka* are filled. In Karak, cisterns (figure 4) are *biyâr mujammi'* or collecting wells, while wells filling from underground are *biyâr hayy* or living wells. *Thumaila* (pl. *thumâyil*) are wide pits dug to a rock layer at three to five metres depth filling from underground and runoff (figure 5), and it is a version of these that are dug in *ghadîr*s. *Mushâsh* are like *thumâyil*, but rather than utilising layers of rock depend on a layer of clay soil that rapidly becomes impermeable. *Qalîb/zalib* (pl. *qulbân*) are deep hand dug wells in the *hamad*, but in the Middle Euphrates are shallow seasonal wells (d'Hont 1994: 29). A *qalama* (pl. *qalamât*) is a machine drilled well or borehole. Musil (1928: 676–684) gives a list of physiographic features, many concerned with water, used in the *badia*.

An alternative small scale water collection system was used in al-Labbah, an area used by Rwala when there had been rain and grazing was abundant. "We lived largely on milk. If we wanted water, we built a *jiba'* (pl. *jiba'iyyât*). This was a pit in the sand, about a metre to a metre and a half wide, and a couple of metres deep. You needed a bucket to draw the water up, so the *jiba'* was about a couple of metres deep. We lined the bottom with stones, and sometimes the sides too, although if the sand was wet, this wasn't necessary. The stones we took out when digging we put in a circle round the top to stop the sand blowing in. The *jiba'* filled up by water moving sideways through the sand. We made these when it had rained, towards the bottoms of the sand slopes. Everyone knew how to do this. We would come back to them and clean them out another year."

Qusur Sattam is a *hâwiyya*, a sort of *bir mujammi'* which collected runoff onto a rocky surface, and caught by stone or mud walls; each is ten to fifteen metres across, and between one and two meters deep, with plastered walls. Two good downpours are enough to fill them, and the water lasts for three months. The Emir Sattam Sha'alan, like many Rwala, liked to spend high summer there.

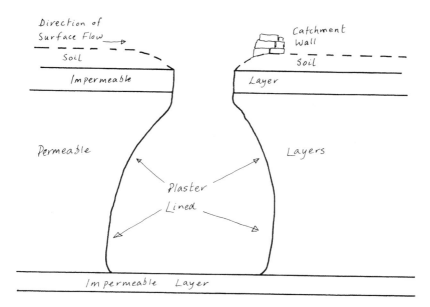

Figure 4 Diagram of a Cistern

Moqr/modhr (pl. *muqwâr*) are enhanced natural cisterns in rock, where the rock is cut out further to make a bottle-shape, and five to ten metres deep; there are thirty to forty *muqwâr* east of Jabal Anaiza on the Iraqi/Saudi border. These particular examples belonged to Sulaib, "who used to cover them over, partly to stop the water evaporating and partly for concealment." Sulaib also owned wells of a type known as *'akl* or *hesyân*, in the extensive gravel beds of the Wadi 'Ar-'Ar. These wells were four to eight metres deep, very wide, and filled by the seepage of percolating rainwater collected within the sand and gravel of the wadi bed and underlying impervious rock strata; water lasted for three to four years after heavy rain.

Small wells imitating *thumaila* in wadi beds were a Sulaib speciality, but others also made them. "You dig out a pit to one side in the bed of a *ghadîr*, and you make a channel so it can fill from water trapped between the gravel and the rock underneath. Then you almost cover it over, you do it carefully, and then you put a big stone over the entrance. You can tell where they are because as the sun sets you can see a haze over them. When we spent

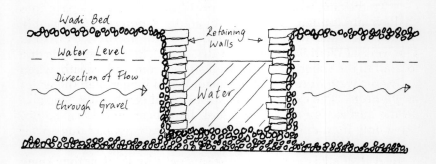

Figure 5 Diagram of a Thumaila

the summer, thirty years ago now, in Wadi al-Cadillac there were some and that was the water we drank." Another man recalled his grandfather building one in the *ghadîr* near Wadi Jawa in the *harra* during the thirties; this, together with another *thumaila* further down the *ghadîr*, and an enlarged and deepened *ghadîr*, provided household water for the years they did not go to Jabal al-Arab, while the animals drank from deep pools further up the Wadi Jawa and the *Biyâr* al-Jawa. This family had also built cisterns, fed by channels from the Wadi Jawa a kilometre and a half distant; this channel goes slightly uphill against the gradient, but "the *sayl* was strong enough to push the water uphill, and you banked up the channel or dug it deeper to help the flow at various points each time," while a second branch went off to another cistern further east.

Wells in and around the Nefud are usually called *qalib/zalib* (pl. *qulbân*). These are either stone lined and deep, as in the *hamad*, or unlined and wide. The usual soil formation for *qulbân* is a mixed area of clay, sand and rock, lying between ranges of deep and looser sand. Wallin (1854: 160) quotes Yaqut's description of the wells of Shakik in the Nefud as being possibly named after this type of soil formation, called *al-shakîka* (Lane [1863] 1984: ii, 1579); he adds "at every place where I have met with wells and water in this Nufood land, the soil has been of this same character." The *qulbân* at Mughaira, south of Sakaka are in this type of formation, stone-lined and "twenty five *ba'* (fathoms) deep" i.e. about thirty metres deep. More wells, among others, in the area are at Murut to the west, northeast of Sakaka, at Hdaib and Swair, also northeast of

Sakaka and at Shuwaitiyyah. (Modern water provision of the oasis towns is discussed later).

Traditional wells in the *harra* fill from rainfed seepage into shallow and limited water tables or aquifers; none supply reliably permanent water with the possible exception of Biyâr Ghussein. The wells at Wadi Salma are now dry, which local users say is due to a fall in the watertable; "the wells filled from underneath, when the wadis were in *sayl*, and the water lasted longer than the water in *ghudrân*". The existence of a former well can be told "because on the surface are stones that look like sugar, and these stones come only from underground. So earlier someone had dug a well at the place." In the dry years of the mid-fifties, Sharafat families of the Ahl al-Jabal re-excavated the Biyâr al-Khudheri, as they knew of old wells in Wadi Khudheri. "We knew where to start digging by the kind of stone on the surface. The nearest the water is to the surface is about nine or ten metres. When I was let down on a rope to go on with the cleaning out, and I looked up, I was so far down the opening looked like a mirror" recalled a woman who had been a young girl at the time. At Biyâr Ghussein, young men from ash-Sharafat dug two wells in the summer of 1988 because "we get water by hand and the water level has dropped so much in the government wells that hauling by hand is too difficult. It's all right if you have a pump and a tanker." The wells were three or four metres deep, and were pits into which water seeped. The following year, these two wells had been lined with basalt blocks. Water in wells at Mahdath was some four to five metres deep in June 1991, when a Sharafat was pumping water into barrels on a lorry for his sheep. The Mahdath wells are groups of shallow wells in an extensive *sha'îb* cut through by *ghudrân*, and almost unnoticeable unless they are in use or have been cleaned out so a spoil-heap is visible. The Wadi Salma wells are in two groups at one side of the wadi, against the hill; there are the remains of a small mosque and mediaeval inscriptions nearby. Wadi Khudheri wells in three or four groups lie in a shallow valley in the uplands, north of Biyâr Ghussein. The Biyâr Ghussein follow the line of the wadi which, as well as carrying floodwater, is also at the lower end of a wide basin surrounded by hills; the numerous wells are associated with Safaïtic rock carvings, and mediaeval and modern Arabic inscriptions. Al-Khudheri and Mahdath wells have dates from the 1960s inscribed.

Similar shallow wells are at Jawa, in a group of three going from west to east on an east facing slope; the top two are old, while the third was re-dug in the thirties by members of the Mesa'id section of the Ahl al-Jabal. The Jawa wells are *bir mujammi'*, filling from channels dug to carry flood and runoff water. Unless water is flowing in the channels, the lines of stones removed when digging the channels are more noticeable. One channel started some two kilometres distant bringing water from the south bank of the Wadi Rajil, and collected runoff from the slope it traversed, while a second channel brought runoff from the slope to the south.

All these systems of water storage provided drinking water for people and animals; the water was not used for irrigation, with an exception for a very small garden at *Bir* Ghussein. How do people decide where they will water? What does ownership of a well, a developed water source, mean? Is water a scarce resource that causes disputes? Customary law permits only developments by a known person to be owned, so water in *khabra*, *birka* or rock pools, *ghudrân* and *mahâfîr* (except for the few recently dug ones) is open to all. Wells known to have been dug by particular groups or families and constantly maintained by their descendants are owned by them, with preferential access. The wells at Mughaira southeast of Sakaka are all owned by different ibn 'amm of the Mur'ath section of the Rwala, and each ibn amm also owns wells at other places; for example, the Zuwayyida Nsair own wells at Mughaira, Murut, Shakik, Shuwaitiyyah and northeast of Sakaka. The wells at *Biyâr* Ghussein, *Biyâr* Khudheri and *Biyâr* Mahdath belong to Sharafat three generation *ibn 'amm*. The two old wells at Biyâr Jawa are free to all, but only Mesa'id of the Ahl al-Jabal use the third as they re-dug it in the thirties. The *thumâyil* in Wadi al-Cadillac was not known to have owners by those using the water in 1968. Ghayyath who came to Jordan in the early eighties use Burqu' or government wells. Water is seen by *badia* users not so much as a scarce resource, but as a varying and unpredictable resource. Disputes over water are avoided by knowledge common to all that groups have preferred areas given where rain has fallen, of which wells are owned by whom, and by the principle of first come, first served ameliorated by the assurance of the quenching of thirst.

Water conservation has always been important in the grazing areas of the whole region. Tristram, travelling in al-Balqa, remarks (1873: 186) "everywhere is some artificial means of retaining the occasional supplies of water." Cisterns in the western rainfed arable zone are for animals and people, although household water for families living in tents in season is now often fetched by tanker from government pumping stations. The northern Karak plateau has permanent flowing water from its springs in Wadi Mujib, Wadi Nukhaila and Wadi ibn Hammad. Every modern village is situated on the plateau with a piped water system. Traditionally in summer people grazed the flocks on the harvest stubbles of the plateau. It was only when the cisterns were dry that people took the flocks down to the springs to water.

The villages and arable land have owned cisterns. Many men have dug them, when an existing cistern became unusable (breakage at the top being the commonest reason), because additional water storage was needed with increasing human and animal numbers, or a spring had dried up, as at Fagu'. Earlier, men dug with pick and mattock, now they hire pile-drivers. An area is chosen "where there is a layer of hard rock about a metre thick, and five or six metres of soft rock (*thihân*) under it, followed by another layer of hard rock or bedrock. The difficult bit is digging the first metre. Using a pick, a man might take a month over this, although he wouldn't be working on his cistern all the time. Hiring a pile-driver costs five Jordanian dinars a metre for this part, but forty piastres a metre for the rest; it costs between three and five hundred dinars to build a cistern (at 1992 prices). The top metre is roughly square cut, about a metre square. Then, in the soft rock, the cistern is widened out to become six metres deep and wide, more or less. With luck, the dug out soft rock could be used to make lime for the plaster. Sand for the plaster was brought from the wadis, and sieved, washed and cleaned of earth. You made your lime, mixed it with sand, and you plastered the walls. A cistern should last at least fifty years with only yearly cleaning. They usually break at the top if the rock is not hard enough, rather than the plaster deteriorating." Pile-driver dug cisterns are finished with a cement top, usually incorporating a drinking trough for animals, and the opening to the cistern is closed by an iron door, with a padlock. Otherwise, large stones are used to close the opening.

In addition to needing the right strata of rocks, cisterns had to be positioned to fill by gravity from runoff. The runoff area needed is surprisingly small; about 0.5 of a dunum (0.05 of a hectare) is usual. The runoff area should be rock, as the water is cleaner; some cisterns, for household use, have settling tanks. In late summer and early autumn, debris is cleaned out of the cisterns, and new channels leading the runoff from two or more directions to the cistern areas are dug with a mattock. The first rainwater filling a cistern for domestic water is pumped out for watering young olive or fruit trees, thus washing out the cistern. Each family, or rather the *khâna* which is the land, house and cistern owning unit, has at least one cistern in the village, and more on its arable land. Various people said that one to four cisterns, two in the village and two outside, and two in the village and seven outside were what was necessary for a *khâna*'s requirements. Cisterns were not necessarily used for storing water, some were used for grain storage, using the principle of anaerobic respiration; they then "become stores (*khazna*) rather than *bir mujammi*." A cistern holds enough water for one hundred sheep for four months. Who uses which cistern for what is arranged within and between families. Cisterns belong to the land in which they are dug, and are known as 'the cisterns of that particular piece of land – for example, *Bir* Umm al-Habaj' – while an individual cistern is known by the name of its owner, or his father or grand-father. Cisterns cannot be sold away from the land, although the water contained in them can be sold. The current price for a cistern of water is thirty or forty dinars; purchasers are sharecropping families working the arable land, and who have animals. Buying water for their animals is more convenient than going to collect water by tractor-drawn tanker from a government pumping station in a nearby town, although domestic water is obtained in this way.

A Majali family who used to spend the summers in tents at Azzur said that the cisterns on the upper slope of the area were for household water, while the lower held the water for animals – cows, sheep, goats, horses, donkeys and camels. The first rains were diverted past the upper cisterns to flow over the rock around these cisterns so that everything was clean for the following rain to fill them. Some cisterns were linked by underground channels near their tops, explained as either overflows, or transferring water to other cisterns. No-one could remember cisterns not being filled every winter.

These pear shaped cisterns are characteristic of the limestone and *terra rossa* upland plateaux of Jordan, although names shift somewhat. At Buraigha in the north of Jordan, on the southern Hauran plateau, the cisterns are called *bir jâmi'*, and are cut down square for two metres and then widen out into a pear shape. There are also large shallow pits, sometimes stone-lined, called *birka*, while small shallow pits for water storage are called *siyh*; small natural or man-made basins in rocks are called *samâ'*. In this area, water in the wells and cisterns is for animals, but also for fruit or olive trees which have to be watered in the summer months if winter rainfall has been low.

In the Hishe part of the Shera mountains, around Qadisiyya and the Wadi Dana, springs provide year round water at places in the wadis, although flows lessen significantly in the summer, and further fluctuate through day and night. The springs such as Ain Dana and Ain Ladhha continue to flow, as does the smaller spring of Ain Taraq on the western edge of the plateau to the east. There are cisterns, either at the *khirba*s used as stores and summer living places for the arable harvest on the plateau, or at *khirba*s in high places in the mountains. Qadisiyya itself used to have large stonelined cisterns, but after the provision of a piped water supply these fell into disuse. As most of the rock formations in the mountains are sandstone, with some limestone overlay, people said the construction of the pear-shaped cisterns, the *bir mujammi'*, typical of Karak and the Balqa', was usually impossible. Exceptions are Bir Harbe at Kula, one at Sila' west of the road to Tafilat, and a third near Khirbet Samra on the road to Shobak.[19] A more common method was to enlarge natural basins in rock formations, sometimes building walls to complete a basin, and then to roof over with corbelling, "but when people don't use them, then the roof falls in"; an example was seen at Khirbet Dakhlet on Maqta. On the plateau, cisterns "get covered with stones and forgotten – sometimes people go on using them as stores for *tibben* (chopped straw)," as at Khirbet Talaya. Water storage continues for animal use, and in Wadi Dhalma a herding family was using water from a seep, tapped by a rock cut channel. A funnel took the water from the rock channel to a jerrycan for domestic supplies, while below the basin two circular troughs had been dug and edged with stone on the downstream sides for animals. Hudaira, a nearby plateau village formerly seasonally used by herders and farmers, has a number of disused cisterns. A little further east, Umm Hwait has two modern cisterns, and several old ones, as well as four shallow pools near the top of a ridge; these pools were constructed by building

---

[19] Water catchment and storage in sandstone, including rock-cut bottle cisterns, at Petra and Humaimat from the early Bronze Age and in the Nabataean periods is discussed by Oleson (1995).

walls around the deepest areas of bedrock, and have channels directing runoff water to them, reminiscent of the *barbak* in the *badia*.

The preceding examples of water storage have been for the use of domestic herds and people. The following section is concerned with the storage and channeling of water for irrigation.

It is often assumed that irrigated agriculture is directed to the market, and that intensive vegetable and fruit crops are its main focus. In the Bilâd ash-Shâm this is not necessarily so, although much modern development of irrigated agriculture has this purpose. To produce crops soil needs certain levels of moisture throughout the growing season. These levels can be achieved from rainfall, probably aided by a bare fallow in the previous year; or from directing water to the soil by a variety of means. The simplest method is by using gravity to channel water to soils, and this can occur naturally. Areas in the *badia* that would rarely, if ever, have sufficient moisture levels purely by rainfall, achieve them through gravity-fed runoff from a wide surrounding area; this is how crops of barley, wheat, sorghum and water melons are grown in some years at places such as the Rahba in Syria, Mahdath and Shabaika in the *harra*, or around the ar-Rishas in the *hamad*. To increase the possibility of a crop, earth dams have been built in two wadis around the ar-Rishas; one built in 1968 was designed by an engineer from a company. When the *sayl* ran, flood water backed up for three or four kilometres behind the dam before the water overflows round the edge of the dam. This dam works well, and a crop of sorts can be grown in most years; the second has been less successful owing to the underlying shallow soil formations, so that the water percolates through the dam. In the Wadi al-Harth, west of Azraq, rainwater runoff is directed onto the low-lying better soils. Various development projects such as Point Four in the fifties and the *Hamad* Basin Project in the eighties have also developed water resources in the *badia* for agriculture; the intentions of such projects rarely reach fulfillment. The dam built by the *Hamad* Basin Project was said, in 1992, to be for agriculture; in 1995, the reason for the dam was "to encourage recharge of shallow water resources."

An interesting example of channeling can be seen at the former farms or gardens in the Wadi Jawa which used rain runoff and snowmelt from the Jabal al-Arab. These were developed in the thirties and abandoned after 1968 or '69 when the Syrian government built dams across the Wadi Rajil just north of the border so that amounts of water coming down the Wadi Rajil and entering the

Jawa system significantly diminished. One farm was started in 1931 by Ghayyath al-Khlaif ash-Sharafat who chose the location because there was an existing spring a little way upstream, and three branches of the Wadi Jawa cross the land. (The spring was blocked by the army in the late sixties to stop *fedayeen* from Syria using the water; when Ghayyath's sons went to unblock the spring it never revived.) The summer crops were rainfed, achieved by building complexes of channels and stone walls which directed flows of snowmelt water to parcels of land or to cisterns (figure 6). "To make a channel you have to move the stones, and this is the reason for the walls." Snowmelt could last for up to six weeks, but usually less. Neither the crops or the water storage was intended or used as a permanent system; "we were never here all the time, or all of us together; we were out herding in the *harra*, this was a place we liked to be in the summer. But we weren't here every summer or all of a summer. Sometimes we went up to Mlah or other places in the Jabal al-Arab. But we always got something from the farm even if it was only grazing the crop we'd planted, and we had the stored water for us and the animals."

The channeling of flood water to irrigate land for arable and vegetable crops in the Hauran was noted by Burckhardt (1822: 218). Hauran systems were examined by Braemer (1990: ii, 453–78), who considers some to date possibly from the sixteenth century and definitely from the nineteenth, while others are recent. Of the northern Ghor Merrill (1881: 179) wrote "Between the Menadirah and the Yabis, there are a number of wadis which have living springs and streams. These streams are carried in numerous canals over the plain for the purpose of irrigation." He noted similar systems at various places in the northern Ghor and along the Wadi Zarqa for arable crops, and describes the system thus (453); "the dams which the natives make are always slight structures. They go far upstream, where they make a small one, and expend their money on a long canal. This makes less expense for a dam, gives them greater head of water, and enables them to carry it to fields which otherwise would not be used." In the southern Ghor, Tristram (1873: 47) saw the "*sayl* tapped by little conduits on its left bank, so that the whole Ghor can be turned into a watered meadow."

The *wudiyân* (sing. *wâdi*) of the Karak plateau, like those of al-Balqa and Jabal Ajlun, and the Aghwar, have seen changes over time between irrigated agriculture for commerce – the sugar of the Mamluk period or the indigo mentioned by Burckhardt – and cereal

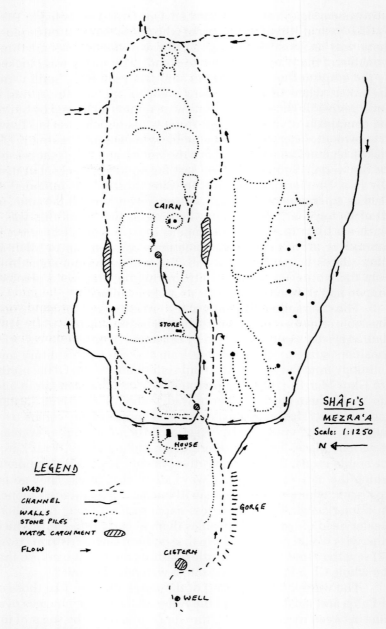

Figure 6 Shâfi's Farm – Irrigation System

cultivation as part of a multiresource economy as described by Burckhardt or Musil. In the Aghwar, irrigation was needed at all periods, while in the *wudiyân* it was often though not always used. Irrigation depended on the channeling of water from springs or from *sayl*s, the flood water in the wadi beds (figure 7).

Water is taken out of the *sayl*, or from a spring, by a channel constructed along the contour line; from this main feeder channel, subsidiary channels are taken off at an angle to lead water to parcels of land; here smaller channels carry water to each garden, which are sometimes leveled. The water cannot be sold away from the land; the garden, its channels and its rights to water are inseparable. The amount of water each parcel of land receives is determined by the group of people, the *jamâ'a*, using the land off the main channel; over time, these amounts become more or less fixed, so that each parcel gets so many hours of water a day, and the amount of water a parcel gets decides what crops can be grown. This is like the "well arranged system by the riparian proprietors" at Safi noted by Tristram (1873: 47), and in many other locations. Disputes over water allocation are sorted out within the *jamâ'a*, who collectively own the water rights because they own and work the land. Share-croppers take over the water rights for the land they are working for that cultivation season. New channels can be

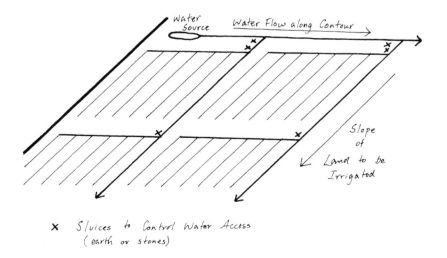

Figure 7 Diagram of Irrigation System at W. Ibn Hammâd

made by anyone if he takes new land into cultivation, either using water from an existing source with the agreement of all *jamâ'a* members (and this land and its user would become part of that *jamâ'a*, to which he probably already belonged), or from a new source if a new spring was found on the owner's land. Uncultivated land on the slopes of the wudiyan belongs to the government, as it is 'dead' land, and anyone can ask the government for the right to cultivate a piece. However, all land is associated with named groups by use and history, and the group has to agree to someone from outside taking up 'dead' land in their area. Such agreement is extremely rare, unless there are existing close ties through women to the traditionally owning group, such use is seen to establish access to that land by the user's descendants.

The history of irrigation techniques in the area is obscure, since waterchannels and watermills are difficult to date without documentary evidence, but it is clear that there is a long history of such installations as at the Nabataean sites of Petra and Humaimat. Some *qanat*, now destroyed, are said to have been 'Roman', "because we had no name for the man who built them" like those at Umm Namasa in the Wadi Nukhaila. Water channels (*qanat*) in the Wadi Fuwwar (below Ain Jubaiha) were built by the present owner's grandfather at the beginning of the century; in the drought year of 1915, crops grown on this irrigated land was the only grain he produced. Channels in the Wadi Mujib were built by the Majali using Christian workmen early in the century; others belonging to the Mu'aita in the Wadi ibn Hammad were said to have been built, or possibly renovated, in the fifties; the majority are described as old, to have always been there.

Access to water gives rise to continual minor disagreements. One group centres around the right to drinking water for people and animals, the other around amounts of water for cultivation of commercial crops. There are constant complaints that the amounts of water for irrigation from springs are declining. A series of relatively poor rain years are blamed for this, together with the concreting of water channels, encouraged by agriculture officials to reduce water loss from evaporation; people say they in fact get less because water no longer soaks right into the soil along the line of the channel and so the effects of the water no longer trickle on during the hours when there is no water flow. A major cause of declining water amounts is due to a perceived lack of recharge of springs, blamed partly on the series of poor rain years, but also on the building of dams by the government to the east at Qatrana and to the activities of the Jordan Valley Authority who are said to pump out too much water in the Ghor below, thus lowering the water table and/or depleting aquifers. The dam building at Qatrana encouraged the development of Beni Attiya small farms at Qatrana, while the JVA action increases the production of commercial crops in the Ghor below Karak. Complaints about the decline in water for irrigated agriculture in the Karak wudiyan often focus on the complainants' 'right to livelihood' being interfered with through actions of the state as owners of 'dead' land, or through the enrichment of one set of people at the expense of another.

Complaints about access to drinking water also focused on 'rights' and 'ownership'. The ownership of particular areas is associated with specific tribal and/or family groups, but there can be discrepancies between ownership by registration with the state, and ownership by customary use. Some groups have 'always' used particular areas for household arable crops (which in customary law did not confer ownership of the land but only a preferential access since rainfed arable cultivation required no development of the land itself) and for grazing (which does not of itself confer ownership if there was no development of cisterns; if herders relied on springs, then there were no legal claims to ownership through development but only through use, and rights to livelihood). Specific parcels in such areas had been registered with the state by other groups who had developed the land with water channels for reliable cereal cultivation. These groups also claimed the area as a whole by virtue of their political leadership of the northern Karak plateau, either through conquest or through protection of weaker groups. A small herding group was one of the political neutrals, outside the arena of Karakiyin politics; they accept the ownership of developed parcels and the water that feeds these, but not of all springs in the area by members of a politically dominant family who, they claimed, were trying to force them to pay for water for their animals. In the opinion of the *mutaserrif* (senior government administrator), springs are registered in the name of the group who own the surrounding land; if the land round the spring is not cultivated anyone has the right to that water; if the land is cultivated, people still have the right to water for animals and people, to the capacity of the spring, although in reaching it, they must not damage the crop. Water from a spring cannot be sold because there has been no development of it, whereas water in a cistern can be, because the cistern owner has built and maintained it.

The same summer, there were complaints about Beni Sakhr herders coming in to the plateau and using the water at the government pumping station. (Although there are Beni Sakhr herding families who regularly summer on the plateau, these were a family who had never used Karak before.) The *mutaserrif*'s response was, "the Beni Sakhr have the right to come here and negotiate with whoever they please for grazing. The only grazing is on the harvest stubbles, which have to be paid for, and that is between them and the owner of the crop. Then, because they are here, they have a right to water. They can use government water because they are citizens, or they could buy the water in cisterns, through negotiations with the owner."

As well as irrigation by channeling water by *qanat* from springs and *sayl*s, since the early seventies diesel motor pumps and plastic pipes take water from springs in the wadi bed gravels to new gardens on the slopes high above the springs. The water is pumped into artificial ponds or *birka*, made by excavating a large pit with a bulldozer and lining this with heavy duty polythene; the pond is situated at the highest point of the garden, and the water is gravity fed through plastic pipes to trickle irrigation piping along the crop rows. These new gardens are on formerly 'dead' land on the *wudiyân* slopes, registered by members of the tribal group associated with each area and grow vegetables, especially tomatoes, for the urban market. This use of water also causes disputes, centering around accusations of over-pumping or of the use of water rightfully others'. The owner of a farm at Umm Namasa on the left bank of the Wadi Nukhaila accused sharecroppers working on the right bank of "taking our water without having title to the land" and threatened to call the police to confiscate their pump and pipes. The sharecroppers were growing tomatoes on a small series of steep earth terraces on land on the right bank of the Wadi Nukhaila, which is outside the lands of Karak. (Boundaries between tribal groups tend to follow the main wadi systems.) The share croppers' response, after a shouted exchange in which 'our right' was mentioned frequently, was to turn the pump off, shift it a short way nearer the bank on which they were cultivating, and re-start the pump. Information received later revealed that the sharecroppers did have the right to this water.

Springs may dry up but may also burst out in new places. A Fuqara was moving rocks with a bulldozer in his field, and "we found a spring! Because the spring does not have much water, just enough to cover my garden, I own it outright." The same principle of the amount of water relative to land conferring ownership is seen in a garden on the slopes west of Sirfa, which uses water from a seep. The owner of the land built walls and concreted around this small seep to collect enough water to irrigate his land. The following winter, he dug a traditional cistern for watering animals. The enhancement of seeps for irrigating small fields or filling cisterns is called *qattâr* (Oleson 1995: 710), while Musil (1927: 680) records "qattâr; a spring the water of which does not actively flow but merely trickles drop by drop".

Safi, in the Southern Ghor, had springs, fed by rain and *sayl* water from the plateau and the Wadi Hasa. Musil (1908: i, 160–161) describes the irrigated crops grown by the Ghawarna, and this pattern of irrigated wheat, barley and dhurra crops continued until the fifties. The development of the potash industry and citrus groves meant that all springs were taken over by the Potash Company or by the JVA, and farmers get water from the JVA. Vegetable crops for urban markets and for export also grew in importance from the late seventies, and modern irrigation methods of pipes and ponds, called *barâbîsh*, came in 1984.

In the Shera mountains there is a long tradition of orchard and garden cultivation using water from springs, as there is in the mountains southwest of Karak (Burckhardt 1822: 396–7). The example given is Dana, mentioned by Burckhardt (1822: 410) as having "fine gardens and very extensive tobacco plantations". The springs for the gardens, now growing fruit and vegetables, rise at different levels on the hillside. Each spring's water serves a set of gardens, and people have land in two of more blocks of gardens. "All the gardens have rights to the water from the spring (associated with that block of land). There are no problems because there is enough water for everyone, there is plenty of water here. Nobody is responsible for seeing to the distribution of water between gardens because there is no need. It's common sense. If you make the channels and keep them clean, you have the water." This attitude may reflect the current use of the gardens at Dana by the 'Ata'ata, which is for the supplying of household needs with fruit and vegetables rather than for a commercial market. The gardens are one strand of a multi-resource economy, and the labour available for the gardens is also tied up with demands on labour by the other economic activities of household members.

Westwards down the mountain slopes and valleys towards Wadi Araba are more developments using stored or permanently flowing water. At Ghuwaibah, where water recharged by *sayl* water is stored in the gravels of the wadi bed and comes up to the surface in pools called *'ayûn*, a small garden has been developed by channeling water to the plot of earth. The plot lies between the main path of the *sayl* and the gravel beds where permanent water bubbles up as *'ayun*. Part of the *sayl* water is diverted through the garden and in the summer, water pumped from the pools is used to grow grapes, figs, pomegranates, olives and some vegetables. At Fainan, where Wadi Ghuwair joins Wadi Dana, a group of the Rashaiyida irrigate gardens using gravity fed water piped for some nine kilometres from the Ghuwair to land on the south side of Wadi Fainan. The piped water is stored in artificial pools excavated by bulldozer and lined with plastic, and then fed by trickle irrigation to the crop. Like the wudiyan around Karak and many other areas of mountain Jordan, the area has the remains of old field walls and water channels leading from the Ghuwair, a Byzantine aqueduct, and a mill. Rashaiyida say that earlier they grew only cereals on the land, using water via the now broken aqueduct, and point out that the present use of plastic piping is only a cheaper and more convenient substitute for stonebuilt *qanat* and an aqueduct needing major renovation. There are stories of water channels from springs at Fainan, Fid'an and Ghuwaibah leading to long disused areas in wadis. In the recent past, channelled runoff from cliffs to small fields was common, with the water being held on the fields by little earth embankments.

Guraigira is a larger scale development whose recent history is hard to establish. The area is a wide depression of relatively good soils where the drainage of a large section of the mountains naturally assembles; in addition, the Wadi Fid'an (the name for this section of the Fainan) brings *sayl* water which recharges the springs of Ain Fid'an. The land belongs to Amarin and Sa'idiyin families as tribal land, but because it was uncultivated or *sahara* (lit. sand desert) belonged to the government who could make it over to anyone who would develop it. The area attracted the attention of Sharif Jamil ibn Nasr, a well known entrepreneur, who made an agreement with the members of the two tribal groups for the development of Guraigira. In 1973, the Sharif started the Co-operative of the Wadi Araba, and the land was made over by the government; he took 30% of the land outright and agreed to supply all the infrastructure, and the rest was between the people living there, the Amarin and Sa'idiyin. The Sharif dug, over time, eight or nine wells which supply water to the farms, and supplied the pumps. The later wells were found by a water diviner, with water between thirty and seventy metres down; according to the Sharif's agent, the rate of recharge of the wells is unknown as no studies have been made. Further expansion of the farms was seen as unlikely because profits did not allow for the costs of more wells, pumps and pipes. In the early 1990s, a series of continuing disagreements between the parties ended in the Sharif withdrawing from the agreement, and removing all the pumps, motors, pipes, even the pipes from the government solar well for agricultural water and the government diesel well for household water. The removal of the pumps, motors and pipes is accepted as right and proper as he had paid for them. By 1995, some pumps and motors had been replaced, and farming was continuing. Commercial gardens at Tlah and Faifa in the Wadi Araba have been developed in similar ways; both locations have remains of earlier water storage systems.

Qanat or *foggara* are water systems using groundwater resources. *Qanat/foggara* are subterranean or subsurface channels that collect water by tapping an underground water table, and carry it through very gently sloping levels to surface channels which then provide water to crops (figure 8). This system was widespread in the Islamic world (and other regions of the world); in Oman, Yemen and southern Arabia they are known as *falaj*, in the Hijaz *qanat*, *'ain* and *khaif* are used, in Syria *foggara* and in Jordan *qanat*. *Qanat* also applies to any channel for water, as is the use at Karak and the *badia*, for example. Nasif (1988: 157) gives the meanings of these various terms; "the plural form fuqqara..... pronounced foggara......refers to a chain of wells linked one to another....... The literal meaning of the word 'ain is a source of spring. The word khaif, in a classical context, refers to 'the part that slopes down from the ragged portion of a mountain and rises from the channel in which the (flood) water flows...... falaj and 'ain ....describe the process of an uninterrupted flow of water over the surface of the land." He says, too, that *'ain* and *khaif* refer not only to the *qanat*, but also to the land irrigated by the *qanat*, at least in the Hijaz. *Qanat* appear to have originated in northwest Iran in the seventh century BC (Lambton 1989), while the first technical work to describe their working is al-Karaji, mentioned earlier (and see Landry 1990: 279–81).

Lightfoot (1996a) discusses *qanat* in Jordan, finding a total of ten sites with 32 *qanat* in the country. Many of the supposed *qanat*

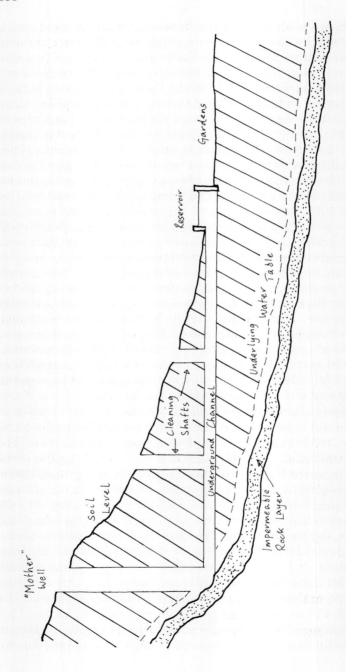

Figure 8 Diagram of a Foggara

he was shown in his survey were (1996a: 3) "short, rock-cut spring tunnels, (which renew or increase natural spring discharge), or slab-covered surface canals. Both resemble qanats in form and function but are usually much shorter and far more common than true qanat." The cutting back of springs is reminiscent of the cutting back of seeps noted above in the *qattâr*. As we have noted, form and function rather than typological features are the defining qualities in Arab terminology. Spring flow tunnels are examined by Ron (1989), who sees their development for irrigation as a consequence of following falling water tables further into the aquifers. *Qanat* in this section describes those systems of tapping underground water for irrigation defined as *qanat* or *foggara* by users and scholars.

The limited number of *qanat* sites – ten with thirty two *qanat* galleries – in Jordan, compared to some 130 in the Hijaz (Nasif 1988) and Syria with 67 sites and 239 *qanat* galleries (Lightfoot 1996b), is associated by Lightfoot (1996a: 7–11) with specific physiographic features; the greater depth of groundwater at below 100 metres, and the less transmissive groundwater flow characteristic of Jordanian aquifers. That is, favourable conditions for constructing *qanat* are in piedmont regions with shallow aquifers in alluvial fans, synclinal beds, or along the margins of large wadis coming out of mountains, and with highly transmissive groundwater flow. The few sites in Jordan with such features, the northwest and Ma'an and Udhruh (Killick 1987) in the south, are those that have *qanat*. In the areas of al-Karak and al-Hishe there are none.

In Syria, conditions are more favourable, and in the region of the Bilâd ash-Shâm, the oasis of Damascus (Thoumin 1935; Thoumin 1936; Bianquis 1977), Qalamoun (Thoumin 1936; Haj Ibrahim 1990), and Qariatain (Musil 1928b: 35) had *foggara* in the present, recent or historic past (Safadi 1990: 287; Lightfoot 1996b: fig.2). The *ghuta* of Qalamoun follow two south-west – north-east lines (Thoumin 1936: 34), one connecting Ras al-Ain, Yabrud, Deir Attiya and Sadad, the other Qutaifa, Mwadama, Jerud, Utna, and Nasiriyya. Haj Ibrahim (1990) discusses the differences in source of water and rate of flow between the *foggara* of the various Qalamoun ghuta, while all depend on access to an underground water table fed from rainfall and snowmelt of the high mountains. He describes the *foggara* of Deir Attiya, starting from two shallow springs then uniting and flowing in a partly open, partly covered channel. This differs from *foggara* that tap into an underground source, and carry the water in underground channels, with shafts at intervals for removal of debris and for cleaning, until the channel emerges into the open. Since its builders and users call the Deir Attiyah system a *foggara*, tapping into an underground water source and carrying water by channels to irrigate land seems the essential factor; that is, the role of a system is more relevant than its features.

The first known date for the *foggara* of Deir Attiyah is 708 AH/1308 AD, when a daughter of a commander of Salah ad-Din restored a channel. The next known restorations were made at the end of the first World War, followed by two in the thirties. Two new *foggara* were also dug, one called The Municipality, the other The Maksar. One channel goes north for four kilometres. Its first section was half-underground to allow the building of the first of three water-mills; downstream of the first the channel was open, facilitating cleaning and maintenance; some water could be taken off for semi-irrigated crops of vines and cereals, and for watering sheep and goat flocks. *Maskûr* are the places where water is taken off for irrigation. Directly after the most important of these, the course of the channel went underground through the village; its water was reached by wells (*nejmât*) for household use, the bath, mosque and grape presses. North of the village, the channel divided into three, one of which fed the screw for the second mill, and irrigated the fields.

The ownership of water in Deir Attiyah was totally separate from that of land; water-owners had their own register (Haj Ibrahim 1990: 301), with officials responsible for water distribution, and who ensured the free flow and cleanliness of water, as well as preventing illegal use. Following pollution of the old channel and population growth, the Municipality had a small *foggara* built to provide clean drinking water. This *foggara* started nine metres deep, and ran for some hundreds of metres until it came to the surface at a watering-place in the extended village street; surplus water flowed into the old channel. In 1943, villagers formed the first Agricultural Co-operative Society of Syria, digging a five kilometre *foggara* to use 2,500 dunums of dry land; this took three hundred work days. In 1944, the society agreed with the municipality to divert the water of the Municipality *foggara* into that of the Co-operative by digging a linking channel, building watering places in several village areas, and setting aside 750 cubic metres of drinking water for the Municipality reservoir each day. The drought of the fifties lowered the yield of these two *foggara*. It was decided to revive the *foggara* by building a small dam across the wadi bed a kilometre upstream of the *foggara* sources to hold back the occasional floods. By chance, the digging of the dam added to the rock fractures followed by flood waters, so increasing rapid subsoil water storage and reducing evaporation losses; the heavy autumn rains of 1987 yielded a fourfold increase in amount of water delivered by the *foggara* over several months.

Foggara in Jerud and Qastal, Qariatain and Tudmor, and in the Ghuta villages, were seen by Porter (1855: 159, 221, 252–3). How cultivators gained access to water delivered by *foggara* in Qalamoun is discussed in Thoumin (1936: 40ff) Each village had its own cycles of distribution and customary law, and he uses that of Nebk, which had one of the simpler systems. Nebk had a twelve day cycle, in twenty-four divisions of twelve hours each. These 24 divisions bear the names of the land-owning families owning the 12 hours of water. Each division is further sub-divided into *shaya*, worth one twelfth of an hour or five minutes. In 1930 the value of one *shaya* was 20 Turkish Lira or 2,200 Fr.francs. Generally, a *shaya* is five minutes of the total water flow, but in Nebk was five minutes of a third of the total flow; additionally, as was general, successive take-off points had lessening amounts of water available. In practice, these difficulties are reduced by rough compromises by water-users. At Nebk, right to land implies right to water, and one contract was sufficient, at that date, in Deir Attiya as well (Thoumin 1936: 48), although at Yabrud, people needed one contract for land and a second for water. The association of land and water in sales was common in the Damascus Ghuta in the eighteenth century (Rafeq 1981: 673–4), who also notes the system of water regulation, using law court registers. In Qalamoun, as in the Ghuta, peasants took complaints over water to official arbitration only in exceptional circumstances (1936: 104). In Thoumin's view (108), "the irrigation systems of Qalamoun and the valleys of Mount Hermon represent very nearly the best distribution that one could get with the flows of their springs", although that in the Ghuta of Damascus did not.

Wallin (1854: 15) mentions that al-Jauf had *qanat*, which have been seen but not investigated, while al-Sudairi (1995: 50) says Sakaka also had *qanat*, the mother well of which was examined by Nasif (1987: 127–33). He was told that in the last third of the 19th century, two Sakaka clans decided to renovate the system; having settled contested claims of ownership, the work was carried out, but the small amounts of water resulted in the abandonment of the project. Musil (1927: 329–333) saw the ruins of *qanat* at al-Haditha in the Wadi Sirhan, and Doughty ([1888] 1936: ii, 220) saw the *foggara* of Khaibar. Nasif's (1988) survey of the *qanat* of al-'Ula, the surrounding villages, Medina, and Yanbu' is the most comprehensive for the southern end of the region. The Hijazi term for the underground water-systems and the land they irrigate is *'uyûn* (sing. *'ain*), but Nasif uses *qanat* firstly because it is generally used by Western scholars to signify the *'ain*, and secondly, because *qanat* expressed the exact meaning of underground water channels. The al-'Ula area has sandstone and sandstone rock strata, which necessitates mother wells, deeper tunnels and shafts as opposed to Qalamoun practice.

For the working, renovation, and construction of *qanat*, Nasif (1988: 160–208) begins with the series of shallow holes seen on the surface, considered as recent work by the *qanat* owners for renovation or maintenance, enabling workmen to enter a blocked or collapsed tunnel. A new opening is called a *baqîra*, while the original well-shafts are *mutla'* (pl. *mutâli'*). Work on *qanat* takes place during the summer months, between the cereal and date harvests when there is no agricultural work. The alluvial sediments and sandstone rocks are friable, so tunnel walls and roofs must be stone-lined. Roof collapses were quite common, as were blockages by clay or tree roots. Most *qanat* at al-'Ula have stones, called *qurra'a*, the same width as the tunnel with an opening at the base through which water can flow, and leaving a space at the top for a man to crawl through; these control the volume of water passing through the tunnel and mitigate the effects of increased flows after heavy rainfall. Another device, the *mahsar*, is a narrowing of the tunnel in certain short sections to the width of 30 cms. and a height of 60 cms.; this protects the tunnel where the surrounding material is soft, or possibly, where one *qanat* crosses another. Yanbu' *qanat*s have a 'plunge', a *ghattâsh*, where a *qanat* goes under a flood channel. Medinan *qanat* have a series of small cavities, called *maghâtîs* (sing. *maghtas*), dug in the tunnel floor which trap water borne debris.

Several of the forty al-Ula *qanat* renovated from the last century to 1957 were damaged in the violent storms of 1972, when most were blocked. Instead of renovating, *qanat* owners dug new wells operated by diesel pumps. Diesel pumps have similarly superseded *foggara* in many of the Qalamoun *ghuta*, especially with a falling water table.

Nasif gives details (1988: 217–8) of *qanat* construction in Medina which he received from Shaikh 'Ali Hafiz, who was informed by Yusuf and Majid Madani, *qanat* owners. Successful *qanat* construction depends on the positioning of the main well, and on the rate of slope and direction of the tunnel. The well is usually sunk in an alluvial fan between the bottom of a mountain and the plain or valley. One, two or three wells are dug, depending on the water supply, to fill the underground water reservoir. From these wells the tunnel is dug towards the land to be irrigated. A vertical shaft is dug every ten metres for cleaning and linking tunnel and surface. Diggers work by candle light, and keep the tunnel, large enough for a man to stand in, in a straight line by soundings with a crowbar at the bottom of the shafts. This description of *qanat* construction is similar to al-Karaji (Landry 1990: 279–81), but differs from Goblet (1979: 310) and Safadi (1990: 293), both discussing Syria, where work starts downstream and proceeds upstream to the watertable.

In al-'Ula, the owners of the land irrigated jointly own the *qanat*, and *qanat* did not necessarily have similar maintenance arrangements. The owners of one *qanat* carried out maintenance on a voluntary basis after a public announcement; owners of a second bought their water one day out of seven, the money paying for maintenance and repairs (Nasif 1988: 172). In Medina, *qanat* construction (Nasif 1988: 217–8) was financed by the builder/s, either according to the amount of water to be used by each individual, or the costs were shared equally and the allocation of water was made later. Each town had different customary law concerning the division of land, water distribution, and maintenance and repair. The villages around al-'Ula also had *qanat*, with varying systems of management: one village has the same as al-'Ula, since the *qanat* headmen and most of the *qanat* owners live in al-'Ula. The other villages and farmed areas were very small, and had "no customary law governing the maintenance and water distribution of the *qanats*" (Nasif 1988: 215).

Legal ownership and claims to the water of the Tid'il *qanat* in al-'Ula is presented by Nasif (1988: 173–4). The Tid'il supplied the domestic water of all or most of the inhabitants, and this water was available to all; rights to water for irrigation were limited to the Hilf section. Before the Tid'il reached the lands of the Hilf, the tunnel crosses the lands of the Shuqaiq section who needed the water for irrigation but held no shares in the *qanat*. The Shuqaiq made an opening in the *qanat* tunnel in its passage across their land, which the Hilf, although angry at the decrease in their own supplies, were unable to prevent. A tradition developed by which the Hilf would enter Shuqaiq lands and block any openings, while the Shuqaiq neither objected nor immediately re-opened the tunnel but waited until they needed water. This continued until King 'Abd al-'Aziz b. Sa'ud sent his first governor to al-'Ula, who proclaimed that any persons with a grievance not more than a hundred years old could bring forward their cases for his consideration. The Hilf elders came forward complaining that the Shuqaiq stole their water, and the case was referred to a court of Islamic law at Medina. Both sides presented their cases; the Shuqaiq said they had no shares and recognised the Hilf as the owners of the *qanat*, but they had become accustomed since time immemorial to taking water for irrigation from the *qanat*. The court decided in favour of the Shuqaiq, but the Hilf were dissatisfied and took the case to the Court of Cassation, who decided the Shuqaiq should have a fixed share of the water of Tid'il of a *wajba*, the hours from sunrise to sunset, or from sunset to sunrise. This right to a share was given in exchange for the passage given by their lands to the *qanat*. Other disputes eventually reached working or legal arrangements because of the importance of water to land.

Similarities between water customary law in al-Ula, Qalamoun and al-Karak may be noted; small communities manage without written contracts and customary law, compromises for the benefit of the parties are general, and the ownership of land carrying channels is as important as the ownership of water.

Deep wells were the other traditional source of water in oases. In Jauf, Wallin (1854: 140) says the wells supplied additional water to the gardens, with "every orchard contains generally one or more wells" whose average depth was less than twenty metres. Musil (1927: 472–3 with photographs) notes many wells in Jauf were from four to twenty metres deep, the deepest wells having tepid and somewhat salty water; "where the four gardens meet there usually is a well from which camels and cows draw water for the irrigation of the gardens." A *hâdhr* family of al-Wadiman of Jauf showed their *qanat* that carried water from a group of three wells five hundred metres to the gardens; this water irrigated the wheat and barley. The family had other wells, one of which was "very deep because the walkway was one hundred metres long, and this distance needed a camel to pull the wheel; cows only did the short distances. Someone had to walk with them, and the ropes were made of palm fibre and cut your hands. We had our own camels and training them to walk properly was quite difficult." At Qara, a well called al-Dda'y was shared between two gardens, one of which had belonged to one of the al-Tayyar who then sold it to a Rwala and the other to Mohammad Hamdan al-Bnaiyyir of the Rwala. Al-Dda'y was very old and deep, and was renovated by Mohammad Hamdan's grand-father, Hamad ibn Bnaiyyir. It was worked by "camels who walked up and down pulling a rope which ran over a (pulley) wheel, and this hauled up a leather bucket full of water which emptied into the *birka*, and then the water was distributed into the main channel. And the water was led to the trees by smaller channels. We put diesel pumps in about fifty years ago, and that was when this steep sloping shaft was dug for the pipe (to go from the well to the channel)."

Doughty ([1888] 1936: i, 335–7) describes the great well of Taima in the Hijaz, with its sixty *sawâni* wheel frames which allowed water to flow to a pond in each garden. Irrigation by *sawâni* from wells was more typical of Qasim, as in Hail, Buraida and 'Unaiza, and Nejd than the Hijaz. Altorki and Cole (1989: 40–2) describe the old wells at 'Unayza worked by one to five *sawâni*, wooden pulley wheels, attached to the trunks of *ithl* (tamarisk) trees. Each pulley wheel needed a camel (or person) to pull up the bucket. Irrigation by *sawâni* needed a high input of labour, and the camels were expensive to buy and to feed. Doughty ([1888] 1936: i, 593) says that landowners of Taima bought or hired camels from bedouin. The usual arrangement on larger farms was for the cultivator to hire a family to do the work of walking with the camels and overseeing the emptying of the water into the *birka*. Altorki and Cole (41) quote a woman, the wife of a sharecropper talking of forty years ago; "although my husband distributed the water to the fields, he hired a family to work the *sawâni*. We owned twenty camels, which were used to draw the water out of the well. The family my husband hired consisted of a man, his mother, and his wife and children, and they lived on the farm. They worked throughout the night and into the morning the next day. The man and his wife and mother would rotate in guiding the camels back and forth as they brought up the water. The camels were used in four shifts of five at a time. They had to be harnessed and this was usually done by the man. The women prepared a mixture of alfalfa and of straw or thorn, which was like a sandwich, and fed this to them. It normally required two people to feed the camels – the woman and one of her children or her mother-in-law. Occasionally, they also had to level the ground where the camels walked back and forth, and this was done mainly by the women. One person also supervised the pouring of the water from the buckets into the *birka*, where it was collected for later distribution to the fields."

Like the digging of *qanat* (Nasif 1988: 160) and *foggara* (Haj Ibrahim 1990: 312), well digging in 'Unayza, like Jauf and Khaibar, was not a specialised occupation. It was the most difficult and dangerous craft specialisation, but "anybody, whether qabili, khadiri or 'abd, who had a strong body could choose to do this work. It brought in good money, but the men who did died young. This was because of the smoke [dust] of the clay of the wells..... There were not many who did this work and they could work in other things as well." (Altorki and Cole 1989: 35–6). Opening a new well was common when Doughty was staying at 'Unayza in 1876–77; both deep stonelined wells that went through sand to a blue clay layer and then down to sandstone, and wide wells lined with brushwood to the clay layer were dug. At Buraida, he saw ([1888] 1936: ii, 355) a new well that had taken three skilled men twenty-five days work, and in Hail ([1888] 1936: i, 665), a well fifteen fathoms deep (c.25 m) dug by fifteen workmen in twenty days.

Following state development of physical infrastructure, financed from the oil economy, water for irrigation has been transformed, particularly in Saudi Arabia. *Sawâni* irrigation, especially, was costly both in terms of labour and camels, and limited agriculture. The first changes was the introduction of pumps. By 1961, there were 523 pumps in 'Unaiza (Altorki and Cole 1989: 98, quoting Sharif 1970). The newly established gardens around Sakaka from the early 1970s depended on pumps to deliver water from the new wells. The pumps have now been replaced by deeper wells, where the water is pumped by electricity, and delivered to central pivot irrigation systems, stationary irrigation machines or drip systems. The greatly increased use of land for agriculture has been effected by the reduction in costs of reaching underground water at greater depths, and carrying it to plants. Oil prospecting has had the additional benefit of revealing sources of geological or fossil water, as well as deep aquifers. The provision of electricity grids enables the use of deep pumps for raising water in the newly drilled wells. Plastics technology has developed piping to carrying water from artificial *birka*, lined with polythene, or tanks to the crop.

The first engine-powered water pump was installed in Jauf in 1368/1948–9, the first artesian well in Dumat al-Jandal in 1373/1954, and in Sakaka in 1390/1970. By 1404/1984 twenty four water supply schemes to villages of the region had been completed, and more have been installed since that date (al-Sudairi 1995: 173). Although electricity generation started in al-Jauf began in 1389/1969–70, generating capacity increased with a new plant in 1404/1984–5 and by 1992 all villages and many farms were connected (al-Sudairi 1995: 174). Each farm, large or small, has its own well and water tank; the larger farms use a number of central pivot overhead sprinklers, the smaller farms have one or use stationary irrigation systems. Water depth varies; at Sultanah outside Sakaka, wells are drilled to a hundred metres depth, while on a farm at Busaita the wells are at sixty metres. "Our farm here at Busaita is six hundred hectares, and we began it four or five years ago. We have five central pivot wheels, *mahwir rashâsh*, and five wells that go down sixty metres, but we got water higher than that. The wells are drilled deeper than the present water table to allow for the water that we will be using. It isn't fossil water, but an aquifer that comes from Turkey, we've seen maps of the water from the company."

On the western edge of the Jordanian *badia*, many people developed gardens using deep water to grow irrigated fruit and vegetables, initially for export to the Gulf and Saudi Arabia, now for the home market. A member of the Sardiyya, east of Mafraq, talked about his garden; "I have four hundred dunums of which two hundred are irrigated at the moment. The government put in my well, but I had to get a licence and pay for the well. It's very deep – four hundred and fifty metres, but the water is at three hundred and sixty metres. The well is deeper because the pump has to be under the water. The pump is electric, and it pumps a hundred cubic metres an hour. If there was no electricity, I couldn't have the well. To get the whole farm going, it cost one hundred and twenty thousand dinars – the well, the pumps, piping, clearing, ploughing, everything. The first year, we did well and I paid off most of the money. Then the border was closed into Saudi Arabia, and the profits aren't like they were. You need a lot of financial capital to do something like this." A Druze farmer at Umm al Guttein is developing an orchard on two hundred and fifty dunums; his well is deeper, at five hundred and ninety five metres deep. In the oasis of Azraq, the pumping of water to Mafraq, Irbid and Amman, as well as to local villages, has transformed Druze traditional farming based on cows, grain and fruit trees. "Water pumping meant *sayl* flows were lighter and shorter, so there was not enough *halfa* (the grass *cylindrica imperata*) for the cows. We always had wells but they were little, light wells, and we used the water for drinking and for some olive trees, pomegranate, dates and grapes. In the fifties and sixties, diesel motors for wells came, they were British motors. In the sixties we grew vegetables using water pumped from the wells, there were more people here. The water table is very irregular, no one knows why. Sometimes it's a few metres from the surface, sometimes it's a hundred metres down. People say the water really comes from Turkey, deep underground. But it's difficult to know, and it's difficult to plan. When I dug my wells in the fifties, I dug seven metres down. In the late eighties, they drilled down one hundred and twenty metres for me to have water; now it is going dry."

Water management systems in various parts of the Bilâd ash-Shâm have parallels with those of other parts of the Arab peninsula and the Middle East. The *foggara* of Syria and *qanat* of the Hijaz are like the *aflâj* of Oman (Wilkinson 1977) and of Iran (e.g. Beaumont, Bonine and McLachlan 1989). d'Hont (1990; 1994), Geyer (1990) and Northedge (1988) give accounts of water systems on the Euphrates in Syria; these systems included various types of wells where water was raised by hand or animal drawing power for supplying the needs of animals and people, and systems of raising water for irrigation by animal traction either in a straight line or in a circle; both wells and animal powered irrigation systems were practised by Agaydat, while a *nûria* system, raising water by hydraulic power, was the property of townsmen on the Khabur river. *Nûria* also existed at Ana. Mundy (1995), among others, describes a water management system in Yemen, and Vidal (1955: 114–148) the system of the al-Hasa oasis. Bruins (1986) discusses the viability of rainwater – harvesting agriculture in the Negev over history. Water use by Bedouin in Upper Egypt is shown by Hobbs (1992), while Costa (1991) gives a brief indication of water management in Musandam.

CHAPTER 5

LAND USE; THE PRACTICES OF PRODUCTION SYSTEMS USING LAND AS A PRIMARY RESOURCE

HUNTING AND GATHERING

These oldest food production methods continue although, for some thousands of years, rarely and solely for primary subsistence. Their importance for subsistence and for exchange or trade was greater for groups like Hutaim and Sulaib and for individuals from other groups up to and including the nineteenth century. In the present, hunting and gathering are for seasonal or occasional consumption and pleasure, with a small market component for certain products. Rock carvings show the use of dogs, nets, bolas, bows and arrows, traps, spears, and sticks for hunting. Falcons have been used since the pre-Islamic period (Allen 1975; Jabbur 1995: 132). Guns became the main method by the early nineteenth century. Doughty reports ([1888] 1936: ii, 116) a distinction between *gannâs* (a stalker of large ground game with a gun), and *sayyâd* (a hunter with hawk and hound). While many hunted on foot, falconry took place from horse or camel back. From the 1920s sport hunters used cars; from the sixties, the very rich hunted gazelle and oryx from small planes in the south of the peninsula. The section concerns hunting for use, not for sport (for falconing, see Musil 1928a: 31–35; Allen 1975; Jabbur 1995: 135–40).

In the extremes of hunger, anything can be hunted and eaten. Hunting is usually opportunistic, depending on which animals or their tracks are seen, or on what animals are known to be in the area in a particular season of a year. The usual prey for hunters in the *badia* are rabbits, quail, partridge, sandgrouse, rock pigeons, or wading birds on migration; bustard, gazelle and ibex are now protected, while ostrich and oryx were rare by the early years of this century. Shooting is the prevalent method, following stalking or a chase. Wolves and hyenas are shot when thought to be attacking flocks. Hunting is enjoyed as a skill, and in addition to the meat, the skin and/or horns are used for bags and knife handles, or sold.

The *dhubb*, a large lizard, is tracked and killed for food in sand areas of the *badia*.

The sale of ostrich skins, feathers and meat is frequently mentioned. Burckhardt (1822: 351 and 1831: 219–20) records sales of feathers in Aleppo and Damascus by traders who had bought from Bedouin, and that Shararat sold eggs and meat in al-Jauf but traded the whole skins in Damascus. Hutaim hunters sold ostrich skins and feathers to traders accompanying the Pilgrimage (Doughty [1888] 1936: i, 173–4), and Sulaib sold to merchants in cities (Jabbur 1995: 127). In the 1870s the value of two ostrich skins, the number a skilled hunter could expect to catch in a year, was equivalent to that of two riding camels (Doughty [1888] 1936: i, 325). Musil (1908: iii, 74) recounts an old story in which a shaikh of the Beni 'Amr wears a cloak of ostrich feathers. Ostrich feathers were used for decoration of lances (Burckhardt 1831: 53 for illustration), women's litters, and in the trappings of riding camels and horses, and the Markab of the Rwala is adorned with feathers. Ostrich feathers decorating lorries and taxis are now imported. Ostriches were hunted in a variety of methods; from horseback (Jabbur 1995: 127), by throwing a snare of ropes and stones to catch their legs (a Rwala), or on foot by Sulaib hunters hiding behind high piles of stones and approaching them against the wind when the birds migrate (Musil 1927: 255).

Gazelle hunting was particularly associated with Sulaib (Burckhardt 1831: 14–5). Doughty ([1888] 1936: i, 613) saw Sulaib coming into Taima with their donkeys loaded with gazelle which they sold to the villagers for meat. Musil (1927: 216) quotes Sulaib families as owning hunting territories, on which they also grazed their herds of white donkeys. Gazelle were also hunted by peasants from the desert edge villages like Qariatain (Jabbur 1995: 361–6). Gazelle move northwest from the Inner Desert in early summer to drink from springs or pools; near these are *masiyyida*, very large open spaces on the plain with a wall surrounding the space on three sides. In parts of the wall are gaps, and on the outside deep pits are dug. "When the hunting is to begin, many peasants assemble and watch till they see a herd of gazelles advancing ...... towards the enclosure, into which they drive them; the gazelles, frightened by the shouts of these people and the discharge of firearms, endeavour to leap over the wall, but can only effect this at the gaps, where they fall into the ditch outside, and are easily taken, sometimes by hundreds. ..... The gazelles thus taken are immediately killed, and their flesh sold to the Arabs and neighbouring fellahs. Several villages share in the profits of every masiidah (*sic*)" (Burckhardt 1831: 220–1). In the *harra*, a Sharafat told of his uncle's use of masiyidah. Here these are lines of stones extending for two or three kilometres that push the gazelle towards a hide where the hunters are waiting; "the gazelle were herded towards the hidden hunters. The last time my uncle hunted like this, he got twelve gazelle." He claimed his uncle was using this method until ten or fifteen years ago. The Ma'aza of Upper Egypt are not granted firearms permits (Hobbs 1992: 41ff) but hunt gazelle using dogs, wheeltraps, or stealth and stone-throwing. Gazelle numbers dropped fast when hunting from cars with automatic weapons became popular with individuals from outside the area. Attitudes to hunting change so that remarks like "I used to shoot gazelle, but now I get more pleasure from watching them" are common.

Oryx (*mahâ'*) live in the sands of the Nefud, while in the *harra* there are many enchanting carvings of oryx, some with young. Shararat and Sulaib hunted them for trade. Doughty ([1888] 1936: i, 613) was told that this meat was the most esteemed of all game, and the skins were used to make sandals; he saw oryx skins at Ma'an in 1875. The long straight horns "were common at Taima; the most are brought in by the Sherarat and bestowed upon their town friends, who have them, in their ware-rooms, to break up any hard clotted store of dates: I saw that the Taima Sulubba (Sulaib) families used them for tent-pegs."

Ibex (*badan*) live in mountains and need water more than gazelle, ostrich or oryx. Summer is the usual hunting season, since their need for water is at its highest, and the males are in peak condition before mating. Hunting is on foot, with a gun, and with or without dogs. Tristram (1873: 288, 292 and 306) met ibex hunters from Beni Hamida and Beni Sakhr who hunted in the mountains between the Balqa and the Jordan Valley; the Beni Sakhr hunter (306) owned agricultural land personally, cultivated by slaves. Carruthers (1935: 82) "witnessed the unusual sight of two Rwala youths hunting the wild-goats on foot; with the aid of falcons and long-dogs (salukis) they literally ran them down." Hobbs (1992: 42–3) watched ibex hunts by Ma'aza in Upper Egypt in the 1980s. Using a dog to attract the attention of the ibex while the hunter aimed a stone at the animal was one of the two principal ways of hunting ibex; possibly a third of these hunts succeeded. The other technique is the wheel-trap, a ring fifteen centimetres in diameter made from palm fibre, with a circle of inward pointing palm leaf-ends or plant spines. The hunter digs a pit ten centimetres across and thirty deep in the ground on an animal path or near a food or water source, and places the trap over it. On top he puts a noose with a slipknot, attaching the other end of the line to a notched stone or heavy stick, and buries the trap in shallow sand. The ibex or gazelle steps through the noose and ring into the pit below, and as it tries to free itself from the spikes, the slip knot above the ring tightens, and the stone hinders the animal's escape. The heavy stick, preferred by the Ma'aza, has been blackened in the fire; the animal can run but is slowed down by the ring and stick, which leaves marks on the rocks and eventually becomes lodged amongst the boulders and the hunters catch up. The Ma'aza used these techniques after the loss of their guns, having studied ancient remains of old hunting sites along ibex paths. An elderly Menaja'a described an ibex hunt at Fainan in the Shera mountains when he was a young man; "there were five or seven of us, all young men, and we hid ourselves before dawn downwind of where the ibex would come to drink. Seventeen ibex came down. One of us had hidden separately, and he moved so the ibex got wind of him, and so they moved down towards where the rest of us were hidden. When the ibex got in range, we shot one with a rifle, only one because there weren't enough people to eat more than one." The same man described how he and his nephew had watched an ibex drinking at the spring below their garden in the mountain foothills that morning. Like the Ma'aza, the tribespeople around Fainan used to rear ibex and gazelle kids with the goat herds; "the ibex kids often died, but the gazelle did very well, and when they were adult, they returned to the wild," unlike the Ma'aza where the ibex kids often did well but left the herds when adult, while the gazelle would remain (Hobbs 1992: 95).

    Hunting for falcons is a popular and potentially profitable late summer activity in the *hamad* by parties of men from the tribes, the Qalamoun villages, and the new settlements along the main roads of eastern Jordan and northern Saudi Arabia. Falcons are caught by flying a pigeon, wearing a snare, as a lure; when the falcon strikes the pigeon, its feet are entangled in the loops of the snare, and both birds tumble to the ground and can be caught. The falcon is then trained for hunting (Allen 1975: 117–20) and sold or given as a present to those known to have a passion for the sport. Presents of a good falcon yield valuable return gifts, such as pickup trucks, cars or money. The falcons are used for one season, and then returned to the wild.

The collection of wild products was important for food, craft and in supplying products to urban industry. At present, wild foods collected for sale are truffles (fagu') from the *badia*, and mallow (*khubbayza*), a thistle (*akûb*), and fennel (*shaumar*) from the farming areas to the west. Medicinal plants such as citrullus (*hanthal*), teucrium polium (*ja'ada*), thyme (*za'atar*), achillea (*gaysûma*), artemesia herba-alba (*shih*), camomile (*babûnij*), sage (*marmariyya*) among others are collected for personal use and distributed along family networks, and are on sale in Amman and other cities. A wider range of spring-time wild foods are picked for family use and used as flavourings for milk, bread and cereal dishes, eaten raw, or baked. *Hamsis* (sorrel), *jarjîr* or *arûga* (senecio glaucus), *rishâd* (lepidum sp), *bakhatri* (erodium sp), *dha'lûk* (scorzonera tortuosissima), *kusaybra* (pimpinella), *hamaitha* (rumex sp), *khafsh* (brassica tournefortii), *tummair* (erodium ciconium), *kharrît* (allium sp), *shahhûm* (gagea lutea), *roghol* (atriplex leucolada), *rubahla* (scorzonera papposa), *mahrûta* (unknown) are some plants of the *harra* and *hamad* used in these ways. *Fagu'*, a sort of truffle, were and are collected in the *hamad* after rain for household consumption and sale to the towns. Although the trade in gums and *kilw* was reduced in Sinai and Southern Jordan by the late nineteenth century, it continued in Palmyrena until at least the thirties. Weuleresse (1946: 307) mentions the regular sale of *kilw* to Aleppo, Antioch and Iskanderun by traders from Sukhne, who returned with articles for sale to tribesmen; they also took wood and charcoal to the Euphrates villages, and sold truffles, gums, liquorice, salt, skins, saltpetre and antiquities to other traders.

Semh (mesembryanthemum sp) grows on sand and gravel plains between Ma'an and al-Jauf. It was used as a flour substitute, and mixed with dates into a long-keeping paste tasteing like chocolate. From al-Jauf *semh* was traded to Najd (Wallin 1854: 126). For a good crop, *semh* must have October rains. After harvesting, the seeds were put into sacks and left in rainpuddles to loosen the husks, then threshed. The seeds of the grasses *thummâm* and *hawwa* (launea nudicalis) were also harvested as flour substitutes.

Wild tree fruits are collected. In the *badia*, the fruits (*tel'*), of the *msa* (nitraria retusa) were boiled down into a syrup. This shrub grows in the Wadi Sirhan, al-Azraq and the Wadi Araba. In the Wadi Araba, the fruits are known as *ghurqâd* and used to sweeten bitter water. The *za'arûr* (cratageus azaroleus) grows in the Wadi Rajil, in the wadis of Jabal ad-Druze, and Qalamoun as well as in a few places in the wadis of Karak; its fruits (*nebk*) resemble like tiny apples. Wild figs (*hamat*) and wild dates (*balah*) are common in the wadis of al-Karak and the Shera; children eat the fruits, which are also fed to animals. Acorns were eaten in the past, certainly around Ajlun and as-Salt, when the cereal harvest was poor. The fruits (*dawm*) of the *sidr* (zizyphus spina-christi) were eaten raw or ground into flour. Another was the fruit (*khzhama*), of the terebinth (*butm*) tree, widely used for oil or ground to mix with flour. Musil (1928b: 34) mentions the *butm* trees on the eastern foothills of Jabal al-Ruwak, whose fruit was picked by the 'Umur and peasants from Qariatain and Tadmur for the oil. De Boucheman (1939: 79–81) says that the people of Sukhne collected and processed the fruit for oil, used at home and sold among the tribes and in the towns. *Butm* oil was noted by Behrnauer (1860: 382) as used by the proprietors of cooked food stalls in urban markets. Many older people around Karak recalled *khzhama* oil saying it was more delicious than olive oil. *Yâsir* (moringa peregrina) grows in only a few places in the western Shera, but is an important tree in the mountains of Upper Egypt for collection of its fruits (Hobbs 1992: 40). A Marazga at Fainan, where there are some trees, commented that *yâsir* seeds have very good oil or "*samn*; people crush the seeds between stones and the *samn* comes out." Locusts were also collected for food; a trench was dug, filled with green and dry brushwood and set fire, so that the locusts fell from the clouds of smoke. Dried, and ground into flour, they taste not unlike spinach.

Gums from trees and plants were collected. Musil (1928a: 95) writes that the Rwala collected a sweet juice from the *rimth* (haloxylon) plant and boiled it into a syrup. Burckhardt mentions the collection of gum from acacia trees in the Wadi Araba for eating, as it was highly nutritious, and sale (1822: 446), and of 'manna' from tamarisk in Sinai (1822: 601). Fifty years earlier, Hasselquist (1766: 250) says that "gum (is) gathered in vast quantities from the trees growing in Arabia Petreae" and sold to dealers in drugs in Egypt and called by them frankincense. *Butm* (terebinth) and other gums from various *badia* locations were sold in Damascus, Seetzen was told by his guide Yusuf, who had been a trader with the tribes (1854: i, 270). Gums from acacias, pines and terebinth are mentioned by the Arab writers on the duties of the market supervisors collected by Behrnauer (1860: 391; 1861: 6, 15), and used as additives to expensive and imported aromatics and perfumes, and in the manufacture of candles. The same author records other plant products, both cultivated such as cotton seed and apricot kernel oils, and wild like oak galls, rocket, peppergrass, wild lettuce, and horned poppy, that were added to imported aromatic oils, spices and medicines; *akûb*, a thistle species, was mixed with haricot beans and sugar to make a popular sweetmeat. *Kilw* (salsola sp) was gathered and burnt for its ashes, and sold to traders for soap production in Nablus and Jerusalem before the use of imported caustic soda from Europe (Abujaber 1989: 135, 137: Burckhardt 1822: 354; Doumani 1992: 203–4, n.76, 300: Seetzen 1854: ii, 360 by Beni Hamida to Jerusalem, Nablus and Hebron). *Kilw* was also collected by villagers of Qalamoun (Seetzen 1854: i, 279–80) who sold to Damascus, and of Sukhne (de Boucheman 1939: 77–79), who describes (88–9) the *kilw* caravan from Sukhne to Aleppo and Antakya. *Kilw* was taxed by the Ottomans in the Pashalik of Damascus (Mantran and Sauvaget 1951: 22–4), used locally and exported to Europe. *Kilw* was the general cleaning agent in markets (Behrnauer 1860: 379). Tanning agents included acacia bark, the roots of rhus triparta (*'irn*), and Burckhardt (1822: 351) mentions the collection of summac leaves by the people of Salt for sale to the tanneries of Jerusalem.

Wood was collected for craft work. Burckhardt describes coffee mortars made from *butm* and oak wood, and Doughty ([1888] 1936: i, 324) the well pulley wheels, pack and riding saddle frames, and milk bowls made for sale by Sulaib from acacia wood. Some Beni Hamida recalled making mortars from *butm* wood. Agricultural implements, such as ploughs, and handles for mattocks, hatchets and saws, used local woods. In the heavy soils of northern Jordan, locally growing oak was the preferred wood, in the lighter soils of southern Jordan, local willow and juniper were used (Palmer and Russell 1993). Woods used for beams, doors and roofing are local, and change through ecological areas. Tent poles and attachments for ropes are made from local tree stems where there are suitable trees. These trees, such as tamarisk, juniper, sidr, acacia, and willow, are coppiced. Tamarisk is fired to get the old wood for firewood, and to produce new growth for browse for animals and future use. Trees and shrubs used for building wood were also coppiced, as described by people in Karak villages, in the Shera, and around Fainan. Trees in the *ghutas*, such as white and black poplar, and cypress, were grown for packing cases, joinery and building use

(Wetzstein 1857: 478); in the oases, tamarisks were and are grown as windbreaks and for building wood.

Charcoal, firewood and dung were used for cooking and heating before the introduction of primuses and paraffin, and then gas and gas stoves. Damascus was supplied at least partly from the Jaulan tribes, using oak, and the tribes of the Ledja supplied charcoal to Hauran villages, using desert shrubs (Burckhardt 1831: i, 17, 19). The gunpowder and saltpetre manufactories in the Ledja and Jabal al-Arab used wood from the desert for firing. Wetzstein (1857: 480–1) describes the contigents of peasants from the eastern villages bringing in donkeys laden with bundles of hempstalks for the city's bakeries, and the enormous loads of *shîh* brought in from Qalamoun for restaurants and domestic households. In the *badia*, *shîh* is regarded as inferior firewood, *ajram* and *ghadâ* having a much higher reputation. Charcoal burning was carried on in the mountains north of Sukhne (de Boucheman 1939: 80), and continues to be made on a small scale for sale to restaurants and coffee-houses for use in *narghilas* (waterpipes or 'hubble-bubbles').

It is generally assumed that the production and sale of charcoal and firewood must be destructive. Much of the wood or woody material is a by-product of other activities. Prunings and the stalks of *dhurra* crops are obvious examples, but offcuts from carpenters' shops are sold as firewood, as are trees from landclearance for building or a change of crop. Local people coppice tree species such as acacia and oak, and use dead wood from suitable perennial desert shrubs, or take wood from species that shoot again when rain falls. There are considerable amounts of dead wood, from lack of rain or old age in the *badia*, or from winter storms and floods in the mountains. A young man near Ajlun said charcoal was more profitable than selling firewood; in this area, charcoal making is a by-product of clearing land for gardens for grapes and fruit trees, for which approval has to be granted, and then a license obtained to fell a tree. Most village and tent dwellers use gas or paraffin for cooking and winter heating, with collected firewood used for baking bread, or coffeemaking, if used at all. Some *badia* tent dwellers buy firewood, builders' offcuts or old olive or fruit trees, while some families from *badia* edge villages go and collect firewood from *badia* shrubs. Dung from camels, cows or sheep and goats is used as fuel for cooking and heating. It should not be used for *sâj* bread (bread cooked on a thin metal sheet over the fire), but is used for baking *tabûn* bread, where the dough is inside an oven.

The use of firewood is careful and economical. For cooking a kid in yogurt, thirty-five loaves of *sâj* bread, two lots of tea, and heating water for hand washing before and after eating, a herder's wife east of Qadisiya used three branches of oak wood, two about 120cm. long and 7cm in diameter, the third 30cmlong and 20cm in diameter. The cooking used 30cm of each of the longer logs and 15cm. of the shorter. As cooking finished, the logs were carried out of the tent, water (previously used for washing the woman's hands after kneading dough and the kneading tray) poured over the burning ends, which were then pushed into loose earth to completely extinguish them. A fire for making a pot of tea in an orchard used five twigs of dead apricot wood, each 10–12 cm long and 2.5cm in diameter, and a handful of dry grass. For making *farîqa* (roasted green wheat), the fire was a bundle of dry thistles and weeds about 30cm in diameter.

The Ma'aza (Hobbs 1992: 39–41) continue to gather plant products for sale. Their views (Hobbs 1992: 103–9) on the importance of trees for their environment and way of life are echoed by people using the Shera, with a similar ecology; "the trees support all the life; if the trees go we have nothing, there would be nothing. Why does the government think we destroy trees? It is for us that the trees are important." A Beni Hamida in Sirfa on the western Karak plateau recalled scattering seeds of useful trees and perennials. An Azazma near Fainan said his family used to plant *yâsir* seeds to increase the number of trees for their seeds and leaves for feeding goats "before there were all these gardens everywhere". A Rashaiyida at Fainan said he and his family scattered seeds of certain trees and perennials. Hobbs (1992: 46) found similar practices by certain Ma'aza. In the Ajlun countryside the local economy used to focus on goatherding, fruit cultivation for sale, and wheat grown for the family and sold only to bedu from the east. The local trees of oak and *butm* were regarded as what the local flora and fauna depended on; "even if there was no proper rain for two or three years in a row, there would have been stored grain or people used acorns from our oak forests, or they would move to relations in other areas. If there was no real rain over three or four years, young tree would die, but the mature trees would survive because their roots go so deep. It's getting difficult to keep goats because of the decrease in wild grazing for them, with more farming and reserves. But goats don't stop young trees growing if they are herded properly, because they eat only the leaves and pass on quickly; the stems regenerate. They don't eat acorns and *butm* seeds when the seeds are in a condition to germinate, only when they have been blown off the trees in the winds or fall because of galls." The same speaker said that *za'arûr* (cratageus) is now used as a rootstock for pears, plums and almonds as they are hardy and suit the environment. Another informant said that *butm* was used a rootstock for pistachio. *Butm* fruit is sold in the Ajlun countryside as animal feed, while *mallûl*, the acorns of the sindian oak, are sold for tree seeds.

The destruction of trees is associated with the demand for wood for building by the late nineteenth century Ottoman expansion, the Hijaz railway, and finally with the expansion of agriculture that coincided with a replacement of local tree products by imported foods, increased commercial olive oil production, and imported building materials of steel beams and cement. This view is drawn from the countryside of al-Karak and the Shera, but is also voiced in the *badia*. Imported sugar replaced local wild plant syrups and gums, imported cooking oils and olive oil replaced *butm* oil, brass mortars replaced wood mortars, and metal and plastics replaced wooden (and leather or pottery) containers. Expansion of agriculture, central government demands and increasing populations all affected a decreasing use for wild trees and shrubs.

The traditional oasis agriculture of the Arabian peninsula, based on the cultivation of dates, cereals, and some fruits, has examples in the southern extremities of the Bilâd ash-Shâm in the Wadi Sirhan, al-Jauf and Sakaka. Dates were grown in the Aghwar of the Jordan in the mediaeval period, but since then have declined and are not mentioned as cultivated in the nineteenth century. A few gardens exist now for fresh eating dates. Wallin (1854: 140) says that the gardens of al-Jauf grew mostly dates, some figs, apricots, peaches, oranges and grapes, a few vegetables, and "between the trees in the orchards the inhabitants sow corn, the produce of which generally suffices for their wants." Visiting seventy years later, Musil (1927: 472) mentions the excellence of the dates; figs, oranges, lemons, apricots,grapes and vegetables grew luxuriantly in the gardens. As well as wheat and barley growing in the gardens, there were small corn fields outside the town.

A similar but more detailed picture of traditional farming in al-Jauf was given in 1995 by Hamad Uthaiman al-Wadiman of a *hâdhr* family from the Bani Tamim. "We grew dates, and they were what we lived from. We also grew in the gardens wheat, barley, *dhurra*, lentils, chick peas, *tranj* (knobbly lemons), oranges, grapes, apricots, olives – yes, we've always grown olives – and all kinds of vegetables except onions, they came down from Damascus. We could never grow enough wheat as it all had to be irrigated from the *'ain*, and the wheat and barley had to grow in full sunlight, otherwise they don't ripen. ...... All the gardens were attached to a *qasr* (a unit of stone houses), and each garden had a well. Our *qasr* was fifteen families descended from a male ancestor, and a few craftspeople. We didn't have slaves, we did all the garden work ourselves, and we built the houses. The *qasr* was a bit like a little town, each family had its house and there was a common mosque but there wasn't a market. Each garden was owned by a family, but the work of the gardens was by members of the whole group. People could rent or buy individual trees, or you could buy a hundred riyals worth of a particular tree's crop, you could rent or buy a whole garden. Share farming was very common, because the numbers of the men of the qasr went up and down a lot. It was possible to swap trees, so you could get your trees altogether, and you could swap bits of land; you could buy a tree with the land it grew on or just the tree. Sales of trees and of land were nearly always between close *ibn 'amm*. ......

Dates were our food, and each variety has its own qualities – flavour, keeping qualities, thickness of skins, time of ripening and so on. *Hamra rashîd* has very thin skins, so the very old people like them, and people with gastric trouble. *Hulwa* are the big black sweet ones, and *mari'ayya* are like that except that *mari'ayya* are very late; *siatiyya* goes on ripening after they're cut. We like *buwaitha* and so do the bedu because it dries well and isn't sticky; there's another variety, *saur al fagîla*, that's similar. We used those two for putting at the bottom of the date stores because the juice of the dates from other varieties on top would trickle through and you could collect the syrup that was made. We sold the dates to the bedu at the house, and they paid in *samn, bagal* (dried yogurt), lambs and wool, and we'd keep some and sell the rest on; we did this until very recently. ......

"Dates need a lot of care. They don't really have diseases, but they do need careful management. Young dates, until they are five or six years old, have to be watered every five to six days. After that, once a month is enough. The level of water in the wells goes up and down, so water can be a bit short sometimes. A few of these trees are a hundred years old, some are forty, some are fifteen years and a few are new. They all come from offshoots. The ones from self seeding are called *digla*, and we feed their fruit to camels and sheep. It's possible not to fertilise the dates properly; the male organ (*abar*) has to be tied into a bunch of the open female flowers. You can leave too many bunches so they are over-crowded. If you don't harvest the trees, they will bear the next year but they bear very badly the year after that; and dates that aren't harvested go bad, so if they are left to be done with the next harvest, the whole lot rot. Some varieties of dates have to be dried; they have to be washed in clean water and then dried in the sun. Another kind mustn't be washed or get damp from rain or dew, or it won't keep. Dates will keep for three to four years with no problems; and you tasted the dates in our house that are seventeen years old! Dates were profitable until about twenty years ago; we got six hundred riyals a basket. But then the development of agriculture came, more wells and pumps and pipes, all that, and Taima was developed early, and expanded their date gardens enormously. Because Taima is nearer to Jidda, they could undercut us. Jauf dates are a bit late, and people always want the early ones more.

"We keep the land clean under the trees to keep down snakes and scorpions. We had cows for ploughing in the gardens, and for their milk, and to eat up the grasses and weeds. And they ate rushes round the channels, and *digla* dates, and datestones and stuff. Now we have only one local cow left, for milk for the house. That barley under the trees, that's for fodder for the sheep; we cut it green and feed it to them. We can't make a living from dates any longer, the price has gone down so much, and we have jobs – I work in the Post Office – so we have Syrians to work in the gardens."

The basic implements were the *ard* (plough), an iron spade for digging the shallow water channels, and to close and open these up for the control of water flow; a mattock (*fass*) for clearing and hoeing; a leveller for the irrigated beds of earth; a curved knife was used for pruning, and cutting bunches of dates and date fronds.

People in Sakaka and Qara agreed that it was not now possible to make a living from dates. Dates are no longer a basic food for the majority of families since wheat production from irrigation systems using wells and central pivots has been developed, and encouraged by the above market prices paid by the government. The owner of a garden in Qara said that their gardens provided dates, pomegranates, oranges, peaches, plums, grapes, and a new fruit from Turkey, called 'cows' eyes', vegetables for the family, and fodder for the house sheep and camel; they sold grapes. They had developed commercial farms on new land. One farm grows wheat and barley with a centre pivot system, the grain being sold to the government, while the other has forty plastic tunnels growing tomatoes, cucumbers, *cousa* (a type of courgette), aubergines and sweet melons, using Dutch seed. They saw commercial vegetable growing in Jordan and employed a Jordanian to set up the farm. The Dutch varieties of vegetables are universally acknowledged to be tasteless compared to local traditional varieties, but the latter do not all ripen at the same time so that there are not the quantities to make up a lorryload for the wholesale market in Sakaka. From the wholesalers, the produce is sold to local shops and to traders from Turaif and 'Ar-'Ar.

There are numerous farms on patches of suitable soils in al-Jubah. The soils vary; Nathayim earth is said to be poor, Swair and Hdaib earth is better, while the earth at Mughaira is better than any of these. People register land by applying to the Ministry of Agriculture and Water for a parcel of land they think suitable; checking by the Ministry that the land is unowned and that the applicant's development plan is reasonable takes two or three years. If approved, provisional ownership is granted for five years, and provided that some of what was proposed is achieved, deeds are then issued. This means the land can be sold or, what is quite common, the farm is turned into a pleasure garden, providing some fruit and vegetables for the house, growing fodder for a camel and some sheep, perhaps a market crop of onions or grapes, and a place for picnics. To make a farm profitable takes capital, time and knowledge. When the government was pushing for the development of agriculture, partly as a means of settling former nomads, the subsidies available and high guaranteed prices for wheat and barley meant that the initial costs of putting in wells and irrigation could be paid off with two good crops of wheat. This is no longer possible as costs for diesel, electricity and water have risen steeply, while prices for wheat and barley have fallen with the removal of subsidies. Many people who developed farms now say the farms hardly pay their way, and farms would never have been developed if people had had to pay for the necessary infrastructure themselves.

People say that farming, to be profitable, has to be either on a big scale with professional staff or at a garden scale where one or two people do all the work without machinery. An example of serious farming may be seen in the farms of Mashur az-Zaid, described by his second son. "We have four farms, altogether we have one thousand, two hundred hectares. There's a small farm at Mughaira which grows only olives, a farm at Sultana, one in Busaita, and one at Mugwa towards Tebuk. The Mughaira farm is run by a Jordanian who understands olives. We took everything step by step, trying things out and building on results. The first olives we planted were too close together and they didn't do well. The later olives do well, partly because when the young trees come from Jordan, the truck stops outside the border post and the men take the plastic pots off the trees and lay them on wet sacks, so that the customs search is over quicker. The trees are planted within twenty four hours, at the most thirty six hours. If the pots are left on, the journey can take three days. We put a mix of sheep and camel dung on the soil before planting, and later, when the trees are established, we use a little chemical fertiliser. The olive trees are irrigated. The olives go to the press in Sakaka for processing.

"The Sultana farm started as a place for entertaining and is the home farm. There is a small sheep flock brought in from the main *badia* flock for milk for the house. I've made a new enterprise of egg production by hens in deep litter housing equipped with cooling fans. The birds are bought locally as day old chicks, fed on imported and subsidised feed, and looked after by Indian workers. It is profitable and successful. We use the manure for fertiliser on the olive trees at Mughaira and Sultana, and for the grapes at Sultana. Grapes grow very well, but much of the crop is left to rot on the ground as the market for fresh fruit is over-supplied, and we cannot get a licence for a juice factory. You know, everything we had for breakfast came from this farm; two sorts of olives, sheep's butter, *labna* balls, maqdus (pickled aubergines), camel's *laban*, whole apricots in syrup, and the apricot paste. And we supply our families (the families of our relations, of former retainers and current employees) with *laban*, *labna*, meat, olives, olive oil, apricot preserves and so on.

"The Busaita farm is larger than Mughaira or Sultanah at six hundred hectares, and we started five years ago. It has five wells and five centre pivot irrigation systems, and we grow commercial cereals and we have one thousand, seven hundred sheep in three flocks. Wheat was the most important crop, and the sheep were fed on subsidised *alaf* (bought in feed) as well as farm straws and *barsîm*. Growing wheat year after year lowers soil fertility fast. After two or three years of wheat, we used to leave the land fallow for a year or two. Now we grow a water melon crop and this works well, it's better than turnips for a break crop to improve the soil. Barley is less demanding. *Barsîm* takes a lot of water as it must be watered every day; one crop grows for six months while yielding regular cuttings. Wheat and barley take three months to grow, and then we leave the land bare until the next season. We put on herbicides and pesticides through the overhead pivot lines, and we do use chemical fertilisers, but not much and I'd like to try other break crops to lower our fertiliser use. We have two big tractors here – the older one is going to the Mughaira farm and we'll have a new one here, two combines, a big baler, a big deep spring-tine harrow, and a big set of disc harrows. The baler is for the *barsîm*, we cut, dry and bale it for hay. And we bale the wheat straw, and use it to make the wall round the edge of the farm to try and stop sand blowing in. We have to harrow the soil (sandy clay) three times before sowing, the tilth has to come over your ankles. Our big problem is sand getting in the rotating parts of the pumps.

"Now the government has reduced the price they pay for wheat, and the subsidy on *alaf*, we'll grow more barley and oats – and maybe *barsîm*, except that is almost cheaper to buy *barsîm*. There's a quota on how much wheat the government will buy, and they don't buy barley any more. The barley and oats would feed the sheep, and they are profitable. The sheep are in flocks of between five to six hundred ewes, and every hundred ewes has four rams. The flocks are in matrilines, and I'd like to close the flocks. Inside the flock, they're herded in age groups. They lamb twice one year, and once the next. The lambing rate is now 125%, up from 75% – we had brucellosis. We replace about 10–15% each year. At the moment the sheep are feeding on wheat stubbles and volunteer barley – it drops through the combine at harvesting and then grows during the winter. And they get chopped barley straw and a kilogram of barley a head a day, as they've either got lambs or they are about to lamb. The vet is here today to inject the bigger lambs for foot and mouth, and for clostridial diseases. The chief shepherd is Syrian and so is his deputy, they are very good and knowledgable. As well as looking after the sheep, they and the other workers do the crop work on this farm.

"We've grown potatoes at Busaita; the soil's good for them. We did thirty five hectares with varieties from Holland, Lebanon and Syria; the profit was 1.5 Saudi riyals on every five kilograms, which is alright. The Mugwa farm is all cereals by centre pivot irrigation so far, and there are no sheep yet. There are profits in agriculture if you know what you are doing and you employ good people who know what they are about."

Shifts in the patterns of farming described from al-Jauf and al-Juba are similar to those reported by Altorki and Cole (1989) for 'Unayza. The changes are seen as directed by government programmes. In Jordan, farmers of traditional irrigated systems and rainfed agriculture perceive changes to be responses to shifts in market demand and imports of technologies, and partly related to government actions. The traditional agricultural economies were three-sided. In the mountainous plateaux, staple cereals, commercial crops of dried fruits, and sheep and goats were noted by Burckhardt (1822: 350–1, 403–6, 418) and Musil (1908) eighty years later. Irrigated valleys produced grains, commercial vegetable or industrial crops like indigo, and had cows and goats. Rainfed plains grew grain, had commercial sheep and goats, and provided commercial services with camels and donkeys. Commercial exchanges took place through the Pilgrimage (Burckhardt 1822: 405, 418–9), to the Bedouin (Burckhardt 1822: 403; Wallin 1854: 123; Abujabr 1989: 277), and to merchants who came out from towns (Burckhardt 1822: 350, 391, 405; Hill 1896: 36). Exchange has been a longstanding component of agricultural production in the Bilâd ash-Shâm whether or not regions were administered by a central government (Lancasters 1995), while the producing family had first call on the harvest. The current agricultural economies of the three landscapes continue to be founded on grain, sheep and goats, with a shift to fresh fruit or vegetables. The commercial aspect of agriculture varies. While some farms are virtually exclusively commercial, many of the smaller have as their main function supplying the families of its owners (and/or sharecroppers) with supplies or shares of cash sales that supplement wages or pensions.

Agricultural practices in irrigated gardens are described from four areas of Jordan – Fainan/Wadi Araba, Dana, Safi and the Wadi ibn Hammad/Karak valleys – and the *ghutas* of Qalamoun in Syria.

Traditional farming in Fainan and Wadi Araba grew cereals around lowland springs, combined with animal production and movement between the valleys and the uplands where grapes and figs were cultivated around springs; people saw the animals as providing profits and major subsistence. Present farming at Fainan and Guraigira in the Araba developed around production for the market. The main crops are winter tomatoes and cucumbers; spring spinach, broad beans, and cucumbers; and summer tomatoes, melons, watermelons and peppers. Barley is part of the rotation, and often grazed green by sheep and goats. With the high temperatures barley dries before it ripens unless it is watered; it is more cost effective to graze the crop green.

People have various motivations for developing gardens. One is the need to take up tribally owned and favourably sited land so that outsiders are blocked from moving in; this was the case with Rashaiyida gardens at Fainan. Developers expect to recoup their capital outlay eventually, but profits in tomatoes and melons are low since the market is oversupplied and the gardens are distant from Aqaba, seen as a potential market. The produce is now sold in small towns nearer the gardens and to passing herders, and supplies local families. A Rashaiyida plan is "to move away from the tomatoes and vegetables, we'll grow enough for the *jamâ'a* but we'll grow more barley and maybe *dhurra* as feed crops for the sheep and goats. We should put trees around, useful trees like *sidr* (zizyphus spina-christi) and *tlah* (acacia), as the animals do well on the fruits and fallen leaves. We could buy in kids and lambs for fattening." A Manaja'a just likes gardening; "I'm an old man, it's peaceful, and I can't get around the mountains like I used to." His garden at Ghuwaibah grows fruit trees, pomegranates, olives, guava, figs, grapes, and vegetables for the family (who live at Garaigira some fifteen kilometres away); the garden was extended to take in two little fields by the water on which he was (December 1995) planning to grow barley and tomatoes. His herding neighbours poured scorn on his efforts; "the olives aren't any good, there's no oil in them, the pomegranates and the grapes aren't bad but he can't sell them. It'll be the same with the tomatoes. How is he going to get them to a market?" The old man acknowledges this is a problem; "Last year it cost me thirty dinars to get a pickup load of pomegranates to Aqaba, and I got sixty for them. And the grapes are difficult, they bruise bumping around on the dirt tracks, but they're really good grapes." A nephew helps him, and takes the fruit to Aqaba in his pickup. Whether the garden will continue after the old man dies is unknown; but gardens are developed and then abandoned (or sold).

The developer of Guraigira, Sharif Jamil bin Nasr, moved out of that enterprise because of problems with his junior partners. A much smaller garden in the Wadi Dahel was started in the midfifties by "a Kurd, who dug a cistern, planted gum trees and made fields and water channels, and built a house. He and his family grew their food and had goats. And then they left it, and went to Zarqa. I don't know why, perhaps they got fed-up." Sharif Jamil's Garaigira farm was on a completely different scale, intended for commercial profit. The investment of wells, pumps, tanks and piping was large, and was one farm among at least two others at Hallabat and Dhulail, east of Mafraq, owned by Sharif Jamil. His *waqîl* (agent) said "the farm has two growing seasons, autumn and spring, and the main crops are (in 1993) tomatoes, cucumbers, potatoes, watermelons and sweet melons. Where the potatoes are there used to be orange and lemon trees, but they didn't do, so we took them out. Potatoes do well, we use a Dutch variety. We grew peanuts one year and they did very well, but we couldn't sell them; the nut merchants wanted them cleaned and packeted, and we couldn't find a nut processor for our crop. Two years ago we grew wheat, and it did well, but tomatoes made more profit. We sent our crops to Middle Eastern markets, nothing goes to Europe. But the Middle Eastern markets are either closed to Jordanian produce at the moment, or they now supply themselves. At the moment, there's really no profit after costs." Tomatoes continue to be grown at Garaigira on a large scale – a girl was said to have earned herself seven hundred dinars from picking tomatoes in 1994. Guavas are also grown, but many plots of the trees have been allowed to die; they are difficult to sell as they bruise easily, and they are not popular with urban buyers. Garaigira farming has continued after the Sharif's withdrawal. Many of the problems concerned transporting and marketing of produce, controlled by the Sharif as part of the original agreement, since he had the knowledge and contacts for this, while the Amarin and Sa'idiyin did not; the latter never felt they got their proper returns. How irrigation and marketing is managed at the moment is unknown, except that farming continues, and a centre pivot system has been installed.

The Southern Ghor of Safi and Mezra'a al Ghor has an older farming system from archaeological and documentary records. Burckhardt (1822: 391–2) mentions the production of indigo, *dhurra* and tobacco; Seetzen (1854: ii, 363–5), saw wheat, date palms, bananas, lemons, sugarcane and all sorts of vegetables; a naturalist (1885) observed wheat, barley, oats, *dhurra*, indigo, tobacco and Indian corn, as well as white grapes on trellises; while Musil (1908: i, 161–4) praised Ghawarna farming highly. Present-day Safi is described by a farmer from 'Ay, a Karak mountain village; "Safi is divided fairly evenly between people from Karak, most of them from 'Ay, the Ghawarnah, and the Jawahin from near Hebron who came in 1948 because of the phosphates factory. Before fifty years ago there was so much empty land here. We, the people of 'Ay, started using land here for winter cereals because we had no more land on the plateau, our numbers had increased. In October we'd sow barley, *dhurra* and wheat, we flooded the fields with water from the springs. The harvest was May, and we did that with our neighbours. A barley crop was followed by tomatoes. People from Amman and Irbid used to come and buy land here too, but after 1975 it was forbidden to sell land to outsiders.

"Thirty dunums (three hectares) is as much as a family can manage. A Jawahin family work their land as a family co-operative, they don't employ anyone or have share-croppers, and they do very well. Most people have sharecroppers, and the share is fifty-fifty, unless the sharecropper provides the pipes and stuff, and then he takes 75% and the owner gets 25%. Or people work their own land, and many employ labourers – Egyptians, Sudanese, Pakistanis or Indians. About half the farms are sharecropped. What we don't do is employ casual labour at harvests, we use family members from all over the country, our neighbours help, everyone. It's difficult to try and organise your crops to the market, because it's all so volatile, it's impossible to predict the crops from other areas, temperatures which affect the harvest dates, things like that. A lot of people have a job as well, even if they're sharecropping, so it's mostly part-time working. Some people from here work in Amman and Irbid and come down at week-ends. There's not a lot of profit after all the costs, water, transport, labour if you're not doing it yourself. But you have all your fruit and vegetables, and most people have a job or land somewhere else or animals. In the summer the whole place closes down, it's so hot it's like an oven, and everyone goes up to Karak or Tafila, or to work on gardens east of Mafraq or round Amman." In November 1995, field crops seen were tomatoes, french beans, aubergines, sweet corn, *cousa*, onions, *colocassia* and broad beans; citrus groves were being harvested, and banana trees were in flower.

In the Wadi ibn Hammad, the north side has commercial tomato and melon crops, belonging to Majali from Rabba and Yarut, and the southern side has groves of olives, figs and grapes, belonging to the Mu'aita. A Majali said tomatoes were grown as a summer crop in the time of Ibrahim Faris Majali, around the turn of the century. A few Beni 'Amr and Fuqara tribesmen have title to less well-watered land in the Wadi ibn Hamad, while Beni 'Amr own the Wadi Shghayfat and springs on the south side of the Wadi Mujib. The western springs on this side of the Mujib are owned by Beni Hamida, as are those in the Wadi Jarra. Since Majali, Mu'aita, Beni 'Amr, Fuqara and Beni Hamida – that is the tribes owning the plateau around the Wadi – have networks of relationships established by marriages over a long period of time, access to land and its produce along these networks is less restricted than would appear from the general description.

The commercial fruit trees are figs, grapes, and olives, other fruit trees are mostly for the family. A Mu'aita family resting under a tree said that "a few gardens, like that one there with the huge olive trees, are really old. That one, and that one and some others were planted in the thirties, and this was planted in the sixties. Before they were gardens with trees, they were fields for grain. We're selling grapes now (July), my nephew comes and collects up all the fruit from the wadi and takes it to the wholesale market in Amman. We've been picking since seven, and he'll be along soon. The figs aren't ready yet, they will be soon. Have some lunch" – offering bread, grapes and melon – "the only food in the Wadi ibn Hammad now is bread and tomatoes."

A Fuqara from Dimneh regularly went down to the Wadi; on one such trip, he stopped first at a garden growing sweet melons and tomatoes belonging to his *ibn 'amm*, worked by Azazma sharecroppers. Then he visited his own old garden under weedy fallow, with a clump of five *sidr* trees; he recalled "there used to be *sidr* trees like olives here." Next, he called at a Mu'aita garden where a Majali from Rabba was also visiting. This garden grew grain, grapes and olives. Their grain harvest was finished (July), and an elderly Ghawarnah woman was cleaning the grain before it was taken to Rabba to be ground in the diesel powered mill; this would be flour for the family. The grape harvest was in full swing, they were sending off fifty boxes a day. A box of top quality grapes fetched 900 fils (0.900 Jordanian dinars), down to 600 fils. At the Fuqara's new garden, his barley was harvested, and his nephews preparing the ground for the November tomato crop. A contractor with a tractor ploughed and harrowed the ground, the nephews were laying out the pipes and polythene. The water for the crop would be piped from a tank he had built by the side of his spring. "It's possible to grow three crops of tomatoes, followed by a sweet melon crop. Then the land must be rested. Then it's ploughed to clean it, and fertilised with manure from (deep litter) hen houses or white fertiliser (commercial fertiliser 20:20:0) or urea. The first time the land is used, the crop is barley, and barley is grown again after melons and before tomatoes. Each tomato crop takes about two and a half months. I buy them as seedlings, and plant them through the polythene. My nephews help with the work, their fathers' gardens are next to mine." Laden with boxes of tomatoes, sweet melons, grapes and hot peppers, and bunches of mint from the spring and *ja'ada* and *shîh*, he returned home.

Irrigation is used for the vegetable crops, melons, grapes, and young fruit trees. The older trees and grain crops are rainfed. Since the amount of water for irrigation is affected by rainfall which recharges the springs, a dry year limits the amounts of irrigated crops that can be grown. The total availability of water from each spring limits the extent of each garden. Vegetable and grain land is ploughed by tractor and plough. Land between trees and vines is often ploughed by horse, mule or donkey and plough. The shallow ploughing between trees clears weeds and allows a dry crust of soil to form which keeps moisture in the underlying soil. The development by the late sixties of fruit and olive varieties grafted onto dwarf root stocks, so that trees were hardier and matured more quickly, encouraged the planting of orchards. Spades and mattocks are other basic tools. Curved and serrated long knives or saws, *shabriyya* (the local curved dagger), and imported secateurs are used for cutting, pruning and trimming.

In the Wadi Shghayfat, owned by Beni 'Amr, grapes are grown. Wheat and barley used to be the dominant crops until early seventies, when the water channels were improved and grape vines planted in response to the rapidly developing markets in Amman, Saudi Arabia and the Gulf. The vines are a special variety, large and very sweet. A Balqawi (a man from al-Balqa) and his nephew from a village near Madaba were spraying vines and supervising the picking of fruit. "I got this land when my daughter married the former owner, and I went on with the grapes. It fits in well with our other work, growing tomatoes, melons and barley on our land on the slopes west of Madaba. It isn't possible to live from grape growing alone. We come here at times in the winter for the ploughing, and the pruning – we do that. And we employ workmen, Egyptians, in the summer. The grapes have to be sprayed every three or four weeks against mildew. So far, we've taken 1,300 boxes to the wholesale market in Amman this season, but the price is low, because exports to Saudi Arabia and the Gulf are forbidden (1992)." The nephew delivers the grapes in his pickup.

The Dana gardens are high in the mountains, so stone fruit trees are more important. The fact that the Dana gardens have been cultivated by their present owners since at least the eighteenth century also has a bearing on their current use. The enmeshments of land and tree ownership, together with an increased population and a perception that the gardens are a major tribal resource, has resulted in little production for commercial purposes. "Selling fruit is dishonourable, it must be given. The produce of the gardens is for the family, the community. We live from our salaries and our enterprises." That neighbouring villages sell fruit from their gardens, and that tobacco grown at Dana in the early years of the nineteenth century was sold, appears irrelevant. The only crop grown in the Dana gardens that is sold are walnuts, bought by a merchant from Tafila. New trees or crops are planted because some one at Dana has a fancy for it; someone sees a young tree or a vegetable in a market or in a garden when visiting in another village or town. Merchants from local towns come visiting acquaintances in Dana at times of year when it is expected people would be in their gardens, and when future crops could be estimated; the merchant suggests purchase at harvest, and returns then, when amounts and prices are agreed. It is up to the merchant to find marketable commodities, not for the producer to find a market. "I don't sell tomatoes, and I don't take them to the market; I grow them, and I grow them for the family, all the jama'a who want some. The fruit on the trees that our father or grandfather planted, all of us have a right to it; so how could I sell some?" Such feelings are common among the middle-aged retired who are the majority of Dana gardeners.

Those who work the Dana gardens do so by expressing an interest in cultivating land to which they have a claim through descent, and by doing the work. A married woman, living in Qadisiyya, the new 'Ata'ata village on the main road, runs her family garden; "I thought I might as well use the land; the men are all busy, and I wanted vegetables and fruit for my household. So I asked around and none of the others wanted to work it, so I cleaned out the water channels with a spade and a mattock, and started gardening. My brothers did the ploughing, and helped with the pruning, but I've done all the rest. I've got onions, *cousa*, sweetcorn, a sunflower plant for *bizhr* (seeds for nibbling), and a big patch of tomatoes – half from seed and half from bought seedlings. I had to sow the *cousa* three times – birds or mice ate them. And there's *bamia* and I'm getting cabbage plants from a neighbour. I grow spinach, radishes and broad beans in the winter. We've got a beautiful apricot tree, but the birds peck holes in the fruit, and apple, quince, and plum trees. It's useful that the quince fruit stay on the tree until October. The grapes grow round the bases of the fruit trees. Last year, porcupines came and ate the *cousa* and tomatoes, and my brother shot two of them. He wasn't sure if they were lawful meat, so he took them to the Imam, who cut their throats and said they were. We ate one fresh, and salted the second. This garden produces enough fruit and vegetables for all its families (from the grandfather, a three generation *ibn'amm* group). We women dry figs and make raisins, and dry apricots, and we make jam from apples, quince and apricots. And we candy grapes, plums, quince, apricots; my sister's married to a Syrian, and she's learnt things from his family. And we dry tomatoes, *cousa* and *bamia*. We have enough grain from the arable land to make all our own *farîqa* (dried green wheat) and *burghul*. Because I planted the vegetables, they are mine but I give supplies to members of the family, especially if I know they haven't got a house garden in Qadisiyya or if they can't do the work. The fruit belongs to all the family, even though I do the watering, and the pruning; but as it's me that is here most, I get first choice although every one in the family comes down to pick the fruit when it's ripe."

The garden calendar starts in autumn, with the cleaning of water channels to the terrace gardens and each tree. New tree are planted from January to March, when the tree are leafless; these are field grown bare-rooted trees, container grown trees can be planted all year round. Pruning starts with grapes, which are cut back to two leaf buds, followed by apricots, figs, olives and the other fruit trees; pruning opens up the tree so that air and sun reach the fruit, and makes the tree a good shape. The terraces are ploughed in spring, when the grass and weeds begin to grow; some people use little tractors, others use horses or donkeys. Broad beans, spinach, radish and onions are sown in March and April, and some bitter vetch and *barsîm* is grown as forage for cutting. In May, cucumber, *cousa*, watermelon, tomato, sweetcorn, and *bamia* seeds are sown; later in the month, cabbage and tomato seedlings are planted out from sowings made in trays and kept in shelter. Watering and weeding continues until harvesting.

The owner of the walnut trees owns the trees and the land on which they grow with his brothers and the family as far as second cousins, that is his third generation *ibn 'amm*; "but my share of the land and my trees are registered in the names of my sons, I did this when it was clear my wife and I wouldn't have any more children. I did it because not everyone contributed to the care of the trees while they all claimed an equal share of the produce. I hope I will avoid problems now. All of the joint owners think of themselves as a *sharika*, a partnership. There are always problems, we expect it. Most get sorted out, a few don't, and that's when a garden stops being used."

A young man is a builder and part-time farmer and gardener. In 1993, he grew onions on family land with their agreement; the onions will be shared around the family with himself having extra shares as recompense for his trouble and time. This year, no one else in the family was interested in the land which is owned by his father and his two uncles, with eighteen male offspring between them, and registered in the name of the dead grandfather. Next year an uncle is retiring from his job in Amman, returning to the village and wants to garden. "So when my uncle returns, we will see how matters go between him and me, and when we have sorted out the working of the land, and the claims on its produce, we will call in the Dept. of Lands and Surveys, and re-register the garden." Families have claims on many gardens, from inheritance, marriage and use; when a man says "I own four gardens" he means "I have claims on four gardens along with all or most of the male descendants of the man from whom my own claim derives." If everyone who had a claim on a garden enforced it, some people would have very little – " a square centimetre." But many people do not push their claims, because they are busy with other activities, the desired produce such as fruit is there for the picking, or the share is so small it is not worth bothering about. Eventually one or two claimants put forward a proposal for using the land and a redistribution of produce; there is discussion, and a settlement and re-registration (or not).

It is said to be impossible to make a living only from a garden in Dana. A few families manage with a garden, a share of arable land and with a flock of sheep and goats, as everyone did some forty or fifty years ago; but the great majority of families now have income from pensions and/or employment. At the same time, most if not all families have access to garden produce, and to the arable land east of the village on the plateau, and have a few sheep or goats; the proportions of access to these resources varies from inheritance, purchase, shares, labour, and choice, but are available to all members.

The irrigated farms east of Mafraq are established as commercial enterprises to supply a market; the choice of irrigated crops reflects availability of land, water, and labour to the developer, and personal interest. A Sardiyya has forty hectares of land, and sank a well; the development cost 120,000 JD, paid off in the first year, 1989, by exporting tomatoes to Saudi Arabia. In 1994, he irrigated half the land and grew barley on the unirrigated half. The irrigated crop was half watermelons, half tomatoes, with watermelons the more profitable. Cauliflowers are grown in autumn and spring. Women from the Ghawarnah and Ahl al-Jabal handharvested the barley and the melon and tomato crops. None of the women depended solely on casual agricultural work, it supplemented a mixed resource family income. A Druze was developing an irrigated orchard of twentyfive hectares at Umm al-Guttein; "it took a year to set up the farm – bulldozing rocks, fencing, the well, planting, and cost 240,000 JD. It would have taken longer, but my father and my brother had cleared the big rocks earlier. I grow apples mostly, there is a shortage of apples. We have five varieties, Granny Smith, Crimson Star, Fuji, Alette (and one unrecorded); the apples are on a dwarf rootstock from Bulgaria, and come from a Jordanian-Bulgarian plant nursery south of Shobak. I chose the varieties partly because they have different ripening dates, so the harvest is extended. We have some apricots, peaches, and nectarines, a few grapes, and pistachio trees grafted on to local *butm* (terebinth) rootstocks. Wind is a big problem, I've got windbreaks of *casuarina* from the Ministry of Agriculture, on the east, and pines on the other sides. The money for this farm came from the sale of my earlier farm, which grew tomatoes and melons for export to Saudi Arabia from the early eighties. Exporting tomatoes was profitable then, so I repaid the bank loan and I saved money, and I thought I'd expand. The farm is between the family, as far as first cousins, but I'm the senior partner; my two nephews help me, and we have five Egyptian workmen. The trees are each watered every two or three days, and the workers are always clearing weeds. And it is ploughed clean between the trees three or four times a year. We have two small lorries, a tractor, plough, harrow, a plastic mulch layer, and a tanker. The tomatoes in the plastic tunnel go to Bahrain and Dubai. We use liquid fertilisers and chicken manure for the vegetable crops, but sheep or goat manure for the fruit trees."

For rainfed arable farming we use information from al-Karak; we are indebted to Sami ibn Fa'iq of the Faris Majali, Nabil and 'Atir Salim al-Amarin, Abu Auda of the Subhiyin Azazma, and Abu Zaid of the Ghawarna. Although this section is concerned with arable crops, all informants used rainfed land in the wadis of the plateau for new gardens of olives, grapes, figs, plums, almonds, apples, and peaches, patches of tomato, *bamia* and *cousa*, with tobacco in small wadi bottom slopes.

Arable land is *krâb* or *bûr*. *Krâb* is land ploughed ready for a crop and growing that crop to harvest, while *bûr* is land left fallow after harvest until is ploughed ready for another crop. The period may be six, twelve or eighteen months, so *krâb* and *bûr* are a description of the state of the land rather than a system. The principal crops are wheat (*gamh*) and barley (*sha'îr*), with *dhurra* (sorghum) as an alternative if there are late rains; chickpeas (*humus*) and lentils (*adas*) are also grown. Most wheat and barley is sold to the government at their price which is above the market price, although many people keep back some wheat for making *farîqa* and burghul. The *dhurra* is a variety used for brush making, bought by merchants from towns; the seeds and stalks are kept by growers for sheep feed. Formerly, white *dhurra* was grown for grain.

What crops are grown, how much seed is sown per dunum (one-tenth of a hectare), and what land is left under a bare or weedy fallow depends on the amount and timings of the rain. The red earth of the plateau, *trâba hamra*, holds the water and grows wheat, while the white (yellow) soil, *trâba baidha*, of the slopes does not and so grows barley. Wheat takes more water than barley, so it is grown on the better soils. Wheat, barley and lentils can be grown if there are winter rains and snow only. *Dhurra* and *hummus* being summer crops need April rains, as do water melons and summer vegetables grown without irrigation. *Dhurra* and *hummus* are shallow rooted, so if there have been good winter rains and spring rain, these crops can be planted on land that would have been fallow as long as the necessary ploughing is shallow, so as to keep in the moisture in the deep soil ready for wheat the following year. In an exceptionally good year, barley is sown at 10 kgs to the dunum and wheat at 9 kgs, and the harvest should be 300 kgs to the dunum. A really good year is heralded by early rains by mid-October, snow must lie on Jabal Sinina, further rain or snow at the end of January, and more rains at the end of March or early April. The years 1989-90 and 1990-91 were both poor years, and wheat was grown only on the better land after a season of bare fallow to conserve moisture. Barley was the dominant crop, with small amounts of lentils. Both wheat and barley yielded only 100 kgs per dunum, with an unknown sowing rate. Spring rains in 1991 meant that some chickpeas, and large amounts of *dhurra* were grown. 1991-92 had winter rain and snow, and spring rains; a farmer said he sowed wheat at 8 kgs to the dunum, using artificial fertiliser at 225 kgs to the dunum; not using fertiliser, he would have used only 5 kgs of seed. The yield was a disappointing 200 kgs to the dunum, caused by a cold and windy May, and a heavy storm in June. The yield in an exceptional year is for wheat 1:33 and 1:30 for barley; in 1992 wheat achieved 1:25. People emphasised there is never a year without a crop of some kind. Musil's (1908: iii, 306) figures for Karak wheat and barley production at the beginning of the century are for wheat in an exceptional year 1:40 and for barley 1:25; a good harvest for wheat 1:12, an ordinary harvest, 1:7, and a really poor year, 1:2. Causes of lowered harvests mentioned by Musil (1908: iii, 298) are the same as today, except that locusts appear to be absent; cold north winds, drying south winds, unseasonable heavy rains, plant diseases, plant eating insects, and fire.

Arable crops may be sown before or after the rains come. Each farmer makes his own decisions, depending on the estimated amounts of water already in the soil after a dry fallow, whether the land is under a dry or weedy fallow as this affects the ease of ploughing, how much land he has to cultivate, and whether his land is exposed to cold winds and snow. There are good arguments for planting before rain or after. Those with a lot of land plant at least some of their crops, and some of their wheat before the first good rain. Late planting means that moisture amounts are known, but late cold or drying spring winds can adversely affect the crop, while weed control is easier.

Although Musil (1908: iii, 294–5) names cereal varieties grown around Gaza, for Karak he says only that the best wheat was Zraybiyya, a four-rowed sort with large yellow kernels, and Qannari was a tall barley flourishing in the Ghor. Current (1996) popular wheat varieties north of Karak are: "Dir'iyya, the best, with 'red' grains, and especially good for bread; Fi'uthmani, smaller than the above, but better for gleaners as the seeds drop easily; Tihiyya, with black ears and a long head; American wheat is fit only for sheep. The barleys are: Baladi, a two-row barley, it's short and sheep prefer it; and Qannari, a six-row barley from Syria, it's tall so it can be combined, and it drops its seeds so there is a volunteer crop for sheep to graze green. White sorghum (*dhurra*) is now rarely grown. We prefer the brush or *makânis* type, we sell the heads for brushes, and feed the seeds and stalks to the animals. We grow 'red' lentils, the 'white' do not do, and we have two varieties of chickpeas, the Turkish and the Baladi." Some farmers buy government varieties of seed, others – especially sharecroppers with goat and sheep flocks – prefer their own seed; "I never have to change varieties because I am always on different land each year. My wheat and barley are proper *baladi* (which means 'local' as well as a specific variety) sorts, with short stalks and many-headed, at least three heads to a stalk. The seed the government sells is long stalked and single headed, but the heads are bigger. We have goats, and the short-stalked sorts are better for them." There are new short stalked wheats and barleys developed from the local *baladi* varieties, with a range of sub-types known by numbers, available at the government grain centre in Rabba; "these come from our local types of wheat, where we have Dir'iyya, Fi'uthmani, Tihiyya and Almiyya. The first two are the most popular. The barleys are: Iqsad which is really short, Qinari and Rum which are heavy croppers, and Aswad which sheep like. Which seed you choose depends on the kind of land you have, and what you want it for."

Burckhardt (1822: 278) noticed a six-rowed barley called *khishâbi* grown at Hammeh in the Yarmuk valley, because ordinary barley was eaten by wild boar. At Heshban in al-Balqa, Irby and Mangles (1823: 472 and accompanying drawing) saw bearded wheat, with "ears of an unusual size, one of them exceeding in dimensions two of the ordinary sort, and on one stalk." Russell (1995: 696) quotes Bdul near Petra growing local varieties of wheat called collectively Qatma Safra. He found "extremely high yields are possible under favourable ecological conditions" because of the "phenotypic plasticity of unimproved, mass-selected land races of wheats and barleys," and quotes Salim's finding that Qatma Safra has a high protein content. Einkorn, emmer, and spelt wheats all have higher protein contents than modern bread wheats, are relatively less attractive to birds and deer, while emmer and spelt are more easily harvested by hand-picking than by sickle.

Before the early to mid-eighties, people say fallowing was the norm, with little use of artificial fertilisers. Tristram (1873: 121) says fallows lasted three or four years, while Musil (1908: iii, 298) says in Karak land was left fallow one year in every three. Soil exhaustion is one reason for fallowing, although using leguminous crops of lentils and chickpeas, (and formerly bitter vetch (*kersenna*), to maintain fertility is practised. Lentils and chickpeas require hand harvesting, and the areas of these crops are not equivalent to those of wheat and barley. Animal dung deposited by grazing animals is regarded as a longterm fertiliser, as sheep and goat manure are low in nitrogen, while chicken manure from deep litter houses cannot be used at more than 100kgs to the dunum; rain is needed for bioorganisms to break down the dung. The Awlad Salim have recently changed their rotation (*dawra zira'îyya*); "we used to have crop followed by fallow unless there was enough rain to grow a nitrogenous crop like lentils on what would have been the fallow. Now we have very little fallow but we grow *bicia* (vicia – vetch). This takes less water than lentils because it is for grazing, it never has to reach maturity. We buy the seed at the government place in Rabba, and we've been doing this for two or three years. It's to improve the soil, because it fixes nitrogen, and to give a grazing crop as we have more sheep here than we used to. It's not particularly to reduce fertiliser use, although it helps. It's taken the place of *kersenna*, which we stopped when we no longer used cows, horses and mules. We always use artificial fertiliser, and how much we use depends on the amount of rain, time of planting and what you aim to use the crop for. If the rains are only light, so the crop will be for grazing, we don't use very much. Qanari is a good variety because there is a volunteer crop."

The accumulation of soil moisture levels is the reason for bare fallowing, since the clean surface layer prevents loss of soil moisture from evaporation. After a series of poor rain years such as 1989-90 and 1990-91, the good rains of 1991-2 did not produce particularly good harvests because the earth was dry and the rains went to replenish the water content of the soil. An exceptional harvest follows good rains after a series of above average rain years.

Ploughing starts the arable year, after the harvest stubbles have been grazed. This is the autumn deep ploughing with a turning plough (*sâj galâba*). The ideal is to deep plough twice. Before sowing, which would be in early winter following a bare fallow and rain, and in late winter or spring following a crop and rains, the earth is ploughed a second (or third) time with a shallow plough (*sikkat galâba*). After sowing, which is often broadcast, the sown earth is harrowed (*harth hafîf*) to cover the seeds and to get them into rows. The rate of sowing of seed depends on the amount of soil moisture; the drier the soil, the less seed sown, so the plants are more widely spaced and have more water available. Formerly, ploughing was by teams of two, four or six animals, depending on the soil to be ploughed; oxen, mules, camels and donkeys were used. A team ploughed between three and four dunums a day (donkeys ploughed one to one and a half dunums), and four ploughings were normal; the first two in spring made furrows, and opened up weedy fallow; while two more ploughings were made in autumn (Musil 1908: iii, 296–7). Fields were smaller, the size that a team could plough in a day. A *feddan* indicated the relationship between land and the number of teams needed to plough it; a family of the Faris Majali had 200 dunums at Azzur, which required six teams of oxen, so "our land at Azzur was six *feddans*." People often say the land is not ploughed often enough now. Before the common use of tractors, it was possible to sow seeds of tomatoes, watermelons, *cousa* and cucumbers using soil moisture left from winter rain. Tractor ploughing causes panning, unlike animal ploughing, so the earth does not hold the water but dries out too fast. Land for dry fallow is shallow ploughed after any spring rain, and again before the *gaith* (high summer; lit. 'the stop'), in June.

Wheat is often sprayed to kill annual weeds. The first harvests are lentil and *kersenna*, followed by wheat, barley and *humus*, and finally *dhurra*. Many families move out into tents on their land at this time. Some wheat and a little barley is cut by combine harvester, most crops are hand-harvested. Combined wheat straw is baled for use in deep litter henhouses. Hand harvesting is preferred as "it gives a cleaner crop with no weeds, better *tibben* (straws for animal feeding), better stubbles for grazing, and cleaner clearance on the ground (by removing weeds)." Crops used for *tibben*, that is lentil, chickpea and barley straws, are hand pulled, with the whole plant being removed; wheat is cut by sickle, "we just take the ears off as we need the stubbles for grazing. Wheat straw isn't as good feed as barley straw." (Around Petra, the Bdul say they harvest by hand because it is quicker on their sandy soils (Russell 1995: 699–702).) The crop is taken to the threshing floors (*baidhar*) by donkey and mule load, pickup or trailer load and stacked so that the heads face inwards and slightly upwards to the centre, thus minimising damage from dewfall. Each pile is a share of the crop, if shares are being paid in kind. Threshing (*darasa*) is by machine, powered by tractor engine. The Turkish threshing machines are owned by local families who hire them out at 6 JD an hour; the labour is done by itinerant Syrian threshing crews, who may be paid in sacks of *tibben* in some years. A sack of *tibben* in 1990 was 4.5 JD. In the past, grain was threshed by sleds edged with flints pulled round and round over the crop by a team of animals – oxen, mules or donkeys, and a few people continue to thresh in this way if there is only a small amount. Alternatively, animals were driven over and over the harvested crop. Threshing is followed by winnowing with winnowing forks, tossing the grain into the air for the wind to loosen the husks and chaff; it is then sieved, with the grain remaining in the sieve. Combined grain and threshed grain is taken to government centres. Chaff and grain for household use are put into sacks and stored, and the grain taken to local diesel mills for milling or processed into burghul at home.

Grain was always sold as well as providing for household consumption and paying taxes. A Majali said "We sold grain to the bedu, and they used it themselves or they sold it on to the Hijaz, and we paid our workmen from Palestine in grain" and a Beni Hamida said " we grew grain for ourselves and we sold some to the Haj." Most grain is sold to the government at government prices; some grain is kept back for *farîqa*, burghul and for household bread flour where wholemeal bread is preferred. There are diesel driven mills in Qasr, Rabba and Jiddat as-Sajayda; earlier, water mills in the Wadi ibn Hammad were used. White flour is bought back for household use at below market price, as is barley for feeding animals. Subsidies were introduced to stop grain hoarding, but are gradually being reduced. Most households (man, wife and dependents) reckon that they need a tonne of wheat a year, as bread is the staple; if the household entertains constantly, a tonne and a half is necessary. In Karak, a hundred dunums of good land is said to produce 200 sacks of 100 kgs each in a good year, and 100 sacks in a poor year; that is, a tonne of grain requires 5–10 dunums of land. A general estimate was that forty dunums of land, allowing fallow and other crops, are required to produce a family's needs in grain, and animal feeds. The figures of a family needing a tonne of grain a year, and twenty to forty dunums of land to produce this, depending on rainfall, was also given in Ajlun. This was from people who cultivated their land themselves as part-time farmers, having employment or pensions, and members of families with sheep and/or gardens of other enterprises. With sharecroppers providing the labour and the machinery, the partners divide the harvest

50–50 as both contribute to its production, and both sides are also members of their respective families with other productive enterprises. Through inheritance and marriage, there are many claims on the produce of arable land, and families have different ways of accomodating these claims.

Local concepts of land ownership and production produce two important points. Firstly, ownership is a function of claims on and access to resources, rather than a system of control and absolute rights to disposal. Secondly, local social practices of agricultural production imply mobility between ecological areas and within them. As well as the three-sided landuse economies mentioned above, agricultural resources for some included two or three kinds of cultivation. Some using irrigation for fruit or vegetable crops grew arable crops on rainfed land, and on their irrigated land some crops had more irrigation than others. This is particularly so in the descriptions of the *ghutas* of Qalamoun, and of the village of al-Ji at Wadi Musa in southern Jordan by Milne (1971: 33–5). Here irrigated terraced gardens grew olives, grapes, apples, oranges and pears, and a wide variety of vegetables including onions, garlic, cucumbers, melons, aubergines and tobacco. Fields near the gardens grew tomatoes, melons, and cucumbers interspersed with maize and tobacco plants; some cabbages and sunflowers were grown here too. Arable crops of wheat, barley and *dhurra*, with lentils and bitter vetch, were grown on rainfed fields some two kilometres from the village. Antoun (1972: 6–15) discusses the traditional farming of Kufr al-Ma in northern Jordan, which was arable based, growing wheat, barley, bitter vetch, lentils and chickpeas, sesame and maize. Vegetables grown were broad beans in winter; onions and potatoes; in summer, tomatoes, *fagûs* (a variety of cucumber with a pale green ribbed skin), and watermelons. The villagers owned thousands of olive trees in their village of origin, Tibne, in the mountains. People say that no-one ever did or could live from arable farming alone (Antoun 1972: 14). Palmer (fc 1998) records traditional arable farming practices in the hill lands and the plains of northern Jordan.

Arable agriculture in the *badia* is opportunistic and depends on rainfall and timing. Crops are grown for household consumption and for animal feed. In the exceptionally good rain year of 1994–5, less cultivation was undertaken in the ar-Rishas than in other years with lower rainfall. Barley cultivation often attempts to enhance scattered natural grazing rather than increase widespread natural growth. At Mahdath, in the eastern *harra*, the family who prefer to use the *sha'īb* grew wheat, barley and watermelons on half of the area in 1991. The family left the area in 1992, and it has not been used since. At Faydhr and the ar-Rishas, where land has been registered, only barley is grown in suitable years. Two Jordanian salesmen for a tractor and seed-drill company attempted to interest the agent of the tribal owners and other local men in growing wheat using modern irrigation. The scheme was dismissed as too expensive and pointless, since " we don't want wheat, we want barley for the sheep. We only get one harvest in five or six years anyway; the other years it's only fit for grazing or we don't even sow as there's not enough rain." The salesmen countered that at Muwaqqar, with no more rain, wheat is grown in some years. But the agent replied that "although the rainfall might not be any more, and the land is poorer, the land at Muwaqqar holds the water better. It isn't only the total amount of rain; it's essential to have at least two lots of rain, so that the first rain raises the general moisture level in the soil and the second brings the soil moisture up to the amount needed to grow a crop. Anyway, we want barley." The salesmen set out a grandiose scheme, involving an estimated forty pumps along the *sha'īb* from the Syrian border to ar-Risha, with each pump having its own generator, *birka*, fixed spray gun irrigation (*rashaishât*), diesel tankers, mechanic and driver. The agent pointed out that the crop would need water twentyfour hours a day, because of the danger of burning the growing crop if irrigation were not continuous. With one breakdown the whole crop is lost, so spare parts and mechanic are needed. In addition, having made the investment in pumps and irrigation, it becomes necessary to have your own combine harvesters and seeders, especially seeders as when there is an opportunity to sow, it must be taken at once. The complications of intensification were clearly spelt out; too much capital – even if available – is tied up, too much depends on the machinery working properly, there are unacceptable social problems, and it is unnecessary.

A farm east of Jawa started and registered in the 1930s by a Sharafat worked on snowmelt flooding from the Jabal al-Arab. If enough snow had fallen in the Jabal, the snowmelt would be flooding down the wadis over six to eight weeks, and could be channelled to the fields. White *dhurra*, barley and watermelons were the normal crops; occasionally, wheat was grown. No crops were ever sold, but the grain was stored against a poor year. How much of which crops were grown was variable, depending on the levels of soil moisture and from what sources. Rainfall was augmented from the snowmelt in favourable years. Since the building of dams on the Jabal in the late sixties, the farm has not been used for a crop. In the Shabaika depression to the east, moisture levels were raised by natural flooding. In 1992, the Sharafat families who had registered ownership earlier grew white *dhurra* following late spring heavy storms. "But it wasn't a good crop. The germination was poor, and the seed heads weren't full and plump. We sowed four hundred dunums (forty hectares) at a rate of one kilogram to twenty dunums. I bought the seed from a merchant in Mafraq, and we'll sell the harvested grain to merchants there too. I don't know what they will sell it for. We'll use the *tibben* for our animals. White *dhurra* makes nice bread." There were no further crops, and very little use of the area for grazing until spring 1995, when a barley crop was sown on the western end of the depression.

Accounts of large scale arable production in central Jordan resulting from late nineteenth century Ottoman reforms are to be found in Abu Jabr (1989) and Lewis (1989). Seetzen (1810: 44) comments on the fertility of the Hauran and that in 1805–6, cotton was a major crop. Burckhardt's description of agriculture in the Hauran (1822: 295–303) shows the mobility of agricultural families, crops grown and yields, and taxation. The horsebean crop, which fed cows, sheep and camels, has diminished; winter cereals were followed by a fallow year, or summer crops of melons, cucumbers, sesame or pulses; these summer crops depended on irrigation or spring rains. Schumacher (1889b) comments on arable and irrigated farming, as well as herding, in Jaulan. Current agriculture is based on cereals with increasing irrigation, using plastic pools and piping and diesel pumps, for a rapid expansion of summer vegetable crops and fruit-trees.

In the *ghuta*s of Qalamoun, Thoumin (1936: 115–120) reports that these oases are unable to have the layered cultivation seen in other oases, such as at 'Ana on the Euphrates (Northedge *et al*.1988: 21–4), since the winters are too harsh. In Qalamoun, winter crops of wheat, barley and maize were sown from mid-October to mid-November and harvested at the end of May or the beginning of June. In the open fields of the high plateaux in Qalamoun, the harvest was not until July. A tenth of the sown oasis land was down to lentils, peas and broad beans, and onions were grown in profusion. Potatoes were a speciality of Yabrud. The summer crops were

Land Use

tomatoes, cucumbers and water-melons. Apricot and walnut trees grew along the water-channels. The oasis lands were never fallowed, but an empirical rotation of crops was used instead. On drier land, cereals and grapes were grown, with one year in two being fallow, and in dry years no crops were sown. Haj Ibrahim (1990: 301–2) gives the crops of the irrigated lands of Deir Attiya; along the first order channels poplars are grown for building wood, second order channels have walnuts and third order channels grow apricots, and figs. Pomegranates, more figs, summac and a local plum are grown with grapes in land outside the oasis. Winter crops are wheat, green forage crops and carrots. Potatoes are a big summer crop and occupy as much land as wheat, although potatoes need a lot of water, fertiliser and hand-labour. Deir Attiya, like Nebk and Yabrud, is renowned for its potatoes, carrots, turnips and garlic. Yabrud is now famous for its fruit, especially apples.

Métral (1984) demonstrates the ingenuity and application of agricultural workers and their multiresource activities on a government irrigation scheme in the Ghab in Syria. The changing activities of the oasis of Sukhne are examined by Métral (1991); in the 1930s the caravan trade was declining and irrigated agriculture expanded, in the 1980s trade expanded after the opening of the new desert highway, dry farming and herding flourished, but the oasis gardens were abandoned. The riverain agriculture practised by the Agaydat of the central Euphrates is examined by d'Hont (1994). The Agaydat used irrigation, raising water from the river, and falling flood waters before the river waters were regularised by the hydraulic engineering projects of the sixties and seventies.

The diversity and flexibility of farming practice in the Bilâd ash-Shâm over the last two centuries has been demonstrated, with farmers using their knowledge of their landscapes and the rainfall amounts and timings in each rain-year to make their choice of crops, locations, sowing rates and dates, and so on. It is likely that, were written records available, similar flexibility would be shown for earlier periods like the sixteenth century; Braemer (1990: ii, 465, n.19) considers this to be the case for the eastern Hauran, and the probability for continuous agricultural production with a market component on the Karak plateau from the sixteenth century is indicated in Lancasters (1995).

There are invariably integrations between agriculture and herding, either directly as in arable farming, sheep and cows, or indirectly between camel herders (with sheep and goats) and oasis farming. We examine first herding in farming areas, and then herding in the *badia* regions, and we focus on those animals kept as herd animals - cows, sheep, goats and camels. Before the introduction of modern transport, camels, horses, mules, donkeys and cattle provided transport and draught power, and a part of agricultural production was to feed these energy providers. Horses, mules and donkeys were not herd animals, but kept for specific functions. In the present, mules are kept in the mountain areas of Jordan for ploughing small fields and terraces; some horses are plough animals, others are for riding and racing. Donkeys accompany sheep herds, carrying the shepherd's water and food, carry water for sharecroppers living in tents, and in the mountain areas, plough terraces and orchards.

Oxen were the common plough animal, and cows were milked in the Hauran (Burckhardt 1822: 295), in the Balqa (Burckhardt 1822: 369), in Karak, where Majali said "we had twelve oxen for ploughing" and "families had between fifteen and forty cows and oxen", in al-Jauf (Wallin 1854: 148 and reminiscences from Hamad Uthaiman al-Wadiman) and in the Hijaz (Doughty [1888] 1936: i, 194, ii, 205, 374). The Beni Hamida and other tribes of the Balqa were known for their cattle herds (Tristram 1873: 251), as were the Ghawarnah (Musil 1908: i, 164). 'A Naturalist' (1888: 271) describes their cows as "chiefly small pretty black animals with white faces." The cows of the Damascus Ghuta were *asîl*, distinguished from the little cows of the peasants and nomads by their large body size and greater milk yield (Wetzstein 1857: 477).

Salim Sulaiman of the Beni Hamida sold his cows about ten years ago, as he was unable to feed them properly because of a series of dry years and there was less natural grazing with expanding agriculture; he replaced them with goats. He and his family used to spend the winters in the Ghor with the cattle, and then move up through the wadis to the plateau for the summer. "The cows ate all kinds of annual and perennial plants, and we fed them barley *tibben* in the winter, like we do the goats. They needed extra because they calve at the end of the winter. We always grew barley for ourselves and the animals, we have fields in the flat areas on the wadi slopes, and on the plateau. A tent (parents and dependents) needed five or six cows, and a cow produced the same as twenty five goats (this seems an over-estimate on several grounds: Arab goat herding households usually say they need a minimum of sixty animals, while in Britain, a cow is said to equal five or six sheep or goats in terms of grazing, and reckoned to be equivalent to ten goats in milk terms). We had all that milk, and we made fresh butter, *samn* (clarified butter), and *jamîdh* (dried yogurt). We lived on it, and we had our barley. *Baladi* (local) cows are much better than Hollandi (Friesian or Holstein) cows, they don't take so much to feed, they eat more kinds of plants, and the milk has more fat in it. And we always had goats as well, we milked them, and we need the hair for tentcloth." This household were living in their tent while harvesting the barley (late May), and the goats were grazing on the natural vegetation; in a week, they would return to their village house on the plateau. The picture drawn of cattle herding by Salim Sulaiman was like that of Tristram (1873: 257), but omits additional sources of income from services.

The change from oxen to tractors as draught power for arable farming in Karak was said to be a result of the droughts of the late forties. "There was a series of dry years, and animals died. Many of the cattle died, like the camels and sheep and goats. The young men got jobs in the army and with the government. When things improved, we needed sharecroppers to work the land, and Palestinians had arrived, especially the Azazma. At the same time, more tractors came. Cattle for ploughing and camels for carrying water and tents and the grain, neither of them were needed any longer." Cows for household milk continue, with Ghawarnah and, to a lesser extent, Beni Hamida and Ajlun villages in Jordan, and Druze villages in Jabal al-Arab in Southern Syria. Cow-keeping is seen to need a high degree of water fed vegetation, lacking in many areas with increased irrigated agriculture and piping of springs. The Ghawarnah cows are herded along field edges and water channels, and yard fed on weeds and crop residues. Others keep a house cow in yards or tethered in gardens; these animals are frequently bought and sold since their costs are high, they make considerable work for the women, and the returns may be less than expected. If the animal is a Hollandi, she must be fed on barley and *kersenna* hay, and in summer some women spend time washing the cows to cool them. A *baladi* animal eats *tibben*, barley and browses, but the calf fetches less as meat. In some areas, around Jerash and Ajlun, dairy companies buy and collect milk from small cow keepers.

A modern system of cattle management for commercial milk and meat was seen at Dhulail. The impetus for commercial production of fresh milk was the demand from growing urban populations, with electricity and refrigerators in homes and shops. A local business man said that the arrival after 1967 of Palestinian refugees, with their dislike of camel and goat milk, was a major factor for intensive dairy cow systems; "not at first, there was no money and the refugees and most people used imported powdered milk, but later on as Palestinians moved out of the camps and got jobs and businesses. The first large scale commercial dairy in Jordan was a Jordanian-Danish government project, and the private commercial dairies did not start until the beginning of the eighties. The original dairy here was as part of the local co-operative. The government had encouraged the people here to change from camels and goats to cows. The co-operative was people who'd always lived here and Palestinians who had moved here, it's very mixed. There's water not too far down at Dhulail, so there were a lot of gardens being developed, growing vegetables for Amman and Zarqa. Then the present owner bought the dairy itself from the co-operative, and one of his sons, with a doctorate in agriculture, runs it. The dairy buys and processes milk, especially cow's milk produced in Dhulail, where nearly every house has a barracks for cows. The sheep's milk, which goes for yogurt, is brought in by the sheep owners, or by small traders who go around collecting it up; if there is a big source of supply, the dairy send a tanker out."

An owner of a cow barracks described how he worked. "The whole block of land with these barracks and the dairy was bought by the co-operative, and it was set up for a dairy – milk production, processing and distribution. I wasn't here then, I bought my shareholding three years ago (1991). We're from Jaffa, and I live in Amman and come here everyday. The dairy itself by that time was separate, and there was the Hammuda dairy which was always a private company, and the one in Marka. The dairies make contracts with cow owners for milk, and these contracts are negotiated through the co-operatives council in Zarqa so that supply and demand can be evened out. I contract to sell to X company, but if they have plenty and Y company is short, I can switch to supplying them. (Exactly how is unclear, but it is ad hoc and partly on price). All the barracks here work on subcontracting or shares, which again can be bought and sold, or people swap. I'm really a trader in cows, always Hollandiyyas, like the man down the road (pointing at a barracks with a Turkish registered car outside), and producing milk is part of being a trader. I deal in bulls, cows, calves..... This barracks, with its stalls for cows, milking parlour, milk store, feed store, and yards, can hold up to three hundred animals, but at the moment I have a hundred. The feed is straw from Disi and Saudi Arabia, plus *alaf*, and alfalfa. The alfalfa is fed green or as hay. I buy most of it in, locally if I can, and I grow alfalfa in my garden. All the gardens here grow alfalfa for cows. The bull calves are sold for meat, to a wholesaler if I have a large batch, or to a small trader if there's just one or two. I never buy in young bull calves to grow on, and I don't always keep my own to grow on, it depends.

"The cows are injected every month, according to government regulations. Everything has to be kept very clean, because the barracks are close to each other and diseases would spread quickly. I've got a slight problem with mastitis. I employ one worker for every hundred animals, he's Egyptian and he feeds, cleans and milks, and I pay him a hundred JD a month. People have Friesian cows because they give a lot of milk with a low butterfat content. People want a lower fat milk now they aren't as active. It's alright for the bedu to eat *samn* and drink milk, they never stop walking, but most people spend all their time sitting. The milk goes for fresh milk, *laban* (yogurt), *labna* (soft cheese), some (hard) cheese, and *shanîna* (buttermilk); butter is not made locally but imported, which is cheaper. A cow gives twenty to thirty litres a day, and an exceptional cow will give thirty five litres, from two milkings."

"I sell milk to the dairy at 210 fils a litre, while the feed alone costs me 70 fils a litre. Of the profit on milk, I reckon I get 40% and the dairy 60%; there's no great profit, but it keeps us; on the trading, I make a profit sometimes, and sometimes I lose. A top bull costs 1,200 JD, an average bull 1,000 JD; a cow is 5-600 JD. I sell the manure for 25 JD a small lorryload, but by the time it gets to the people who'll be using it, it costs them 45 JD. I deal in sheep a bit, too, sold some to a man from Salt the other day. And I've got a flock of a hundred, which are at Shafa' Badran with a Bedu shepherd; I'm close to one of the Adwan, and so I graze my sheep over there."

There are spinoffs from the basic idea of cows in barracks supplying a commercial dairy for the urban markets. In local villages of Khaldiyya and Hallabat, block factories, building suppliers, and electricians get a large proportion of their work from building and furnishing barracks; metalshops make feed hoppers, feed troughs, water tanks and drinkers; there are also agricultural, veterinary and feed suppliers.

Sheep are the most numerous herd animal in the Bilâd ash-Shâm, and the main breed is the Awassi or fat-tailed sheep, also known as the Na'imi by Aneze tribes. These sheep are white with wholly or part brown or black faces; some have brown legs. A few in al-Juba and in the *badia* have some Najdi sheep, which are black with white faces, long tails, and less curled wool. Najdi sheep are far more susceptible to cold and wet, give more milk, and their lambs are less valued as meat than Awassis. Their neck wool is preferred for weaving the patterned sections of dividing curtains by some Rwala women. Information is presented from sheep management practices in Karak and in the *badia* by Ahl al Jabal and Rwala. These distinctions appear to differentiate types of sheep management practice, but practice depends on the functions of raising sheep and supplying family and urban markets with milk products, wool, and live animals for meat. It varies according the aims of owners and herders, and the various constraints on achieving this. There are long-standing commercial transactions in sheep-herding, recorded in travellers' comments.

For the Karak region, Seetzen (1810: 40) mentions several hundred sheep from Karak purchased by people from Jerusalem and Bethlehem for the Easter feasts, and (1854: ii, 363) the trade by Ghawarnah of Ghor al-Mezra'a in donkeys, sheep and goats to Jerusalem. Burckhardt (1822: 388) says that the people of Karak sell their sheep, goats, mules, hides, wool and a little madder at Jerusalem, and all kinds of provisions to the Haj at Qatrana, while Merrill (1881: 475) saw "large flocks of sheep and lambs, numbered by thousands, driven up from the plains of Moab, designed for market at Jerusalem and some of the coast towns." The numbers and productivity of sheep are attested in Salt by Buckingham (1825: 32) and in Karak by Burckhardt (1822: 385), both commenting on "great quantities of butter" eaten by the population.

In the recent past and the present, the majority of families of the people of Karak own or have access to sheep flocks. The composition of the flocks is flexible, and their herding varies, depending on the grazing and labour available, which themselves depend on the season. What is standard is the opinion that sheepherding is more profitable than grain cultivation, although each production option necessarily implies the other. Most herders say that in most years milk products are more profitable than meat, although this applies to family flocks rather than to flocks bought and managed opportunistically. Most are family flocks, and a shepherd herds up to two hundred and fifty animals; opportunistic flocks are smaller as they are more difficult to herd, as the animals are bought in from different flocks, often as lambs.

The sheep year begins in July/August, when mating starts; one ram to twentyfive or thirty ewes is the norm. Each sheep in the flock is known as an individual animal, and flocks are made up of matrilines, each of which has particular characteristics, such as ease of lambing, regular conception, good milk, nice wool and so on; within matrilines, sheep are grouped by age. Most flocks have a proportion of goats, rarely more than a third, who "eat things the sheep don't care for, keep the sheep moving, and they give more milk and for longer, although it doesn't have such a high fat content." At this time, the sheep are grazing on wheat stubbles, and watering from field cisterns, and continue to do so until the end of September. After this they feed on *dhurra* stubbles, rough grazing and are fed *tibben* (chopped barley straws) and *alaf* (dry feed of barley or other grains, bought from the government at a subsidised price). Lambing is from late November through to February or March, and ewes produce a single lamb. The sheep are often taken down into the shelter of the wadis or the slopes, or if they are on the plateau, return to barracks at night. The lambs have the milk for the first six weeks, and then milking starts and continues for eight weeks. The amount of milk given by a ewe varies according to the quality and amount of the grazing. Milking and milk processing are usually women's work, and when possible, the household, or a combination of members from it, move out of the village into their tent. Men do milk, but never process. In dry years, there is milk only for household and family use, in good years there is plenty for sale to merchants in the towns. *Samn* (clarified butter) and *jamîdh* (dried yogurt) are the main commercial products. These, with *labna* (soft cheese) also dried and stored in olive oil or *samn* and fresh yogurt are made for the family. People expect to move as far east as possible in the spring, to use the fresh grazing of annuals; Karak sheep rarely go much further east than Qatrana, although an Azazma family herding Karak sheep was met in the *harra* south of Burqu' in 1995. Grazing on annuals is expected by January or early February in the wadis, if not yet further east, and to continue until May. Shearing takes place in late April or May, so the sheep have time to grow the beginnings of a new fleece before the summer heat. Male lambs are sold to traders at four to six months old, with great demand at the times of the great Muslim feasts. Flocks of young lambs are quite common in the summer; these are the progeny of ewes which did not conceive earlier. Surplus breeding sheep are sold in September, when owners know what the harvest is, and what supplies of *tibben* and prices for *alaf* will be.

The returns on a flock vary not only from year to year but from herder to herder, depending on skill, interest, time and labour, including that of the women of the household. No household lives from sheep alone, all have land, a job or pension as well. The successful flocks are the household flocks, or those with a paid shepherd and active supervision by the owner, who has access to agricultural land for summer stubbles and winter feeds. Some flocks are intended mostly to provide milk and milk products for the household, meat for household and family entertaining, and wool for the household; the owners of three or four such flocks join together and hire a shepherd, or provide a shepherd from among members on an ad hoc basis. Share partnerships for sheep flocks exist, where one partner owns the sheep but is unable to herd them, often because he is employed outside the area, and the other herds. Share milking is fairly common in years with good springs; a woman who has time will milk and process the milk of a flock where the women have no time, and share the produce. A few owners concentrate on commercial meat production, and only enough milk for the household; a man from Hebron living in Rabba who has three thousand sheep is an example. Another household who focus on meat production has large amounts of arable, and therefore a lot of *tibben* and feed; they feed the sheep all the year and lamb twice a year. For some, selling milk or milk products remains dishonourable while for others, selling milk is now acceptable and an important source of income.

A Majali sheep owner has an employed shepherd to herd his flock of 200 sheep and a few goats. In 1990, he received 3 JD each for his fleeces, which weighed three kilograms each; "I got such a good price becuase we washed the sheep before they were shorn. We took them down to Ain Jubaiha, sometimes we use Wadi ibn Hammad, or if they are really dirty, we walk them down to Dead Sea along footpaths; it's a day down and two days back from here. The wool goes to Turkey, there's no big factories here for wool material. I sell the lambs for meat to merchants from Amman or Karak; a good large lamb fetches 40 JD, like last year (1989), a small one 25. The ones I'm fattening, the ones in the barracks, they'd only get 30–35 now. My wife makes butter and *samn* and *jamîdh*. She has a special building in the garden, and she has three machines (twintub washing machines). We use it at home, and some is for the family, and we sell some, but this wasn't a good spring so there was very little for sale." The following year, he was selling a tank of *samn* in his shop for 100 JD.

In 1992, he saw "more profit in meat than milk this year, because with the drought last year and the poor years before that there are fewer sheep around. And people lost lambs this year, with all the snow and the late rain, and the ewes didn't go into the winter in good condition. A lamb this year fetches 70 JD. I've had to rent grazing this summer, five hundred dunums at 2 JD a dunum; it's near my land by the bridge, where I'm growing brush *dhurra*. This paying for summer grazing, it never used to happen. It was between neighbours. That sheep trader from Rabba started it two years ago; he had three thousand sheep and needed grazing, and he was offering people with barley crops that weren't going to be harvested two dinars a dunum. Of course, they took it. And now we all have to pay.[20] Although lambs are profitable this year, I rely on milk products; *samn* is a real moneymaker, and we go for top quality, and I've sold five tanks to the top *hulwayât* maker (patissier) in Karak at 100 JD each. *Jamîdh* makes money too; even though Turkish *jamîdh* is half the price of the local product in Karak, people prefer to buy the local stuff. My wool sold well, even though the price is down. If I wanted to live from sheep alone, without a pension, rent, the shop, and land, and have a house, car, and a shepherd, I'd need at least five hundred sheep. And that would be tight. If we did everything ourselves, and lived in a tent, two hundred would be fine. I want to buy more land and grow crops, I need the grazing." Two years later, he had bought more land from a cousin, and was growing barley.

An Amarin family have "a sheep flock of around four hundred, the ancestors of which certainly came from our grandfather. We also bought a lot of sheep at the end of the very dry summer of 1991 from bedu further east, because we had a lot of barley, and not much of it was worth harvesting. The bedu needed to sell sheep to buy *alaf* for their core flocks, and because they weren't in good condition, they were cheap. But we had all this grazing. They did well and we sold them on, most of them went to Saudi Arabia.

---

[20] Others say that stubbles used to be free, but since 1990 cost half the price of uncut crop which in 1992 was 4 JD a dunum. Weedy fallows are free but few, and better for camels or goats than sheep.

They were always separate from our main flock, they're herded apart." In February 1995, the main flock was lambing on fallow land at the top of the Wadi Balu'a, with a plastic shelter. As there had been little rain until the New Year, there was no spring. The ewes were being fed a kilo of barley plus *alaf* a day. *Alaf* costs 85 JD a ton, while barley is sold to the government at 125 JD a ton; every registered flock has an allowance of *alaf* at this price, although *alaf* can be bought on the free market at 170 JD a ton. "We have one ram for every thirty to forty ewes, and they're always together. We never use sponges to get the ewes to lamb close together. We don't approve, and anyway, we don't want all the lambs to come at once, we'd need more people for lambing and milking. The ewes mostly have singles, there's a few twins. And we lamb once a year only, it's possible to have two lambings a year, but we'd have to feed *alaf* all the time, and the ewes don't live as long. And the work would never stop. Our lambing goes from December to March, and milking from February to May. In a couple of weeks, we'll be starting the serious milking. It's a month or six weeks of almost twenty-four hour a day work, and by the end of it we are all exhausted. The shepherds do the milking, and the women (of the household) make *jamîdh* for sale, and *samn*, *labna* and *jamîdh* for the family. Some of the sheep have got an E.coli infection, that's why the vet. is here to inject them. He's from Iraq, and it's a bit difficult because our names for diseases are different to his. The government pay for injections against the main diseases; blackleg, clostridial diseases, brucellosis, pasturella, a wide variety of worms and husk. Foot rot is very rare, and so is liverfluke."

Earlier herding was recalled by a number of Beni 'Amr. The exact time being discussed varied, but within the last ten to fifty years. "Before we settled, which was the mid-fifties, we herded much more than we do now. That was how we got our living, although we always grew corn in our fields in the wadis and the plateaux." "In the winter and spring we went far, far east, almost to Saudi Arabia. Some families sent their sheep to the Jeraydat section who used the Wadi ibn Hammad, but some took them with the goats and the camels east. And then we moved back west for the summer. Share workers cultivated the fields in the wadis and the slopes, and we only grew wheat and barley for ourselves, we didn't sell any. We lived from our animals, and things we wanted, we exchanged butter or wool or lambs for them." "We lived so cheaply. Every household had thirty or more sheep, and a horse, a few camels and donkeys, and four or five cows, and we didn't need to feed them *alaf*. When we'd finished threshing, we just put the animals down the Wadi beyond the *hammâm* (hot springs), and they looked after themselves until we needed them." "Of course the sheep and goats had two young a year, they gave birth once in June and once in November. And we never had to feed them. There was so much grazing, in the wadis especially. We stopped having sheep ten – no, more like fifteen years ago, because there wasn't any grazing. The government said each village had to have a forest reserve, so that took up a lot of the rough grazing. And people began to make all these gardens in the wadis, so a lot of the trees, *sidr*, *ballût*, *za'rûr* and *butm*, were cleared. We could buy government *alaf*, but that meant we had to employ a shepherd to look after the sheep while we did the paperwork of registering the flock and claiming the feed, and collecting it, and that needed a lorry. And the young men were employed in the army, and it's difficult to combine sheepherding without a shepherd with children being educated. Education is important now, but sheep are good, we lived from them."

Each of these speakers does have sheep in some way. One has a share arrangement; he is employed in the Hasa Phosphate mines and owns the sheep and arranges for the purchase and delivery of *alaf*, while the other partner does the herding. Another is a sheep trader, with forty sheep of his own, and his wife has share milking partnerships in most years, spending the springs in a tent outside the village. The third has a series of bought in flocks of lambs that are fattened for meat or brought on to be sold as breeding gimmers.

In summer, additional sheep flocks come onto the Karak plateau belonging to members of the Beni Attiya, Beni Sakhr and Huwaitat. Similar movement is recorded by Jaussen (1948: 117), and an Ajalin family of the Beni 'Attiya who regularly spend the summers near Hmud claimed their relationship with Amarin of Hmud had been extant for at least eight generations. On the other hand, a Beni Sakhr family with three households, sheep flocks and camels spent the summer of 1991 west of Balu'a; "We've never been here before, we usually spend the summer around Jiza. We made an arrangement with some of the Majali..... yes, like a *sohba* agreement, we're paying for grazing and we've bought water separately. We've come here because this is a drought year (1991), it's worse than the sixties. Karak is about the only place there are any stubbles. We used to winter in Saudi Arabia, but it's too difficult now after the war and with the sheep. The lamb prices are holding up, but ewe prices have fallen by half. People want to reduce numbers, and we need the money to buy extra *alaf* for the sheep. Part of the low price is caused by so many Iraqi sheep coming into Jordan, everyone there was selling them as fast as possible. We have camels as well because we can always live off them; some years we live from the sheep, this year they're living off us." A Huwaitat lived off his sheep and goats, and was camped for the summer of 1992 near Hmud, preferred to Fagu'a where he often summered, because at Hmud there were more uncut crops. The *jamîdh* they made for sale was sold in Hmud and neighbouring villages, "so we don't bother taking it into a merchant in Karak."

In the *badia*, sheep herding has changed in detail with the decline in camels, the development of the nation state and oil wealth. The information comes from the eastern *badia* of Jordan and northern Saudi Arabia, supplemented by material from Syria. Written sources, as well as conversations with herding families, establish the use by sheep herders of the *badia* in winter and spring (Blunt 1879: 90, 121, 137; Musil 1927: 55, 391, 403–5; Musil 1928a: 99). Burckhardt (1831: 203) indicates a depth of time for sheep trading, writing "when at peace with the Wahabys, many Anezes were accustomed to go every year into Nedjd, loaded with dollars and merchandize, to purchase camels and sheep.........The Anezes set out with them from Nedjd in winter, that they might reach Syria early in the spring, when they immediately sold them to the butchers of Damascus and of the Druze mountains". A Rashaiyida tribesman from Shobak remembered his father buying Nejdi sheep from Saudi Arabia and bringing them up to Damascus for sale, and selling some on the way if there were buyers. The ownership and herding of sheep and goats by members of tribes who describe themselves as camel-herders, bedu or living from the *badia* through camels, was widespread; if these sheep and goats were for household use and the family lived from camels, then this is accurate. But if a family gave up camels and herded sheep and goats for profit, then they became *shwaya*; Musil (1927: 391) writes the Aida section of the Weld Ali and the Hessene had made this switch by 1914. It is now common to hear people say "There are no more bedu, we are all *shwaya* now." The owners of sheep flocks in the *badia* may be their tribal herders, villagers or townsmen; some tribes specialised and continue to do so in herding flocks of particular villages or parts of

Land Use

towns. The herding of Hauran villagers' flocks by sections of the Ahl al-Jabal was noted by Burckhardt (1831: 17–8), while French Mandate records (1930: 53, 55, 59) mention Jumlan and Na'im herders of flocks owned by Damascene merchants, and herders from various Umur sections of sheep belonging to villages of the Qalamoun, Sukhne and Palmyra. Thoumin (1936: 150–4) describes the transhumances of Qalamoun. Some of the higher villages lived from sheep and goats; the lower oasis villages had sheep flocks as part of their livelihood. All flocks summered in the mountains and wintered in the *hamad*. Qalamoun paid *khuwa* to the Rwala for protection in the *hamad*, although French military security lowered the price; the rate of *khuwa* was established in the autumn, but paid in the spring when security had been delivered. The Qalamoun set-up, where the flocks were under the care of family members or paid village shepherds, differed from that of town families from Hama, Homs and Damascus, who had bedouin shepherds. These delivered two and a half kilos of butter per ewe and three-quarters or four-fifths of all lambs to the owner, while the shepherd paid the tax and kept all the wool. Alternatively, the shepherd handed over all the lambs, and kept all the butter and wool. Wool in the thirties was profitable, being exported to the United States for carpet manufacture. This pattern of ownership continues, as Métral discusses (1993) for Sukhne. Tribesmen who had sheep might herd themselves or hire a herder. How herders were paid varied. Burckhardt (1831: 18) says that shepherds received a quarter of the lambs and of butter; Musil (1927: 391) that they were either paid in cash according to the size of the flock, or by *'adâyal* (the use of and profit from a stipulated number of sheep between one and three years old) – for example, he may be consigned all the year old ewes in a flock which he milks and shears for two years to his benefit. Rwala herders confirmed all these methods of payment. Abu-Rabia (1994) gives a thorough account of bedouin sheep management in the Negev.

In the present, sheep using the *badia* belong to tribesmen of the area, from outside the area, and urban owners, in that order of frequency. The sheep within a particular flock are usually the subjects of an intricate network of claims on them and their products, since sheepherding is only a part, if often a major part, of the multiresource economics of a family (three generation *ibn 'amm*). Temporarily invisible members may be employed, have other businesses, work land, with other herds, or trading. Women and children may own animals in their own right, from gifts or by purchase. Women may keep the money they make from selling *samn* and *jamîdh*, or they may put it all or a proportion into the household income; what they do with it is up to them. The majority of hired shepherds come from Syria and are employed; a local shepherd herding an urban owner's sheep may have a partnership arrangement, or be employed. 1992 wages were 100 JD a month, winter clothing and boots, full keep, and often cigarettes.

Formerly sheep used the *badia* from October to April, when they moved northwest to graze on arable stubbles. Now sheep may remain in the *badia* all year round, although herders with access to arable land often take their sheep to these pastures in summer. The provision of wells and *alaf*, together with lorries and tankers, allows sheep to remain; in addition, few camels now use the *hamad* so it is otherwise empty. The prime consideration for sheep herding is grazing and water; these vary according to rains and other users, but supplemented by *alaf* and tanked-in water. Carrying water to sheep was always done. Ahl al-Jabal state that donkeys and camels were used to carry water and Doughty ([1888] 1936: i, 474–7) saw Mwahib taking water by donkey to their sheep in the *harra* of Khaibar; just as camels carry more than donkeys, so tankers carry more than camels. Grazing and water are invariably unevenly dispersed around the *badia*. News of where rain fell, and when and for how long, is a dominant focus of all conversations in winter and spring, followed in summer by news of which wells are open, the whereabouts of grazing, and who is where. Families and households have preferred camping sites for different seasons, but move to less familiar areas when necessary.

The management of grazing at different seasons and in different years is a constant juggling of known grazing and water sources, costs, labour and access along wider domestic networks. Shearing takes place in late April or May, and is the start of the herding year, as it was in Old Babylonian times (Postgate and Payne 1975: 19). Owners of large flocks hire shearing teams, often Syrians, while the family shear the smaller flocks. The lead sheep are left unshorn; rams are left with wool on their backs, as protection from the sun; and ewes are shorn but left with enough wool to give protection from the increasing heat. As summer approaches, those who have decided to move west start to return slowly; those who are based in the *badia* search out and organise water and, possibly, feed supplies. A base is established where the tent is pitched, feed sacks stored, water and diesel barrels kept; this is between 5–50 kilometres from water supply, feed depot, school and shops. The flock, with shepherd, may be another fifty kilometres away, visited daily by the owner with the water tanker and shepherds' food supplies, and grazing circling or criss-crossing an area. In high summer, sheep need water twice a day, and spend the heat of the day resting, often under shade shelters. The quality of feed is less important, as sheep are at the period of lowest nutritional needs after conception. Mating is at its height in July and August; births are five and a half months later. Dickson (1949: 399) gives a spread of mating dates from late spring to late summer with lambing times from October to February, while Postgate and Payne (1975: 19) interpret Old Babylonian herding contracts as having a lambing date of November/December and therefore mating in July/August. A few goats are often run in with the sheep, to encourage ewes to come into oestrus (Lancaster and Lancaster 1991: 129).

From early October, herders watch the sky for signs of early rain; if there were good rains last year, soil humidities and dew in the summer (local informants; Agnew and Anderson 1993: 23ff) produce autumn growth among a variety of perennials. If there is no news of rains and better conditions elsewhere, people stay put, tank in water and feed, and consider the options of selling some sheep, moving the sheep along networks to other areas outside the *badia*, or moving right outside themselves. These are decisions by the owner/s, just as the supply of water and feed, and purchase and sales, are his/their responsibility. This is why Syrian shepherds are employed, not partners, as they bring only their labour and have no share of inputs or responsibilities. Some are able to move across borders to better conditions, as did Rwala from 1991–94. People do eventually move from bases after a long series of poor years like 1989–94, like a Sharafat Ahl al-Jabal family who, having based themselves in the Mahdath area for eight years, moved to the northern Ghor in 1991, and then moved between the north Ghor and Mafraq, via Jerash for the next three years. If there are signs of good early rains, or indeed any rain, herders move as far east as possible fast, to use flushes of early annuals.

Lambing starts at the end of December, although earlier lambs are quite common. As the weather is cold, often windy and sometimes wet, tents are pitched in sheltered places in the lee of hills and slopes. Clean sites are chosen for lambing to minimise risks of infection. Stone corrals or tents made from feedsacks shelter the lambs. Ewes are taken out a couple of hours after sunrise to graze, return mid-morning to suckle their lambs, and may be fed and watered if there is no fresh grazing or rainwater. They then are taken out again to graze, and return before sunset, when the lambs suckle again and the ewes fed and watered. Women and children undertake part of the herding in family run flocks, set out feeding bowls and drinking troughs, and deal out the feed. The sheep are inspected and carefully watched to ensure each is eating and drinking properly, and suckling its lamb. Each ewe and its lamb are known. Ewes are fed at least a kilo of *alaf* a head at this time, unless there is really good grazing. The lambs are kept in age cohorts; on fine days, they are taken a short distance to clean grazing by small children in the middle of the day. If the weather is settled and warm, and the ewes are grazing nearby, the lambs accompany them. Weaning is between six and eight weeks of age, and the ewes are then milked for a further six to eight weeks, depending on grazing. Family flocks are milked by the girls and women, who process the milk into fresh yogurt and butter, *bagal* or *jamîdh*, *samn* and *labna*. Hiring professional cheesemakers is common for larger flocks, or where there is a shortage of women and girls. The cheesemakers (*jabbân*) prefer to make contracts for milk from a thousand sheep, so those with fewer often join together. If grazing is poor, milk is kept for the lambs and household needs only. Herders generally say their flocks have an 80% lambing average (though from observation this sometimes appears optimistic), and of these, 80% survive to maturity; losses of between 10 and 20% of ewes are regarded as normal, similar to those in hill flocks in the North of Scotland. Sudden heavy snowstorms or prolonged very cold and dry periods cause deaths. The working life of a ewe is said to be six years, although many herders have older ewes. Tooth loss occurs, but seems not to be a problem.

Daily movement for a family flock is a series of daisy petals out and back. Flocks with shepherds may make similar movements around a base, or follow a linear movement along a wadi system or a series of *shi'bân*, following the directions of the owner and supplied with water and feed by him or a driver each day. When a family flock moves camp, the herder/s, sometimes children, take the flock out after sunrise and slowly make their way to the new camp site, while the flock grazes along the route. The children have been told where they are to make for, shown the landmarks, and had their way described in terms of soil surfaces, rocks, pools, vegetation, and former campsites, as well as direction. The woman takes down the tent, and it and the furnishings, with water barrels, feed, feed bowls, and troughs are loaded into the lorry by the adults and other younger children. The lorry is driven, usually but not invariably by the man, to the new site, where everything is unloaded, the tent put up, furnishings stowed, and the arrival of the flock and herders awaited. Stopping at a newly erected Mesa'id Ahl al-Jabal tent in the eastern *harra*, we were asked if we had seen two children with a sheep flock. We had, and a girl of eleven and a boy of seven arrived shortly, with the family's one hundred and fifty sheep and some goats, having walked some fifteen to twenty kilometres between dawn and one p.m. A usual day's grazing covers some eight to ten kilometres. Longer moves of over forty kilometres are made by the flock under the charge of youths or older men or, if going to a different region or crossing a border, by lorry.

Shifts in the use of the *hamad* are indicated in the following intermittent and incomplete observations; rainyear evaluation is by herders.

1987–8. A good year in *hamad* and *harra*, with good grazing and tents well spread over the area. Dry feed was used only at lambing. Wells recharged and in use. Rwala, Ghayyath, Zubaid, Ahl al-Jabal tents from the Saudi to the Syrian border, and from the eastern edge of the *harra* to Ruwaishid watered all summer at Burqu', where a hundred tankers a day were filling up, supervised by the *Badia* police. Ahl al-Jabal herding in the eastern *harra* watered at their wells at Biyar Ghussein and Mahdath, those west used Bishriyya, Qatafi and Azraq. Rwala, Ahl al-Jabal, Beni Khalid and Ghayyath herders east of Ruwaishid used the government well at Jisr al-Ruwaishid, or the Sha'alan wells at Faydhr or Fowk ar-Risha. Arable crop harvested at Mahdath and Shabaikha by Ahl al-Jabal, and Faydhr, Fowk ar-Risha (Rwala) and Wadi at-Taxi.

1988–89. A poor year, a little rain in late spring, but good rains in western Iraq. A few Ghayyath used Burqu' and Wadi Milgat. Al-Wudyan in Iraq was opened to non-Iraqi tribes, and many Rwala, and some Ghayyath and Ahl al-Jabal herded there for July to September. Others grazed along the Saudi border with Saudi herders from Rwala, Shararat, Utaiba and Harb. These latter, and Rwala, moved south from late September. The southeast of the Ruwaishdat was used by a few Dahamsha, Amarat and Ahl al-Jabal. From mid-September, Dahamsha and Amarat returned to Iraq, the Ahl al-Jabal went north of the road. Flock numbers were being reduced from before lambing.

1989–90. A poor year, with a little November rain along the Saudi-Jordanian border, and three heavy but scattered rains in mid-February and March, which gave ten weeks grazing, with water from *khabrât* and *ghudrân*. Before January, no-one was moving from their current site; dry feed had been given since late summer, and flocks had halved since summer 1988. Some Ahl al-Jabal considered going to the north Ghor. By mid-March, herds from Ahl al-Jabal, Rwala, Jumlan, Fedzir and Fuware congregated around Anqa, Mahruta and the Ruwaishdat. The summer saw Rwala, Beni Khalid and Ghayyath tents northeast of Faydhr, and *alaf* was fed from July. Some Rwala and Beni Khalid had been to the Jezira in Iraq. The Ahl al-Jabal were scattered in the *harra* herding goats, having sent their sheep to the villages. A crop was harvested at Mahdath, and a poor crop was grazed at Wadi at-Taxi.

1990–1. Another poor year, flocks at base levels, and fed for eleven months. Ahl al-Jabal well spaced in the *harra*. Rwala and Ghayyath using area close to the Syrian-Jordanian border; by mid-September, some had moved south, nearer to Burqu' and the ar-Rishas.

1991–92. The good rains and snow further west rarely reached east. Rain fell east of Ruwaishid in late December, and in late April in the northwestern *harra*. In late January, Ahl al-Jabal were spread along the Jordanian-Saudi border. By March, Rwala who preferred the northeastern *hamad* were considering moving southwest of Azraq; the only Rwala in Jordan were those based there. Ahl al-Jabal who preferred Mahdath had gone to the northern Ghor. In the summer, some Ghayyath were dispersed around Burqu' and Wadi Milgat. The April rain enabled some Ahl al-Jabal to harvest *dhurra* at Shubaika.

1992–3. Little rain in eastern Jordan, but good rain in northern Saudi Arabia, while rain on the Iraqi border which flowed east gave a flush of annuals, used by Rwala, Beni Khalid and Sardiyya. May rain filled *ghudrân* but gave no pasture. Umur used the western *hamad* since Fuware had gone to Saudi Arabia; Ghayyath used the Athaina in the spring, and Burqu' in the summer. Many Ahl al-Jabal wintered around Tell Asfar, while others were scattered in the eastern and southern *harra*. Some moved to the northern Ghor and west of Irbid for spring, returning to villages for the summer via Jerash, Zarqa and Mafraq. All Rwala herds summered in Saudi Arabia, and if their owners returned to Jordan in winter, their animals remained, as did those of some Sardyiyya.

1993–4. Late rain in the northwest *harra* filled cisterns and recharged wells at Jawa, but was not enough for crop sowing. Most Rwala were in Saudi Arabia. Ahl al-Jabal flock numbers were at base levels. Mesa'id and Sardiyya flocks grazed irrigated vegetable crop residues around Dhulail, Umm al Jamal and other villages. Ahl al- Jabal and Zubaid herders were dispersed in the *harra*, and used the northern Ghor and slopes west of Irbid. No wells at Mahdath and only one at Ghussein had water or was in use. Ghayyath summered around Burqu', Mingat and the Syrian border, collecting water from Burqu' and Jisr al-Ruwaishid.

1994–5. Exceptional rains in the *hamad* gave "the best spring for twenty five years." Herding management concentrated on the slopes as this growth withers first, and avoided grazing the *shi'bân* during flowering and setting of seed by annuals, and growth by perennials. Areas associated with locally based families were left, and incoming herds from Huwaitat and Beni Sakhr went to the areas south of road.

Herders describe their management as a function between possibilities of grazing, feed and markets, with a built-in assumption of livelihood in most years. The flocks are part, often the base, of the multiresource economics of the wider domestic group with a core of three generation *ibn 'amm* members. Different market aims for milk or meat are recognised. "We Rwala herd to produce lambs for meat rather than producing for milk, like the Ahl al-Jabal. We never did milk processing for the market. Herders who produce for milk separate lambs from the ewes as soon as they can be weaned. With herding for meat, lambs stay with their mothers because the loss of milk income is more than compensated by the increase in the value of the lamb, whether it's a ram lamb for meat or a female lamb for breeding. Of course, we separate lambs from ewes for other reasons, bad weather or distance. But really we keep lambs with the ewes, and we make just enough *samn* and *bagal* (*jamîdh*) for the household, never for sale; unless it's a really good spring, when the jabban come out and buy the milk for cheese-making. The early spring milk sells at 4 JD a tank, and goes down to 1 –1.5 JD later on, when it's thinner. We use five to six tanks of *samn* a year, and we're a big household (ten to twenty people) and there are always guests. We sometimes have Nejdi sheep because they give a lot of milk, although they don't like the cold. In a poor year, we can be feeding *alaf* for eleven months of the year; there's an allowance of cheap *alaf* and after that it's at market price, and it costs. But it's important to feed the ewes, a wellfed sheep doesn't die, nor does her lamb, and it'll grow well. When they're feeding a lamb, they need up to two kilos each a day, that's what we're feeding at the moment because there is no value in the dry grazing. Some people feed less. Most people feed three times a day, first light, midday and sunset, and that's when the lambs suckle. There are people who produce two lambs a year, but they have to feed *alaf* all the year round, and the ewes don't live as long. And the meat is bland, while if there's a spring and the lambs have been eating natural plants they fetch a good price because their meat has a better texture and flavour. There's one herder, he arranges mating so that his lambs are ready for the big feasts, but I don't like that; it happens anyway, so why bother?"

Two Zbaid brothers, who came from Syria in 1983, herding together in the western *harra* said they had 250 sheep and 50 goats, from which they and their families lived. "We have milk from October to the end of May, beginning of June. Usually we make a lot of *samn* for sale, and some *jamîdh*, but this year (1993) we had very little for sale. When there's a spring, a ewe gives two kilos of milk a day, this year it was only half a kilo a day. But we've enough for the household for the year. I didn't know the price for wool has gone up, that's good. We'll start shearing in the next few days and we've got a couple of Syrians to come. Looking after the flocks is a bit complicated with lambs and kids most of the year, but we know every animal so it's fine. And our wives herd and the children, we all work. The animals are in two flocks, one each. The very young lambs and those being weaned don't accompany the flock but stay in the *haush*. The very young can't keep up, and we don't want the others suckling. The babies suckle at least twice a day, sometimes three times if the flock comes back in the day. It depends where the grazing is, what it is.... This year, we've fed *alaf* all the time, there haven't been many annuals, and the perennials are pretty dry. We move frequently, this year to get clean areas and clean grazing - even if it is only dry stuff. But with the borders closed it's difficult. The Rahba was marvellous, and we use the *hamad* in the winter and come back through the *harra* to Jawa and other places."

A Ghayyath woman herding with her husband and small children in the Ruwaishdat in late March spoke of her care in looking after ewes and lambs and in milking and processing. "I, like my husband, know each ewe and each lamb. The ewes have names, mostly their names refer to a characteristic of each one. We use the *shbâk* (rope and its method of attachment for milking, while the ewes are being fed).[21] The lambs mill around and the children keep them away, or they are in the *haush*, or we could put them on a *rabej*.[22] I milk into metal saucepans, and everything has been well washed and dried in the sun. The milk is strained through well boiled and sun-dried cloths, and then I heat it, keep it at the right heat for the proper time, and then cool it. I have to do this if I am making milk products for sale, it's a government requirement. The children like the lambs, they play games with them, and cuddle them, and kiss them. I have ewes I like especially too, I have my favourites. And the rams.... We all make a fuss of them sometimes, even the donkey and the dog. After all, we couldn't herd without them too."

The perception of the construction of a sheep flock is not something that herders talk about much; most flocks have, as it were, 'always' been there with their herders. Constructing a new flock, by those who have given up herding and then return, is seen as difficult, especially when the animals are bought from a variety of flocks as lambs. The control of young animals ignorant of their herders and of their landscape takes a lot of time and a lot of labour. School summer holidays is a possible time. "We got the lambs from different flocks, and kept them in the barracks with the donkey for a few days so they could get used to us and to each other, and the donkey. They were being fed and we watched their behaviour – some were bold, some timid, and so on. And the boys were here a lot and played with them and gave them titbits. So we got some to respond to us, and a few of these were the ones first at the feed trough and the water, and the ones the others followed. Then we took them out, just a short way. But it was hard work, not like when they have their mothers to follow. And we went on like that, but it's difficult because they don't know anything, they have to learn everything. It's easier to keep them in the barracks." D'Hont (1994: 65) noted that the Agaydat had ideas on the nature of herding, seeing their role as herder as "an interested disturbance of the

---

[21] A loop is passed over the sheep's neck and the free end is passed under the neck and drawn through the loop to form another loop for the next sheep – rather like a row of knitting. The whole length can be unravelled by pulling the free end.

[22] A *rabej* is a series of separate closed loops (sometimes stopped slip-knots) tied onto e.g. a tent rope. The loop is large enough to pass over the lamb's head but fits too closely for the lamb to pull out backwards. The system is commonly used in markets to tether sheep for sale.

life cycle in the course of which a flock assured its conservation, its self-defence. ....In this point of view, the animal born wild, captured to be killed or fattened, is not herded but hunted as a species to furnish an attribute." At the same time, the Agaydat keep milk production as their main objective (1994: 75), and sell many young male lambs to specialised herders who are financed by traders. The Agaydat method of training lead sheep is described (1994: 84–5), based on imprinting young male lambs with people. Out of three hundred lambs, twelve will be taken for training; these are hand-reared in the tent, hand fed with bottles or by fingers in buckets of milk, and returned to the flock at four to five months old, when most milking has stopped and the male lambs are sold. These trainees wear pompoms of coloured wool for quick identification, continue to be handfed by the herder, and learn to answer his call. Those who answer the herder's call without waiting for a line of other sheep to form behind it by early autumn are sold off. The successful are castrated for docility, and are never shorn or dagged, so they never have the energy to go off looking for fresh grazing but rely on the shepherd. Using lead sheep, Agaydat reckon a shepherd can herd five hundred sheep. Rwala and Ahl al-Jabal herders using employed shepherds use lead sheep who follow the donkey supplied to the employed herder to carry his cloak, food and water.

People also trade sheep. Some concentrate on dealing, more deal and have flocks. There are two aspects to trading; one supplies markets with meat, dairy products and wool, the other moves sheep to areas of good or surplus grazing and water from those in deficit. The same herders are taking part in both, and the most productive markets often lie across borders. Herders sell their animals and products to merchants who come out to the *badia*, or at local markets. The main market for sheepmeat is now Saudi Arabia, whereas earlier it was Damascus. Wool goes to Turkey, and milk products are distributed along household networks and sold in the towns and cities of Jordan. The redistribution of flocks and grazing grounds used to be by mobility, sometimes within a tribe's associated areas, sometimes along intra-tribal networks within the wider domestic groupings, and sometimes by arrangements and contracts with payment. Reducing and expanding flock numbers was a standard response to poor years (Glubb 1938), although dependence on *alaf* is said to have made this more marked. The cross-border mobility formerly practised by every tribal group herding in the *badia* has now become restricted to a few, and dependent on government regulations, which appear to change often and arbitrarily. It is said that local crossing points between Syria and Jordan north of Deir al Ginn and south of Tinf are open to local herders and their animals in trucks, but many said crossing was either impossible or difficult. Crossing between Saudi Arabia and Jordan is only possible at official crossing places, by truck and with papers; a flock is allowed to increase only by a certain percentage on returning, but if a herder has been in the other country for some years, this percentage may well be exceeded. In such cases, the herder or owner goes to a well known tribal shaikh to mediate on his behalf. Differences in currency official and unofficial exchange rates provide additional profits for astute operators.

A long, amiable and indecisive conversation between two Rwala in Jordan concerned two hundred gimmers (*garâgîr*) for which Abu R. had a buyer in Jordan, but his were in Saudi Arabia. He proposed his buyer should have Abu A.'s two hundred gimmers which were in Jordan, and Abu A. would be paid with 250 mixed ewes and gimmers that Abu R. has in Saudi Arabia, where A. is presently herding with other family sheep. Complicated transport arrangements and swaps of lorries, none of which were in total working order, would complete the proposed deal. Abu A. is a herder and trader, with his eldest son; the flock fluctuates between 500 to 4,000, often split between Abu A. and A., in the eastern *hamad*, western Iraq and northern Saudi Arabia. Abu A. prefers to herd between Raghban, Basatin and Traibil, while A. herds in Iraq and Saudi Arabia, although in the early months of the bad year of 1991 some sheep were sent south of Azraq with other sons. He sometimes, as in 1992, buys breeding sheep in Mafraq, which are taken to Saudi Arabia by Rwala associates for sale. Since the Gulf War, and as Abu A. is getting old, an increasing proportion of the sheep are in Saudi Arabia with A; "they're my sheep but in his name". He bought agricultural land near the Tebuk turn, with the idea that the sheep would move between this base where he installed centre pivot irrigation and Raghaban, "but they sowed three times and nothing came up." In mid-1994, Abu A. said "I've given up herding, I've handed everything over to A, who has the sheep near Swair. The main reason is the borders, it's got too difficult, the borders are almost completely closed. It's very difficult to take sheep back and forth. I can import sheep once, and there has to be a genuine sale, but I can't herd across borders. My grazing areas means we graze across borders, and this brings endless problems with the army and the police, and it's no longer worth the trouble. Getting decent shepherds and drivers is more difficult, because most of them are Syrian and so I have more problems with the authorities. My sons don't want to herd, they would rather loaf in Saudi than work herding. They don't want to be Bedu, they'd rather take subsidies and be morally corrupt (*fâsikh*). Herding is only really profitable if it's a family concern."

Sheep trading into Jordan from Iraq after the Gulf War increased dramatically. "because exports of sheep from Iraq into Saudi Arabia, which was an important market, were banned by Saudi Arabia. And the Iraqi dinar collapsed in value. So Jordanians would go down to the moneychangers, and change Jordanian dinars or American dollars into devalued Iraqi dinars and travel to Iraq illegally. There they buy sheep for Iraqi dinars, with a quarter of the price in a hard currency. They bring the sheep back and sell them in Jordan, or into Saudi Arabia. Then they take their dinars or Saudi riyals down to the moneychangers and start again. People made a lot of money, but it stopped, for a while, when Iraq changed all the notes." "The borders were closed to stop the sheep trading (1994), because some of it was in stolen animals. It had got very bad with shoot-outs on the border, and people were driving sheep over the sand and earth ramparts on the borders." In late 1995, "the official exchange rate was 3.2 Iraqi dinars to the US dollar, and the unofficial was 2,500 ID to the US dollar. The sheep are smuggled out of Iraq but enter Jordan officially, there's a tax of 5 JD per head to come in. Then they can be sold in markets in Jordan. This year the trade is mostly in lambs from the Jazira, because there was a good spring there. If you want to re-export the lambs, you have to get papers, papers of ownership, health and so on. Papers for import and export are organised by X; he puts them in his name, which means no-one asks questions, and he charges 5% for importing and 10% for exporting. In spite of these charges, it's still profitable, because you can buy cheap Iraqi dinars here for buying the sheep, with a proportion in hard currency. I don't mind paying a percentage to X, it saves me doing the paperwork."

A Sharafat herding family traded sheep, buying in Syria, importing officially into Jordan and either keeping them or selling them on. "We do it only when it's profitable because of the differences in the official and unofficial exchange rates. The unofficial rate varies between 60 to 75 Syrian pounds to the dinar, while the official rate is 58. We don't go to Iraq, we don't know anyone there. We don't import a lot, it's a small business."

Another Sharafat (Ahl al-Jabal) "used to drive a bus, but I stopped to help my father with the sheep. The real money is in lambs now, but if there's grazing and the women are in the tent, milk makes more money. But it's difficult to herd properly now, because the children must be educated and that means the women have to stay in the villages to look after them. It's difficult to be in the *badia*, unless the children are very young, or they're grown up. Lots of people sell sheep to save money on *alaf*, but buy more sheep when prices are right or there's a good spring. My father and the sheep are in the *harra* now in Wadi Rajil; he did go down to the Northern Ghor three years ago, but the sheep got ill. Then he used the slopes below Ajlun and Irbid, that was better. We used to go Jaulan in Syria when there were droughts, and in 1954 we went to the Biqa in Lebanon, or we used to go to the *harra* across the border in Saudi Arabia or to the northeast side of the Wadi Sirhan. We can't pass sheep to cousins in Syria or Saudi Arabia when it's a drought year in Jordan, there are no Sharafat there."

There is general agreement that sheep herding is profitable, even though in a series of bad years the sheep only keep a family, there are no profits above maintenance. The flocks managed entirely by labour from within family networks are more profitable per sheep, or the number of sheep needed for livelihood. This latter aspect becomes virtually impossible to estimate with any accuracy, because of the nature of claims on resources contributed by family members; but a very tentative conclusion is that formerly each person had available 10–15 animals, and now, each person has available 10–50 animals. 'Available' is important. The variation between ten and fifty represents the difference in costs between a poor year provision of eleven months' supplementary dry feed, and a good year's supply of four months' feed. Family labour includes the women to process milk for sale, which in good years adds to profits considerably. "For investment, for what we would do with spare money, it's sheep every time. Nothing else is as profitable. There's not much in building land or agricultural land at the moment (1994), and arms smuggling is over-supplied, every-one's doing it. But there's always a market for sheep meat in Saudi Arabia." "If the family does all the work, then each sheep makes more profit, and the costs are lower and a tent needs two hundred sheep to live on. But if you have a shepherd because you're busy with other things, then the flock must be a thousand to make it work; there's the shepherd and a driver for the lorry and tanker, and all of this has to be paid for from the sheep. But there must be someone from the family to oversee the shepherds every day, otherwise it's a waste of money. It's always your lambs that die" (which implies the shepherd has sheep of his own or as shares; and recalls the disadvantages of hired shepherds mentioned by Musil who sold the owners' sheep at night and the next morning would proclaim the animals as lost). The investment is made up of contributions from individuals within a core of the wider domestic group, men and women (and see Abu-Rabia 1994: 77–8, 96ff; Métral 1993: 208).

A presentation of figures on sheep costs and sales would not be particularly useful for several reasons: costs, especially *alaf*, would have to take into account government prices and quotas; the figures we have are inconsistent, and clear explanations from individual owners are not forthcoming; numbers fluctuate through the year; labour arrangements are flexible; sales are for a variety of reasons, some of them unconnected with herding practice; prices received fluctuate widely from market to market, and from month to month – the price of a ewe with lamb at foot can double between December and March, and lambs fetch far more just before the Muslim festivals. One owner saw the profits in sheep herding "as all about futures and hedging. The core flock itself is capital, and we live from them, but above that, the flock isn't capital, it's hedging. I have as many sheep as I can before the winter. If there is a good spring, or something that makes a strong demand in the market, then I make a good profit. If there isn't a spring, I have to sell sheep to buy *alaf*. How many I sell depends on how many I have to feed, how soon I think the price of *alaf* or sheep will rise or fall, where I can find grazing, how I think the market will go......".

An increase in sheep flock size is noticeable. Ahl al-Jabal herders say they used to have 40–100 sheep and goats, and some camels, before 1960; about 200 sheep and goats in the seventies, and from 200 up to 1,000 by the late eighties, although in the early nineties most Ahl al-Jabal flocks were between 200–500. The simple progression has to be modified by the decline in camels, and by the increase of employment so that a herder is probably now herding for a wider group that would formerly have had more of its members herding. On the other hand, the Ahl al-Jabal like other groups has considerably increased in numbers, and so possibly also in animals. A further factor in the increase in sheep numbers is that they are seen as a good investment, with ever growing urban markets for meat and dairy products. On the other hand, the reduction in camel numbers and other employment opportunities may mean that the total demand on badia resources from grazing may not have changed much. What has changed is the limitations on mobility by the reduced areas of grazing available, with the expansion of agriculture, and the difficulties of crossing borders, which does put pressure on *badia* vegetation, while supplementary feeding both mitigate and exacerbate this. The relationship between plant species and their cover and grazing animals is complex, but flexible and resilient.

Local users are aware of the pressures to increase flock numbers. "Before there were trucks, we used camels or donkeys for carrying water to the sheep and goats. A camel carries two to three hundred kilos of water, a donkey sixty, sometimes eighty, but usually sixty. So not so much water could be carried, so not so many animals could be watered. So there weren't as many people. Now the wells can't water the number of people and animals there are. In the past, when people had 60 or 100 animals, life was much better because there was plenty of grazing and plenty of water. We didn't need so many sheep because we didn't have to buy a lorry, tyres, mend it, and put in diesel. And the children all live now, and people live longer, so there are a lot more people. So we need more animals to keep them all, and that means a lorry and a tanker for water, and buying *alaf*." "There used to be far more grazing and more trees. There were *za'rûr* and *tarfa* in the wadis. They went because people in towns wanted firewood, and for building houses, for roofbeams, when people settled and had houses. The grazing plants, the perennials, get eaten down so much because there are too many people so there are too many animals, and there are too many borders so we can't move, and they take unsuitable land for agriculture. In the past, we didn't need *alaf*, even if there was only a poor spring, because there were the perennials, and when we went to Syria for the summer, we had all the dry grazing to come back to." "But the *badia* can't be used as a feedlot, just keep the sheep here and feed them *alaf*. It's not good for sheep, they have to walk and graze, even if they don't get much food out of the grazing. It makes their stomachs work. And when it rains again, the annuals will come and the perennials grow. It's rain the plants need. There are a lot of sheep, but if there are too many they die. I reckon if there's no rain this year, the number of sheep in the *badia* will at least halve, people will move them out, or sell them, or they will die." "Grazing doesn't matter, it's rain that's important for plants. Some plants have long roots under the ground, or dry right out and look dead, but it rains and they come back. And the seeds live in the dust for years, until it rains. The plants have to be like this, even if there are no animals they would dry up and be broken by the wind. And we graze carefully, we don't let sheep eat the plants when they are in flower and making seeds. Most plants don't taste good then. And the sheep are always moving between the grazing, and there are always places where the sheep don't go, like the islands of plants in the *gi'ân* or everywhere in the rocks. No, when there are good rains, there will be good plants." This view of the natural resilience of *badia* vegetation was upheld after the good rains in 1994–5 in the *hamad*, when *shi'bân* known to have been grazed hard over the last twenty five years bloomed with annuals and recovering perennials. *Shaiban* that had been ploughed in some years were also thickly covered with vegetation. The whole area was not covered, but the *shaiban* made islands of flowers and plants, while for mile after mile the slopes were green with waving grasses and pink or yellow from flowers (see appendix).

Rwala, Ghayyath and Ahl al-Jabal herders gave the following as seasonal sheep grazing species: we give local names followed by botanical names, and there is no absolute correspondence. Identifications have been checked with Prof. Da'ud al-Eisawi of the Department of Botany at the University of Jordan and using Mandeville's *Flora of Eastern Arabia*.

Spring annuals and perennials described as *'aishb* include: grasses – *niza* (*poa sinaica, stipagrostis plumosa*; *khâfûr* (*schismus barbatus*); *sam'a* (*stipa capensis*); *al-hossniyya* (?*phalaris minor*); *sajil, shu'ayyirah* (*rostraria cristata*); *da'a* (*lasiurus scindicus*); plus other unidentified grasses.

Annuals and perennials: *da'a, mlih* (*aizoon hispanica*); *salih, islih* (*gypsophila*), *cakile* (*erucaria,lepidum aucheri, neotorularia*); *humbayz* (*emex spinosa*); *hambaz* (*rumex vesicarius*); *khubayza* (*malva parviflora*); *irdja* (*helianthemum lippii*); *khafsh* (*brassica tournefortii, diplotaxis harra*); *khuzaylan* (*savignya*); *rashâd* (*lepidum sativum*); *khuzama* (*horwoodia*); *shiggâra* (*matthiola*); *hazzâr* (all small cruciferae); *nifal* (*trigonella, medicago*); *nai'ma* (*lotus haplophilus*); *gafa'a* (all annual *astragalus*); *shitâde, khidâd* (*astragalus spinosus* when young); *umm al-grain* (*hippocrepis bicontorta*); *bikhatri* (*erodium*); *tzahil*, *kahil* (*arnebia gastrocotyle*); *graita* (all *plantago* sp); *gutaina* (*plantago ciliata, filago desertorum*); *ribla* (*plantago boissiere* and *coronopus*); *gathgâth* (*pulicaria*); *arbiyyân* (*anthemis melampodina*); *jurrais* (*aaronsohnia factorski*); *hawthân* (*picris babylonica*).

Perennials included in *hamdh*, (all year plants, good grazing in general) are *ruth* (*salsola vermiculata*); *ghudraf* (*salsola volkensii*); *hamdh* (*halothamnus bottae*); *rimth* (*haloxylon salicornum*); *ajram* (*anabasis lachnantha*); *sha'rân* (*anabasis setifera*); *neytul*, *yintul* (?); *shumbrum*, *shibriq* (*zilla spinosa*); *thunaybât* (*caylusea*); *shuwayta*, *shuwaydha* (*helianthemum*); *shitade* (*astragalus spinosus*); *za'atar* (*thymus*); *kalsha* (?*halimocnemis pilosa*); *hamat* (*moltkiopsis ciliata*); *shîh* (*artemesia sieberi*); *firs* (*artemesia judaica*); *amrâr*, *mrâr* (*centaurea*); *mullayh* (*reamuria*). *Shîh* and *gathgâth* are eaten only when very young or dry.

Summer grazing includes the drying and dry annuals, some of which like *graita* and *salih* are said to be better dry than fresh. *Graita* is said to put fat on animals. Summer appearing grazed plants include *gathgâth* (annual *salsola*); *agûl* (*al-haji maurorum*); *roghol* (annual *atriplex*); *labt*, *urzh* (*prosopis farcta*). *Hamdh* perennials are the main component.

Autumn grazing; before rain, as above, but dew from soil humidities may freshen some perennials; some, like *shumbrum*, often flower and put on new growth at this season. *Niza* is a useful autumn grass. After the first rains, annuals begin to appear; *niza* and *graita* are among the first.

Herding in the *badia* in Syria is often within a multiplicity of government directed co-operatives. Shoup (1990) describes the *hîma* reserve system in Syria. A Ghayyath in Jordan had experience of the system close to the border where it is managed by the army. His main grievance was that the range allowed to him was so tightly controlled with only a three kilometre range, and a five kilometre range further away from the border. For him, the *hîma* system meant government control of people rather than conservation of grazing through tribal mechanisms, which is how *hîma* is presented (Chatty 1986: 145ff; Shoup 1990). A Rwala in Syria saw the grazing co-operatives, the means through which the *hîma* system is organised in practice, as providing cheap credit to herders so that they do not have to sell animals to buy *alaf*. A herder registers with a co-operative, then he has access to credit and cheap *alaf*, and he sells to the co-operative who sell to the government.

There are also companies with joint Syrian and Saudi Arabian or Gulf ownership that have sheep. One, based between Palmyra and Qariatain, had 1,700 sheep, and was into milk production, with seven or eight women doing the milking. Each milked about two hundred sheep, but the ewes were milked only once a day. The Syrian partner, a government agency, provides the land and fixed capital inputs like wells, possibly electricity and a road; the other partner provides all the working capital for whatever they are going to produce. In the smaller joint enterprises, the active partner often uses further share partners. Métral (1993) has an important discussion of the inter-relationships between herding and agriculture among the traders of Sukhne, Umur and Hadidiyin herders, and Sba'a in the northern part of the Bilâd ash-Shâm. The flocks of co-operative members from the Sukhne district used the *hamad* in winter and spring, and in summer were trucked to the Jazirah, the Ghab or the Hauran, where stubble grazing had been rented for the summer. This material recalls that of Dickson (1949: 545–9) where Muntafiq sheep herders worked between agricultural land in Iraq, and herding and marketing of sheep and milk products in Kuwait and eastern Saudi Arabia in the thirties and forties.

Other material on sheepherding in Syria includes Chatty (1986) on changes following the introduction of lorries; Hannoyer and Thieck (1984) for the Raqqa region; d'Hont (1994) gives a detailed account of herding practices among the Agaydat; Lewis (1989: 170–193) discusses sheep-herding and farming tribesmen of north-central Syria; and Yedid (1984) on the nomadic pastoral system in the plateau north-east of Hama.

Some in the *harra* prefer goats as the main herd animal. These are black, *baladi* goats, noted for their hardiness in tolerating heat and needing less water than the larger brown *Shâmi* or Damascus goat which gives more milk. *Shâmi*s are found only in villages as house goats. *Baladi* goats are also much cheaper to feed, needing less *alaf* and browsing on a wider variety of perennial plants. Goats cope better with heat than sheep, although they do not like cold and wet, and need shelter in winter. The *harra* is more sheltered than the *hamad*, and has, perhaps, a larger growth of perennials. There is not so much profit in goats, as kids fetch less than lambs and their milk has less fat; but they milk for longer, and there is an urban market for goat cheese and goathair for the manufacture of tentcloth.

Three tents of Sharafat Ahl al-Jabal "always winter at Wisad in the *harra* if it's possible. It's a good place, with the pools, and there's good grazing in the *harra* and on the shaiban. We have mostly goats, each tent has its own flock, but we're one family. And we each have a lorry. We have 120 goats, they have 70, and they have 100 and 60 sheep. We make *samn* for the family, and *jamîdh*, from the milk, and we make goats'milk cheese for sale. We sell it to traders in Mafraq and Ruwaishid. We – the family – have land at Deir al Ginn, but we ourselves aren't there every summer. Two years ago we were at Burqu', and last year we were at the village." A Ghayyath at Zilaf in Syria had arrived a week ago (late December 1995) from the villages north of Salkhad. "We always winter here, all the tents here are Ghayyath, all the land is Ghayyath. We have only goats, there's a lot of profit and we like goats, they're better than sheep, not so much trouble. I'm not in a co-operative, I'm in a *jam'iyya*, a society; I register my flock, and I can get *alaf* and credit. All the water came from last spring. It was marvellous, everything grew and flourished. Now there's still *agûl*, *gurr* (*vitex agnus-castelli*), *tarfa* and *za'tar* in the wadi beds, and *yintûl*, *rimth*, *shîh*, *gaysûma*, *mullayh*, *subaycha* and grasses; the goats eat them all."

In the Shera mountains, black goats are predominant herd animal. An Azazma family at Ghuwaibah reckoned a household needed at least a hundred goats to live from, "for the tent and for guests, and money for clothes, pots and pans, tea, coffee, sugar and flour. We bought 14 tons of *alaf* this year (1991–2), this was us and our eldest son and his family, and together we herd between three and four hundred goats. Goats need less *alaf* than sheep, they get about three quarters of a kilo a day before they give birth, and when they have the kid and milking. This year, we're having to feed them for twelve weeks, maybe longer, there haven't been the *sayl*s (rainfall in the mountains which flood down to the valleys) for the plants. Half the goats have a kid that grows to maturity or sale; 10% don't conceive or die, – they fall off cliffs, and things – and the goatlings don't breed, we don't breed from them until they are in their third year (that is, in their second year in our reckoning), and a few of the kids die. We have the children to herd, and another son will be marrying this summer, and we need some of the goats to pay the brideprice, and some for the feasts, and I'll give him some for himself on his marriage. A new household needs sixty-five goats if they're to live properly, and of course, when all the children are married, that's what we ourselves could live from. But we're a family, and although of course we have separate tents, we live as a family. Traders from Amman, Tafila and Safi come and buy the kids. The price is 30 JD, and we reckon that 10 dinars of this is profit. Then we have the milk for *samn*, and the hair for tentcloth. A hundred goats give the milk that makes one and a half to two tanks of *samn*, and that sells for 100–120 JD a tank. We're not making *samn* for sale this year, just for ourselves. Some people work harvesting in the summer, so they get stubble grazing and gleanings as well as pay."

A year later, another Azazma family, also at Ghuwaibah, said that they had been using the area only for the last two years. "We came here from the Wadi Ghuwair area to the south, and we use a much wider area. Like last year, we had our tent a bit further up the wadi in the winter, and then we moved up in the summer to the Wadi Jamal in the mountains, and to Maghta; we came back here through Kula and down the Wadi Dahal. I don't know why the government doesn't encourage people to grow forage crops, like alfalfa and *kersenna*, and to plant trees like sidr and acacias for animals, rather than all these tomatoes. Then we wouldn't have to spend so much on *alaf*, which isn't very good for the goats anyway, and we don't know what's in it. All these chemical fertilisers, they can't be good for people or animals. Goats need a good mixed diet, that's very important. They need annuals in the spring, and *ajram* is good, and leaves and seeds from trees like tamarisk, acacia, *yâsir*, and sidr, and *butm* in the mountains, and *ghadâ* in the sands, and so on, and the *ghaisalân* (asphodel) on some of the mountain slopes is good when it's dry – all sorts. Of course we like trees, if there are no trees, there are no animals and no people. We live from the goats and they live off the trees and plants....... We don't use too much of the plants; the plants continue, it goes on like this. Using *alaf* doesn't prove anything, we usually feed the goats in the winter because they are kidding. This year we fed some of them before mating (in July), because 'Id al-Adha is earlier every year, and to catch the market next year we needed the goats to conceive the first time. On the whole, we don't sell the kids to visiting traders, one of the boys has a pickup, and he takes our kids to the market in Safi or Tafila; and he collects *alaf* and stuff for us, and the family and neighbours, and they pay a bit for transport." "With our goats, we're free. We have grazing, water, air. The goats keep us, and their hair gives us tentcloth for shelter. Living in a house is disgusting, people living in houses want so many things, and they're not free any longer."

The daily round for a goat herding family starts soon after sunrise, or as the day warms up, when the goats are released from their stone or brushwood *haush* or section of the tent. When there is no separate *haush* for the goats, they have half the tent; of the remaining half, two-thirds is for the men and guests, one third for the women. If there are milking goats, these are milked by women or girls; water or feed is given as necessary. The animals are taken out to browse as a flock, or in separate groups in different directions by household members, depending on numbers, ages, available browse, distances, weather, and other activities in progress. The goats rarely go out further than three or four kilometres from the tent. Animals are watered at springs mid-morning and before sunset; when they are being fed, this occurs at midday and sunset. The feed is put out on sacking or old tentcloth pegged out on the ground, and animals are checked to see that all are present and in good health. Exactly who does what when depends on other activities. Men and women, old and young herd, water and collect firewood; girls and women make bread, cook, wash clothes, spin and weave, and erect and dismantle the tent. Men and boys collect *alaf*, shop, deal with paperwork, entertain guests, find out about fresh grazing and trade. In summer, people move up the mountains, use trees for shade and shelter, and, if the year is poor, move everyday along the slopes and between springs and seeps.

A goat trader using the Wadi Ghuwair region said "I buy up weaned kids, and I keep them for another two months or so. They graze what's around and they get *alaf* as well. When I have enough for a lorryload, that's about seventyfive to a hundred, I hire a little lorry with two decks, and I take them up to Amman, where the slaughterhouse is, at Qawsimiyyah. I nearly always make ten JD a head, sometimes fifteen, occasionally twenty. Depends on the market. Animals are a good enterprise."

Camelherding has been transformed from an essential part of the economic and political networks linking *badia*, countryside and urban centres to being of subsistence livelihood for a minority. The development of modern transport based on fossil fuels, and the associated infrastructures, cut away the raison d'etre of bedouin (in the sense of tribal keeping the peace or rule) camelherding. This was founded on a continuing market for camels for transport and draught, and meat, and combined with the provision of protection or *khuwa* to rural groups unwilling or unable to provide for their own security, together with the institution of raiding. The remaining market for camels in the Bilâd ash-Shâm is for meat. Camel transport and their use in raising water for irrigated agriculture went on although in a continuing decline until the 1970s. Raiding, which had a variety of functions (Lancaster 1981: 140–1), ended because it depended on a need for camels, which no longer existed by the mid-thirties and forties; camels were by this time in over supply. As well as increasing turnover in camels faster than breeding, raiding was also a means of gaining a reputation for daring, generosity and mediation. Many animals were out of bounds, through agreements such as *khuwa*, beni 'amm or sohba, and others, and these animals, if taken, had to returned or the guarantors of agreements had to make restitution.

It has often been pointed that camels are the ideal animal for *badia* herding, being browsers, able to withstand high temperatures and a degree of dehydration, although they do need to drink large amounts every three days or so under these conditions; they do not like cold, and since they give birth in January or February, breeding herds move to warm areas for the winter (Lancaster 1990: 182–5: Schmidt-Nielsen 1964). Camelherding provides for daily subsistence as well as supplying a market, but in comparison to sheep-herding, entrepreneurial activities or employment, does not provide as much income. In addition, camelherding in the *badia* is even more difficult to combine with education than sheep-herding, and education is seen as the necessary condition for participation in the transformed political scene and its economic ancillaries. However, "every section of the Mur'ath (section of the Rwala) has families who are herding with camels, and a few sheep. Camels are profitable, not like sheep, but it's steady. The males go for meat, and there are always people with gardens who want a milk camel or two for the house. I know at least twenty families herding camels, without bothering to think. The problem is education, because education is essential with a state system of politics, and it's difficult to live outside the state. A few do."

The decline in camel-herding was described by another Mur'ath when looking at the old wells at Mughaira; he remembered summers "when the surrounding slopes and dunes were black with Mur'ath tents, and the young men and boys pulled up water from the wells. The camels went out south-east, and browsed on *ghadâ*, *rimth* and *hamdh*. We had around fifty or sixty camels, and we lived on milk and dates. We were really healthy, not like children now. People stopped using these wells when pumps came in and tankers. Look, these wells had new drinking troughs made in 1390 AH (1974). And people moved away from camelherding in the years of *jifâf*, scarcity, in the late fifties and early sixties, there were jobs in the National Guard and the oil companies. We didn't need money, and we depended on ourselves, so we were morally wealthy. Now we are rich in money and possessions, and morally poorer."

Daily herding is relatively simple. Shortly after first light, enough camels are milked to provide enough for the household, and the herd is then driven off by the herder to graze. The herder has a riding camel, which carries his *furwa* (sheep-skin coat), water and food. They may or may not return at night, depending on the general availability of grazing and the needs of the household; at night, notorious wanderers or particularly valued animals are hobbled. Herds are now normally watered by troughs taken out to them with the water tanker, although in winter and spring with fresh grazing, camels do not need water. Longer term herding is concerned with locating fresh grazing, where there has been rain and where other herds are. These are the fundamentals of conversation at every tent, and decisions as to future movement of the herd are based on a balancing of several factors – the whereabouts of grazing and its nature, the source of water if needed, the location of other herds and the relationships between their herders and owners and oneself, the need for shelter in winter or for coolness in summer, access to markets. If natural grazing is poor, herders may buy dry feed of barley straw or poor quality dates for camels; there was a system of subsidised feed quotas in Saudi Arabia in certain years, available at particular centres to registered herds. Rwala do not process camels' milk, while the Shararat do.

Rwala reckon a family (tent) needs fifteen to twenty camels for survival. Musil (1927: 64, 405–6) recounts examples of families with fewer, and the employment young men undertook to add to the family budget. Middle-aged and elderly Rwala brought up in camelherding families remembered having fifty, a hundred or two or three hundred camels. Some Rwala continue to herd camels because they like camels, and like herding, and because living off camels is certainly viable. If a herder, often Sudanese, is employed, people say that a herd of a hundred camels is needed; this is the number a herder can care for, while the owner supervises and delivers water. A family run herd may be from twenty up to two or three hundred; the smaller number would indicate the family has other economic activities, while the larger implies the herd is made up of animals from the wider domestic group managed by certain members. There is a demand for camel meat in Saudi Arabia. Camel dairies have not been taken up by Rwala herders.

Members of many tribes continue to herd camels, although in far fewer numbers than earlier. A Beni Sakhr family, summering in Karak in 1991, had about eighty sheep, twenty goats and twenty-five camels; in that dry year they said "We are living off the camels. They eat plants the sheep won't, and we have their milk. This is why we go on having the camels, we can always live off them." A Huwaitat family herded a hundred plus sheep as well as twenty camels, wintering around al-Jafr and Bayir, and summering east of Qadisiyya; "we have a good life, we have good air and a good countryside. We're here for the camels to graze weedy fallows and the barley stubbles. If people treat the land well, the land looks after them."

Serahin Azazma camel herders use the slopes (*wudiyân*) of the mountains, the plateau, and the Ghor. Two tents of Serahin Azazma had goats and camels on the western slopes of the Karak plateau; "We come up here in the summers, it's too hot in the Ghor and the wudiyan. There is a spring right down in the wadi, but we've bought the water in this cistern from a Hamada in Fagu'a. All these camels (65) belong to our *jamâ'a*, so do the goats (c.130). It is profitable, and they keep us. We sell the two year old males at Sahab, we take them in a truck." Winter camel herding was described by a Serahin Azazma at Ghuwaibah. "Our herd, we have between seventy and ninety altogether, are divided in two at the moment. One lot are away east with one of my sons. The ones here are those near calving (December) My grandsons (aged about ten and twelve) herd them; they go with them, and they can tell me and the rest of us where they are. They're watered twice a week at the moment, sometimes at the Ghuwaibah, below where we get our water, sometimes at Bir Hnaikh or at the new government well below Jabal Hnaikh. This is convenient and it's free, but the water is brackish. The camels come back to the tent every night, so we can see how they are, and they go out every day. There's plenty of dry grazing, *ghadâ* and *thummâm*, but nothing green yet – no rain in the mountains so no *sayl* here. They never have *alaf*, ever. There's no milk because they are all dry. It's the time of year, either they're near calving, or they're coming into heat. To have milk all the year round, they need a much bigger range than they have here, so there's green grazing somewhere. A few of the goats have kidded, so we have fresh milk, and there's the *samn* and *jamîdh*. There isn't a lot of money in camels, but as long as we have the labour they keep themselves, and give milk most of the year. We don't process it. And there's the wool, all our dividing curtains in the tents and our rugs are made from it. We sell the males at two or three years old at Sahab, we take them up by truck, and we get between six and seven hundred dinars. Camels are good."

CHAPTER 6

BUILDINGS AND OTHER STRUCTURES

This chapter deals with the structures people place in the rural landscape for storage and shelter, production and processing, and for social and religious purposes. Only those structures arising out of the productive activities, built, commissioned, and/or funded by their users are considered; water management structures were discussed earlier. The structures in the landscapes may have common functions but the ways in which they are constructed vary according to available materials and preferred techniques; in addition, many carry social meaning. There are, then, variations on the common themes of production, shelter, storage, and the disposition of the living and the dead in the landscapes. Those in the *badia* are discussed first, and lead to the recent development of more permanent structures and settlements. Relationships between these and the older villages of the Balqa, Karak, Dana and al-Juba are examined through encampment and settlement layout, and indications of shifts in livelihood. Examples of field systems and routes are included where possible.

The black tent, one of the four pillars of bedouin life (Jabbur 1995: 241), is made of long strips of woven goathair material, sewn together into a roof (*bait*, a word which also means the tent as a whole). The roof strips may, if long enough, hang down at the ends to form the end walls (*rufa*, plur. *rufâf*). This is supported on a variable number of poles (*'amd*, plur. *'amdân* or *'amûd*) along its length, and three rows of poles across its width. The number of lengthwise strips (shigga, plur. shigâg) depends on their width as well as the overall length of the tent so that the length and width remain in, roughly, the same proportion whatever size it may be. A tent is described by the number of free-standing poles it contains, i.e. not counting those at either end. The smallest size is a one pole tent, called a gutba in the hamad and harra, and muwâsit in the Shera. This provides two spaces within the tent, and may have from four lengthwise strips or up to ten if the shigâg are very narrow. A five pole tent, giving six spaces, may have twenty lengthwise strips. At the front and back of the roof, narrow strips of woven goathair (mukhal) are usually sewn on to provide a bed for the pins which attach the front and back curtain walls (*ruwâg*) to the roof. Some tents have *rufâf* that are not an integral part of the roof, but sewn together to make a section whose seams go across, rather than up and down; this is then sewn or pinned to each end of the roof. Tents in Karak and the Shera have roofs with short side overhangs, with the back curtain wrapping around to make side walls. Strengthening strips of woven material, (*taritza* or *tarika*), sometimes black, sometimes black and white, are sewn on the underside of the roof to take the strain of the ropes; to these strengthening strips are attached rings of leather covered wood, wood, iron or rope through which the ropes go. One end of each rope is tied to a large metal tent peg, boulder, or deep rooted bush; the other end goes through the rope holder and is then tied back on itself with a type of running hitch knot. The main *taritza* strips run across each end of the roof and where each line of poles come; larger tents may have additional strengthening strips midway between poles, and for really large tents, between again with the appropriate attachments for extra ropes. The rows of poles support the roof along the taritza strips. Tents often had only one goathair curtain as a back wall; now a front curtain of sacking or tent-cloth, for privacy and extra shelter, is common. These curtains are pinned to the *mukhal* by metal or wooden pins. The back curtain often has one strip with white stripes woven into it.

    The best tent-cloth is that spun and woven by the women from their goats. A *shigâg* for a three pole tent, i.e. one with five bays, as made in the Shera and Karak, is about seventeen or eighteen metres long. This uses the hair of fifty to sixty goats, and takes a month to spin and a fortnight to weave; it is summer work. The women and girls of the tent spin while herding and in between other activities. Weaving is carried out in front of the tent, on a ground-loom. Crowfoot (1945) and Weir (1970) give technical information on spinning and weaving in Jordan. The ideal is to make a pair of new *shigga* every year, although many make only one. If the tent-cloth is in reasonably good condition, the season's hair may be sold or used to make rugs. Herders without goats have to acquire tent-cloth. In the past, tent-cloth was often a part of *khuwa*; now it is purchased from merchants. Goathair tent-cloth woven in Yabrud in the Qalamoun is regarded as the best quality, but much of the tent-cloth on sale is imported from Turkey, Pakistan and Germany. This has a poor reputation, especially the German material, as "it is too loose, the hairs are so short, they don't bind together properly to get a tight weave." Jabbur (1995: 242–52) gives details of various qualities of commercial tent-cloth and its manufacture at Yabrud; in the thirties, Yabrud had two hundred and fifty manufactories, supplying all the bedouin west of Deir az-Zor (Thoumin 1936: 163). Tightly woven material, although heavier, is more water resistant; when it rains, the hair or wool swells and becomes watertight.

Women of camelherding families spun and wove camelhair for tents. A Rwala recalled "When I was a boy we lived from camels. We bought some goathair tent-cloth from Qunaitra, Yabrud or Abu Smura – that's a village near Idlib in north Syria. These were the only manufactories. Many people had all camelhair tents, but we had the roof of goathair *shigâg*, and the women wove the *ruwâg* (front and back curtains) from camelhair. Some people, like the Dulaim in Iraq, had sheep's wool *shigâg*. All fabrics are equally wind and waterproof, but camelhair is much lighter. Goathair is best, because it weaves a nice, even, black material, whereas the others come out streaked. Different hairs and wools are never mixed, it's always like with like. A wool or hair *shigga* will last for fifteen years, and wool has the advantage that if there's a hole, it can be unpicked, respun and rewoven." Patching is the usual method of mending holes in hair tent-cloth.

Women of a Beni Hamida family were weaving tent-cloth with a goathair warp, and a dark wool weft; "he (the senior man, a sheep and goat trader) brought all this spun wool and goathair home. We had been thinking it would be nice to have a tent for the summer. So when we got the wool and hair, we started. We're using goathair for the warp because it's strong, and it's good hair, nice and long. The white sheeps' wool will be stripes in the *ruwâg*."

Poles in the *hamad* and *harra* are usually bought poplar trunks from Kiswe, the Ghuta, or the Qalamoun in Syria. The centre pole is the strongest, then the centre poles at the short sides, then the front poles; the back poles are usually the shortest and worst. These may be local tamarisk, abandoned building wood, or metal poles. In Karak and the Shera, poles are cut as coppicing from tamarisk, willow, oleander, acacia and juniper trees in the *wudiyân* and mountains. Ropes in the *harra* and *hamad* are bought sisal or hemp; nylon ropes are rarely used, as "if there is rain and wind, these ropes don't break but the material does, and it's worse mending the material than to join a rope." In Karak and the Shera, however, nylon ropes are common. Musil (1928b: 64) reports that the Rwala used ropes made from camelhair and bought ropes from Damascus, where hemp was cultivated and the bast made into ropes (Wetzstein 1857: 480).

Sewing a tent is the responsibility of the mistress of the tent, and she is joined by other women of the encampment. The *shigga* for the roof are sewn together first, either over and over stitches or through and through; the roof is then turned over and the *taritza* sewn in, and the 'ears' (at the ends of the roof) for the main lengthwise centre ropes are put in. The roof is then re-turned, and these 'ears' are given their own strengthening lengths of *taritza* along the centre seam. The curtains and the finishing touches of edgings for pins, sewing on ropes to the lower side edges, and so on are done by the mistress of the tent and a smaller band of helpers. Many tents have idiosyncratic details for the fixing of ropes, or in the kitchen area there may be slots for knives and herbs in the taritza, or side extensions for more space. A tent made by Mesa'id Ahl al-Jabal women from bought, poor quality, goathair *shigâg* had a rope sewn along the lengthway's centre seam for strengthening; this feature, when the tent was used in Karak, attracted comment from visitors as a good idea when using poor quality tent-cloth.

Erecting the tent is women's responsibility, although often the man and boys help. The roof is laid out and tent pegs hammered in in line with the rope holders on the roof. Ropes are attached to the pegs and threaded through their holders. The centre pole (*wâsit*) is inserted under the tent cloth and raised, followed by the other poles along the length of the centre seam. At the same time, the main ropes at centres and corners are gently tightened. Then the secondary poles are inserted and raised, along with the secondary ropes being tightened. Poles are raised and ropes tightened alternately until the tent is up; with good judgement, the prevailing wind helps. The back wall is then attached, and the dividing curtain put up. A one-pole goathair tent, the smallest size, takes half an hour to erect by two people, and a sacking tent takes twenty minutes. Poles are always placed on seams, except by the 'Ata'ata who prefer to place poles in the middle of the *shigâg* – the reason was never made clear. Tents in the mountains and *wudiyân* have shorter ropes than those of the *hamad*, whose use of longer ropes is necessary because of more persistent and stronger winds. The mountain and *wudiyân* tents often do not bother with the *mukhal*.

The smallest size tent covers a living space of about three metres wide by six long; a three pole tent about four metres deep and seventeen metres long, while a five-pole tent may be fourteen metres wide and eighty-five metres long; (figures 9, 10 and 11). Tents are divided into areas for men and guests, and for women with the cooking area and storage; some herders, particularly goatherders, have a part for the animals. In the *badia*, and on the plateaux, tents invariably face the rising sun; this puts the shortest wall of the tent towards the heat of mid-day. In the mountains and the *wudiyân*, tent position is more eclectic. Here tents are often pitched on slopes, and flat floors are built up of boulders, smaller stones, and earth, topped by sacking or old tent-cloth. The men's side is usually to the south; parked lorries and pickups indicate the men's side, while washing hung on tent ropes, pots and pans drying, firewood, ash-heaps, water barrels and feed troughs show the women's. Tents where the women are closely related often are pitched so that the women's side of each is adjacent.

In bad weather poles are lowered, reducing wind resistance and tension on wet tent-cloth and ropes, and ropes slackened. If rain seems imminent, a ditch is dug round the tent to carry water away. Tents, although airy, are warm in winter from the warmth of the sun or from a fire; *zirb*, screens of cane held together by wool threads woven into coloured patterns, may be used especially when the family expect to be in one location, a known cold area, for a long time. In the summer heat, the women raise the front edge of the roof by putting the poles under the ropes rather than under the material, tie the front or back curtain up to the ropes or hang coloured material from the roof to make shade, and tie the side walls to the ropes to let air through. In the *hamad* and *harra*, some have white cotton tents for summer which are made of lengths of heavy woven Syrian cotton. Although cooler, these are expensive and not rainproof. Others change black tents for sacking tents, home-made from hessian sacks split into lengths; sacking tents cost almost nothing, and are light and airy.

Tents in winter are pitched on the leeward side of ridges, in the *shi'bân* and along the *wudiyân* to use shelter from winds. In summer they are pitched along ridges and on hillsides to get the benefit of cooling breezes; an alternative in the black basalt *harra* is to camp on the edge (occasionally in the middle) of a *ga'* where the pale earth reflects less heat than the surrounding basalt. In Karak, black tents seen in summer are usually those of incoming Beni Attiya, Beni Sakhr and Huwaitat herders, or of resident Majali, Beni 'Amr and Beni Hamida out at their harvested fields. Azazma, who live in tents all or most of the year (some rent older houses in

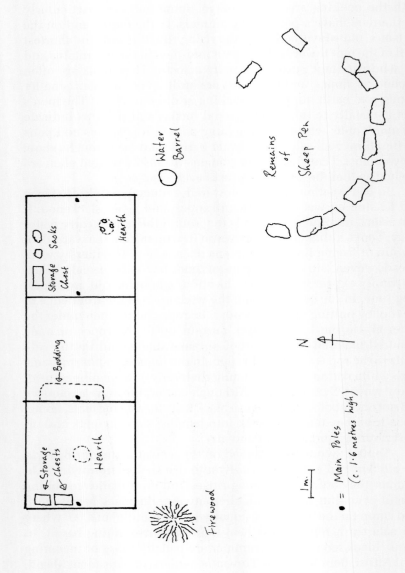

Figure 9 Tent of Mobile Herder

245

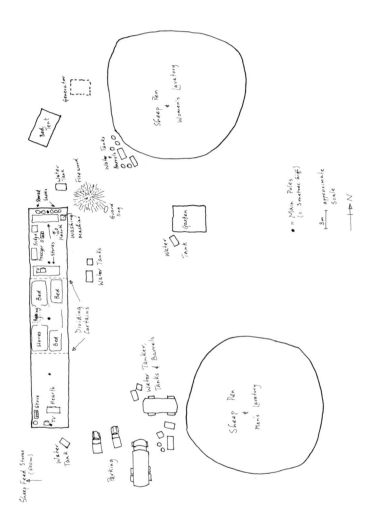

Figure 10 Tent – Permanent Herding Base

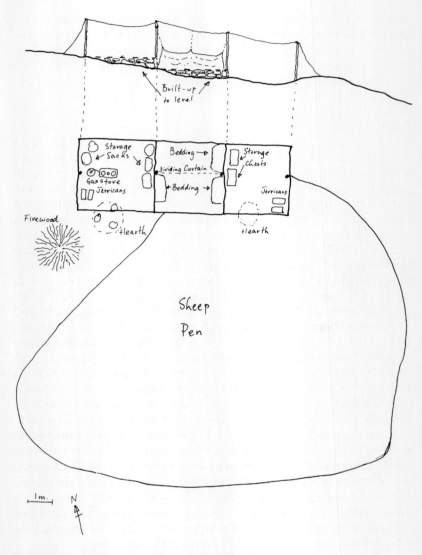

Figure 11 Tent at Annual Summer Site

villages for the winter), use sacking tents in the summer and store their black tents at favoured winter sites in the *wudiyân*. In the Shera, as summer approaches, people move away from valley shelter to headlands on the mountain slopes. Tents are reduced in size and erected near or under trees. The tent is used mostly as a store, and people live under trees, saying "the shade of trees is cooler than the shade of tents." Some store their tents altogether in the summer, and live completely under trees, which are turned into bowers (*ar-rîsh*), by the addition of brushwood walls for privacy.

The care of the tent is up to its mistress. Many women take pleasure in this, devoting time to sweeping the floor of the tent to produce a smooth surface, cleaning coffee pots, cooking pots and pans with ash and sand until they shine, and arranging them in ranks. The dividing curtain has its decorative face to the men's side, where guests are received. Women like their dividing curtain, rugs and cushions, and make patchwork covers for bedding piles. The teenage daughters of the tent often work cushion covers in intricate designs derived from weaving patterns. The pile of bedding quilts is also valued for its visual appeal, as well as its practical symbol of hospitality – like the shining pans, trays and tea glasses. Many tents in the *hamad* in the winter have *zirb* and these may be covered by more patchwork. In tents pitched where people expect to remain for some time, and the family is large, *zirb* are often used to construct bedrooms. An 'Ata'ata tent in the Shera had bedrooms made by hanging rugs from the tent roof for walls, with ceilings of coloured material. Men make hearths in their side, where they make coffee, the symbol of hospitality; the brass coffeepot keeps warm in the ashes. Coffee pots are often ranked and displayed alongside the hearth. In the Shera, the big pots and trays for cooking and serving *mensif* (sheep cooked in a yoghourt sauce) are often visible in the men's side, though in other areas these, like the rugs, bedding, cooking pots and stores, are kept in the women's side. Weapons are kept in the men's side. Women cook using scrapes in the ground for fires, with three stones for the cooking pot and the convex metal sheet (*sâj*) for baking bread.

Musil (1928b: 61–76) gives information on Rwala tents and their contents at the beginning of the century; Dickson (1949: 66ff) discusses tents in Kuwait in the 1930s and 40s; Bienkowski (1985) reports on contemporary Bdul tents and their use of caves in Petra.

In the mountain slopes and valleys of the Shera and Karak plateau, goats and sheep are sheltered from weather and predators in tents, caves, *haush,* or village houses in the past. Sharing living space with the family animals in village houses has gone, and now goats and sheep are housed in a room of a disused building, or in a purpose built concrete barracks. Barracks are always rectangular, with a roofed section at the rear long wall, and a lockable store or two against a short wall; the rest is walled yard. Barracks are always built near to a water supply, often a cistern on arable land. Caves, used by herders living in houses or tents, may be improved with sheltering or dividing walls. Tent dwellers shelter their animals in tents made of feedsacks and pitched in front of the tent, or in circular *haush* built from a base of large stones, and brushwood walls up to four or six feet high. Some use *haush* made of wire netting, old barrels, wood, old tent-cloth and other debris. Hens find shelter among the stones at the bases, or have special shelters built from stones. Small sections in the *haush* or tent may be made for an unseasonable lamb or kid, or a sick animal. Camels are hobbled, and couched in front of the tent rather than being housed; very young camels wear sacking or sheepskin coats in bad weather.

In the *hamad*, sheep and goats are usually provided with tents made out of polypropylene sacks or old tent-cloth, bounded by wire netting enclosures, or combinations of barrels, netting and tent-cloth. Sick sheep or goats are isolated, kept in a pen on their own or tied to a pole in the owner's tent; orphaned lambs or kids are also tied up in the tent. Young camels may be kept in a stone *haush* in bad weather, but *badia* camel herders tend to take calving herds to warmer regions. The *hamad* has little stone for building, but there are in the Ruwaishdat low walls of rough boulders added to low rock outcrops to shelter sheep.

Stone abounds in the *harra*. People build, or more often rebuild, corrals (*sayra*, plur. *sayrât*) for sheep and goat shelters. Those with higher walls are for goats, which jump better than sheep; the smaller corrals are for a few ewes and lambs, or for lambs or kids. These *sayrat* are all over the *harra* and associated with old campsites and graves. Men, women and children build them by putting basalt boulders in a circle of the desired size, and then add stones until the walls are the required height; the walls are of single stone thickness and vary in height from a metre to a metre and a half. An opening may be left, and then blocked with barrels, or a section is built so that it can be partially dismantled for the animals to enter or leave, and then rebuilt to enclose the animals. Winter tent sites in the *harra* are often surrounded on the west and north by stone walls built from stones moved when erecting the tent. The corrals are always in front – that is to the east – of the tent, and often downhill of it. A modern addition is an area of slope, in front of or at the side of the tent, cleared of stones, for the lorry; this is not just for convenient access but to provide a run to start a lorry with a defective battery. The ground covered by the tent is also cleared of boulders and stones; how thoroughly depends on how long people expect to stay there, and on personal choice. Boulders are rolled or heaved out of the way, stones swept. Clearing ground of rocks and stones is quicker than might be expected.

Women may make structures when processing milk for *samn* and *jamîdh*. In the women's side of the tent, they dig a pit and line it with stones; outside the tent they build stone platforms. Such structures were seen on the Karak plateau and in the *harra*. Women kept butter in the pits "while we were getting enough together to make *samn*. We put it there to keep cool and sweet, away from the heat of the sun. I'd rest the butter on clean *alaf* sacks or white cloths, but first I'd put *shîh* or some other sweet smelling herb on the stones under the butter, and cover it with more *shîh* and a few stones to hold it in place. We made pits this year (1992) because there was a spring and grazing to give a lot of milk to make butter and *samn*. The stone platforms outside the tent, I put brushwood on them and then I rest the *laban* in cloths there to drain when I'm making *jamîdh*. A lot of water comes out and attracts flies, and it doesn't smell nice, so it's better outside." These structures are convenient but not essential. If there are nearby boulders on which to drain the *laban*, there is no reason to build a platform. If a storage place for butter can be found by digging a hole on the north side of a nearby rock, a pit in the tent is unnecessary.

Storage only needs structures if the material to be stored is bulky. People stored possessions in the sand, caves, or holes in rocks. In the last few years, people have built a few stone structures (*qusûr*, sing. *qasr*)[23] in the *harra* to store animal feed. A family who used part of the *sha'îb* at Mahdath in some years for growing wheat, barley, and watermelons, built one. Another family who grew crops on their land east of Mafraq and herded that winter in

---

[23] Literally any stone structure, as opposed to mud-brick or other materials. Is meaning can cover 'castles', houses, corrals, store-buildings, etc. It is an example of a descriptive use of a word as opposed to a definitive one.

their favoured wintering site west of Burqu' built a store there. Both stores were constructed as single dry stone walls without foundations, one rectangular in shape, the other circular, about two metres high, and roofed with canvas. Most herders in the *badia* store *alaf*, collected from depots in Ruwaishid, Safawi and Azraq, outside their tents in wire netting pens, in the tent or in the lorry, to stop their animals breaking the sacks. In Karak and the Shera, where arable crops are cultivated regularly, *tibben* and barley are stored in a part of the barracks, caves, disused houses, or in purpose built stores/shops.

The summer and winter placing of tents in the landscape has been mentioned. In times of plentiful grazing, tents are widely distributed in small clusters; when grazing is localised, tent density increases. When grazing and water are in short supply but widely scattered, tents are dispersed, often singly. Springtime location of tents follows the flocks, while at other seasons encampments are often bases from which the distant flocks are supplied. Bases are positioned with regard to grazing, the collection of water and feed, and school (in term-time), together with preferred and associated areas by families. In the eastern *harra*, graffiti of certain family names from the Sharafat and Mesa'id of the Ahl al-Jabal are present and repeated, while names of other families are rare or absent. The same may be noted for Safaïtic names of earlier times (Macdonald 1993). Similar preferences for particular places are known for the *hamad*; a Ghayyath family likes the area around Fawk ar-Risha (Upper or Anwar's ar-Risha), a Rwala likes Rijlet al-Khail, Bowbihiyya and Anqa, another prefers Raghaban, Basatin and Traibil. Tents are pitched alone or close to a tent of someone close to them by descent, links through women, or employment. The white square canvas tent (*khaima*) adjacent to a black tent is often the private quarters of a newly married son and his bride. These tents are bought, and Rwala prefer those from the Gulf and Saudi Arabia, imported from India, to those made in Zarqa, which they say are not as strong, heavy or weather-proof. The professional cheese-makers who come out to the *badia* in the spring invariably use these white canvas tents, and hired shepherds may be provided with a tent of this sort in winter or spring. The discrete clusters of one to five tents may have neighbours at a kilometre's distance who are close or unknown. Between tents with connections there is constant visiting by both sexes; between those without, little.

The stone or mud-brick structures and tents of more permanent encampments, like those at Taht ar-Risha (Lower or Nuri's ar-Risha) in the seventies, were sited using similar considerations (Lancaster 1981). Rwala like to put their tents in straight lines, and as tents face east or southeast, rows are therefore north–south, or northeast-southwest. The tents of close relations are closest together in the line, sometimes with their tent ropes overlapping; family members arriving later start a new parallel line, or fit in where there is a socially convenient space. Household space around a tent extends to the limits of its ropes; many increase this by lines of stones, blocking off throughways for trucks and pickups, for greater privacy. The main track from Ruwaishid to Syria ran between the *sha'îb* and the *ghadîr* on the east, and the ridges to the west where tents and houses stood. Coming from the south, the first tent and *qasr* (in this context a house) was the shaikh's, the house had on its roof neon strip lights for identification, and the tent was large with pickups, cars and lorries outside. Some households had separate kitchens in small black goathair tents or structures made from earth-filled diesel or petrol cans covered with a plaster of mud and *tibben*; a few of these kitchens had additional storerooms. Other households had one or two-roomed block houses, with separate lavatories. Only villagers or townsmen employed as drivers or mechanics had houses with surrounding courtyards. Many people complained that block houses were cold in winter, hot in summer, and less healthy than tents.

The dead are buried on high ground "as they watch over their camping places" (Wetzstein 1860: 26) or, at more permanent encampments, in family cemeteries. Tribes have different practices concerning death and the dead, which are affected by the circumstances of death. Someone who dies at home is buried with the appropriate grieving. Those who die in raids or war may lie on the field, although it is an honourable act to bury those so killed, whether friend, enemy or unknown to the involved. Groups of little heaps of stones, roughly aligned, can indicate such a burial place. Rwala, and Aneze in general, dislike death and the dead, bury their dead simply and do not visit their graves (Musil 1928a: 671; Lancasters 1993: 167), although the places where an ancestor is buried is often remembered for several generations. The Ahl al-Jabal often, but not invariably, construct elaborate tombs (Lancasters 1993) which are visited by members of the deceased's family and by others, who may commemorate the deceased by a sacrifice and prayers. Rwala commemoration of family dead is private, and consists of the yearly dedication of an animal in the name of the deceased for a feast; no-one may know this except for the man giving the feast and killing the sheep. The elaborate Ahl al-Jabal tombs are built of local black basalt stones in a rectangular shape with a high cairn at one end, with extending wings. The cairns are often whitened with paint or lime, recalling Weuleresse's (1946: 228) mention of the whitening of the domes of shrines in the Syrian countryside. Ahl al-Jabal tombs represent the tent as an image of hospitality, but include necessary additions for burial with a platform for washing the body and a mosque for the saying of prayers. The mosque is an outline of a *mihrab* in carefully chosen stones, often with Safaïtic inscriptions, and inscribed with prayers and the names of family and other mourners. Stones are rarely shaped, but selected from the abundant local material for shape, size and surface. These elaborate tombs are not for shaikhs or the politically powerful, but for those held in respect and affection. Some, men and women, are said to be of healers, who healed both physical illness and the social weal of the community; others were "good people", who had sometimes asked to be buried overlooking grazing and camping grounds.

Other tribes have their own expressions of relationships between their dead and social continuity. Dead Beni Sakhr and Adwan shaikhs in the 18th and 19th centuries were given elaborate graves with representations of coffee pots and swords indicating the prized virtues (Merrill 1881: 240; Conder 1883: 313). In his survey of the tribes of Palestine and southern Jordan, Musil (1908, iii) habitually gives the watering places and the cemeteries of each group as reference points. Members of some tribes attach importance to being buried in a particular place, for example near a holy man's (*welî*) tomb (Wetzstein 1860: 26), or a group cemetery.

Builders of mosques in the *hamad* and *harra* have different aims. The outlines in stones of the *mihrab* placed outside the men's side at certain tents express individual piety and are often built at Ramadan. Mosques, whether as walled outlines or an impressive domed building with a tall minaret as at Fawk ar-Risha, are built as a response to wishes of guests and residents, and seen as a symbol of participation in this aspect of the wider society. The mosque at Butme was built to commemorate the builder's father. Firhan ibn Mashur and Fa'iz ibn Na'ur, both Ikhwan, built a mosque in al-Labbah of large white stones in the late twenties or early thirties, and "it is still there." Many tribesmen and women regularly pray, most fast, some have made the Pilgrimage, and all would deny that they are unbelievers, seeing their practice of generosity and protection as the essence of Islam.

Settlements in the *badia*, apart from oases, existed in the past. Notable examples are 'the desert castles' of Jordan and Syria, and the possibly contemporary police post at ar-Risha (Helms 1990; Bisheh 1992; King 1992), and the mediaeval *khirba* or deserted 'village' at Shabaika. Some were established by central government as desert police or customs posts. More recently oil companies built pumping stations and wells, around which small towns have accreted such as Ruwaishid in Jordan and Turaif and 'Ar-'Ar in Saudi Arabia. Others developed around existing tribally owned wells by tribal populations from political or economic motives, such as Swair, Hdaib, Tabarjal or 'Isawiyya. Swair was mentioned by Yaqut (d.1224 AD), quoted by Musil (1927: 199, n. 47) as a watering place of the Beni Kalb; in the more recent past, Swair was the wells of the Ga'adza'a section of the Rwala. Other new settlements may start in response to external pressures to register land, such as villages started by sections or families of the Beni Sakhr, Sardiyya and Ahl al-Jabal on their land at the eastern limits of cultivation, or Rwala villages of Nathaiyim and Rifa'a in al-Juba. The Rwala settlement of Faydhr started because of an association with IPC at Ruwaishid, while the ar-Rishas started as a political and economic response to actions taken by the Ba'ath party in Syria in 1968.

People regard the recent building of stone structures for shelter and storage in the *badia* as part of the opportunistic use of land for arable cultivation and the increasing need of dry feed for sheep. Earlier, crops and stored goods were protected from mis-appropriation by the reputation and authority of *welîs*, the graves of holy men or other sacred sites, as in the Rahba (Wetzstein 1860: 31), and the Karak *wudiyân* (Musil 1908: i, 87 and local memories). A Sharafat described his grandfather starting a farm east of Jawa in 1931. The first building was a store for produce; "we never sold produce from the farm. We stored the grain and straw first in that central storehouse, and later in the largest of the three rooms in the house. The house took two years to build, but we only worked on it sometimes. We put the lintels over the doorways in place by building stone ramps and levering them up with iron bars. We built the house on bedrock, and we had to shape very few stones; there are so many around, we found what we wanted for lintels and cornerstones.

We bought wood for the roof beams for the first house, with bought canes on top and then earth. Before the house was built, we lived in the tent in the *sayra* in the winter. In the summers, we lived in the tent up on the top of the hill when we summered here. But we weren't always here, it depended on the rains."

A Mesa'id Ahl al-Jabal *ibn 'amm* described how the buildings on their land had changed in function over the seventy years they had used fixed structures. Their grandfathers built the first from local stone, mud, and wood as stores and shelters in bad weather, but not for living in; one store was circular, another was square-ish. The second group of buildings were built fifty years ago of stone, but used metal I-beams for roof beams and lintels, bought building wood and bought canes for the roof; these were rectangular, and usually had two or three rooms. At the same time, they built rectangular mud brick stores or kitchens. The earlier structures became kitchens or remained as stores. From the seventies, cement with a framework of reinforcing rods became the usual building material, and these were houses that were lived in; people lived in their tents when out with the flocks. Families whose men were in the security services or employed built most of these cement houses, using paid workmen and providing some of the labour themselves. Former kitchens were used to put *tabûn* ovens in, while the earlier houses were turned into *tibben* stores with ramps built on the outside for filling. From the late eighties, villas became the preferred form of housing, as a result of increased wealth in the region and the general provision of piped water and electricity. Villas are designed collaboratively between the family and the engineer in charge of construction; craftsmen do the skilled work, and the family often act as labourers. Villas are lived in permanently, while many have a tent as a meeting place for the men in spring and summer. Their owners have agricultural land, flocks that move into the *badia* seasonally, and employment, and personnel move between the different spheres.

Sardiyya of Subayh/Sbayha said that when they registered the land under Glubb in the 1930s, "the black stone houses were standing here then. We hadn't used them for three or four hundred years, and we didn't know how to repair them or how to live in them. We had to get workmen from Syria to repair them. At the same time, we cleaned out this cistern, and the big one over there; the water was for animals and people. Most people went on migrating. A few people stayed here more or less all the time, and

Buildings and Other Structures 255

grew barley if there was enough rain. Before, we'd spent the summers in Syria. The black stone houses had roofs of basalt slabs[24] which last a long time, much longer than roofs of wooden beams and canes and earth. We didn't have houses then, we had buildings. That one with 1943 over the door, that was used as living quarters, and we built the black stone lean-to on the back for a kitchen. There wasn't a door through from inside, we went round the outside. Then about twenty or thirty years ago, we built that mud-brick house. It was mud-brick because there weren't any more cut stones from the ruins left[25] and we couldn't get workmen from Syria. At the same time we did a new kitchen of stones and mud. The house with 1943 on it was used for storage, and overflow guests, and the old kitchen wasn't used. We also used some of the houses behind, people used to sleep there. When the 1943 house was used for storing *tibben*, we made the stairs on the outside. Look, there's this very nice hole in the roof (corbelled); when this was a store, we put the *tibben* in through the hole, and when people lived in it, the hole was a chimney......... Now we all have villas. Mine is next to my brothers', and his (a nephew) is with his brothers'. Outside my villa, I have my tent, a four pole tent, for the men. I've put down nice gravel as a floor, and made a garden with rose-trees and flowers outside, fenced against the sheep."

The nephew's villa, completed in 1993, is a variation on the standard house with wide central through-passage as seen at ar-Risha in the 1970s. Whereas the 1970s *qasr* had a bathroom or washroom but the kitchen and lavatory were in separate buildings outside, the villa incorporates all aspects of family living – shelter, hospitality, storage, cooking, eating, privacy, washing and hygiene – and the need to provide these for men and women from outside the immediate family. Men and women use apparently defined spaces for entertaining on different occasions. At a men's feast where the food is served in the salon, the women of the household and female neighbours eat after the men and often use the central passage; but at a women's feast, the hostesses and their female guests take over the salon, and the men of the household visit other houses. The basic villa may be seen, in these *badia* fringe settlements, as a

---

[24] Basalt slab roofs are like those of the nearby town of Umm al-Jimal, discussed with other north Jordanian houses by Azzawi, Salim and Rajjal 1995.
[25] Merrill (1881: 295) records the excavation of ruins near Irbid for building stone.

spatial rearrangement of tent spaces, plus the incorporation of sanitary facilities, in a fixed structure (figure 12). Again, the earlier buildings used as shelters, kitchens, and stores, with the tent for the reception and entertaining of guests and family members, fulfilled the functions of the household previously (or in the *badia*) carried out in the tent itself. The increased space reflects the greater number of people in the household, and more visitors. Furniture and fittings increase in number and have more separately defined functions resulting from more money, more commodities available, and changing patterns of storage and consumption. The permanence of occupation by various household members varies, high for children of school age and their mothers in school terms, much less for others. Regardless of what other economic activities household heads and their sons have, most have flocks or have sheep looked after by members of the wider family, and this means active supervision of shepherds or co-operation with involved family members.

The mud-brick houses of Kaf in the Wadi Sirhan are briefly described by Blunt (1881: i, 85), those of Jauf with the dimensions of the mud bricks by Wallin (1854: 139–40) who saw the buildings as in 'the Syrian character', and the castle at Jauf by Musil (1927: 160–1). Mud-brick was the common building material in the Jordanian Ghor (Politis 1995) and in the Ghuta of Damascus. Nasif (1988: 130–2) describes the old houses of al-Ula and its constituent groups (133–5). Mud-brick houses, now uninhabited, were seen at Qara and at Jauf. The focal point of the Qara house was its reception rooms, "a *maqhwan* or coffee room for men and a *liwân* for women; these rose the height of two storeys for coolness in summer. Behind the women's *liwân* there were two rooms, and one room behind the men's *maqhwan*, and we used these in winter for sitting, entertaining, sleeping – everything. The kitchen was a separate building, next to the women's *liwân*. The walls were made from mud-bricks. The earth for the mud came from the hill by the side of the house and its gardens. The roof beams were *ithl* (a type of tamarisk) cut from the trees round the gardens; then there was a layer of date palm stalks, and finally a layer of palm fronds, not plaited but packed down, with earth on top. And there was a *haush* in front for the sheep and goats. I lived here when I was a child and I'm thirty now (1995)." He and two of his three brothers live in adjoining villas, facing the family gardens; the third brother's villa is diagonally opposite, across the main road. The brothers share

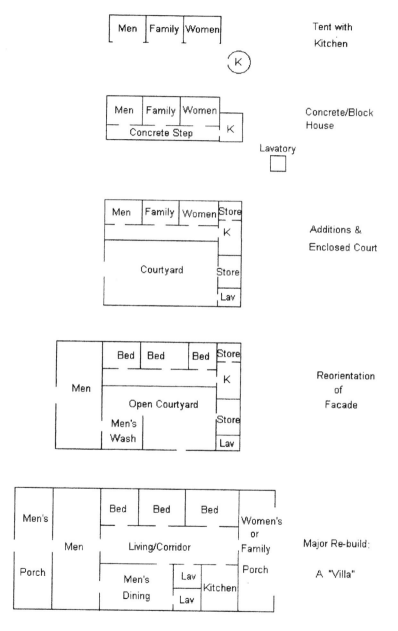

Figure 12 Diagram of "Progression" – Tent to "Villa"

the upkeep and expenses of a black goathair tent, erected over a metal frame against the outside wall of the villa nearest the alley from the street, used as their family public place of entertainment and meeting.

In Jauf the old house was described as 'a *qasr*', (in this instance the group of houses lived in by related families and their dependents: a 'quarter'). The *qasr* was surrounded by a high wall, through which a gate gave entrance to a series of courtyards. Each courtyard contained a variety of buildings, and a family owned buildings in a number of adjacent courtyards. One *maqhwan* faced north, and had a winter room and a summer room. The winter room had four walls, and a part ceiling of *ithl* poles put in at one end over a section of the room for storage. The summer room had one long open wall, the ceiling supported on pillars of palm trunks; small high windows in the rear wall gave light and added ventilation. Both rooms were two stories high, their walls of mud-brick with stone lower courses had been plastered, and the roofs of palm branch and earth were supported on *ithl* beams. The owner, a mechanic and gardener, had had his workshop in the courtyard. Further courtyards had buildings used as kitchens and stores; the stores were a quarter of the room walled off at waist height to make bins for dates, jars for dates, and smaller bins, one of which held chickpeas and another *semh* collected ten years ago. A stone staircase led to the roof over a door-way, with the door made from ithl planks. Rooms had piped water and electricity, which "came about eighteen years ago, and we lived here until nine years ago." Alterations in the function in many rooms were visible, with doorways blocked or put through walls, rooms divided or opened up. The family moved over a period, based around the sons' marriages, to a series of small villas on higher ground above the gardens.

Al-Jauf, like other towns, was composed of groups who derived their origin from different tribes and villages, each associated with a subdivision of a quarter. Wallin (1854: 141–12) lists the quarters, subdivisions and families in 1848, Musil (1927: 472) the quarters in 1909, and al-Sudairi (1995: 163–5) presents the quarters and families of contemporary al-Jauf. The names of the quarters have a fair degree of similarity, while some expand and others disappear; family names are less consistent, indicating physical mobility and/or realignments of alliances and genealogies. The neighbouring town of Sakaka has its quarters listed by Wallin (1854: 151–2), Musil (1927: 279) and al-Sudairi (1995: 162–3). The names in Wallin differ from those of Musil and al-Sudairi; the main families, mentioned by the latter, remain relatively stable.

The establishment of nation states and oil-wealth have increased settlement in the region. Settlement by nomadic tribal families also occurred earlier; two Rwala, one at Qara and one at Jauf, said their fathers or grandfathers had settled after the taking of Jauf by Nawwaf ibn Nuri Sha'alan in 1909. More tribesmen settled in the drought of the late fifties and sixties, with the expansion of employment in oil companies and the National Guard (Lancaster 1981: 100–111), and the growing importance of education. The government of Saudi Arabia has actively pursued the settlement of the Bedouin, some of whom have chosen to settle on potential agricultural land rather than see land go to others. There are now a number of villages and settlements in al-Juba and the Wadi Sirhan, some of which were described by Lancaster (1981: 18–21, 108) for the seventies.

The villages of Nathaiyim, Swair, Hdaib, Zallum and Zubara/Rifa'i were revisited in 1995. All, except Nathaiyim, had grown. Increased infrastructure had provided surfaced roads, electricity from the central grid, piped water, water towers, secondary schools and health centres. In the late seventies, although some households had villas or block houses, people lived in their tents and many were seasonally nomadic. Villas and block houses were not at first surrounded by walls, but then it was held that the building of courtyard walls was either a condition of the government grant, or for the issuing of deeds. All houses are now enclosed by walls, forming a courtyard, divided into inner courtyards for men and women. People say "this is necessary because now there are so many men driving around from the government, and we don't know who they are. Before, we knew everyone who came here, because a stranger would only come with a sponsor, and they knew where to go to see the men." Tents remain the place of general hospitality. Every villa has a tent in the men's courtyard or outside the walls, erected over a permanent metal framework, with a concrete floor, electricity for lighting, air-conditioning, and coffee making, and a side wall of shelving to store coffee pots, incense burners, schoolbooks, and television. People permanently resident in block houses may have the tent over a metal frame, or in the usual manner with poles and ropes. Many block houses are owned by herders, often away living in the tent with the flock, while his wife and children or elderly parents live in the house. New villas or houses are often built for a newly married couple as a part of the wedding gift from the husband to the bride, as part of the bride price, or as part of a son's share of the family property. A youngest son often brings his bride to the home of his parents. As in Jordan, the villa incorporates all household activities under one roof, while a block house will often have a separate building/s as a kitchen, store, or washhouse; tents are invariably used for male gatherings, and private household leisure, while villas have at least one room for formal male entertainment and hospitality. Block houses and villas alike have separate entrances for men and women. Two-story villas, or large one-storey villas, house two units of one household, usually a man with two wives, sometimes two brothers and their wives. Most villas and many block houses have large satellite dishes.

It is not uncommon for men to have two houses in a village and a town, or in two towns, because of their business activities or because they have two wives. In the towns, villas can be luxurious, with much thought and care going into the overall design, its details and furnishings, as well as its purposes. Every villa, like every house and tent, must have a place for entertaining guests, expected and unexpected, and must protect the privacy of the women. The house of a man who entertains frequently on a large scale and has a constant stream of visitors may have two kitchens, and a store for the huge saucepans and serving dishes; in addition, the house may be decorated with a large and freestanding coffeepot or incense burner as a symbol of hospitality. For those who rarely have large feasts or guests outside the family circle, there are catering services that provide, mostly for weddings, tents, cushions, carpets, coffeepots, lighting, cooks and cooking equipment, water and washing facilities, towels and soap.

Village houses in mountain and plateau areas of Jordan and Syria have changed in materials and techniques since the late forties. Before that date, local materials were the norm with construction undertaken by the family and/or by hired workmen. Now materials and techniques are imported (or derived from imports), with construction in the hands of the architect or engineer and skilled workmen. The owner specifies his requirements, decides on details, and produces the funds, or a proportion of them. When a site is agreed, the *muhandis* (engineer or architect) together with the owner and his family trace the plan of the building on the ground. This is particularly so in the rural villages, but not unknown in the larger towns, and part of a long tradition in building;

Carswell (1996) points out that "from the earliest days of Islam, there is evidence that plans for buildings and even whole urban complexes were simply traced on the ground. Al-Ya'cubi (d. 987) describes this method exactly as it is still in use in the Hadramaut to this day." The dimensions and materials for the elevations and internal layout are similarly decided between the *muhandis* and the family. Details of finishes, doorways, windows and window grills, and of the upper story and roof are worked out either from what the *muhandis* suggests or is known for, or by the owner driving around recently completed houses and seeing details he likes. Builders' pattern books for finishes and details are apparently unknown in Jordan (personal communication from Mr. Ammar Khammash).

A *muhandis* is often contracted because the owner has seen and liked his work for a relative or neighbour; "for our new villa, we're having the same *muhandis* as M.M.; I liked M's villa, and my wife visited there and she liked the way it was laid out inside. But we're going to have the top finished differently. We want a series of blind arches to finish off the tops of the walls, to act as protection when we use the roof in the summer for sleeping. The insides of the arches will be painted a different colour to the arches themselves, and the walls underneath will be another colour; we think three shades of brown, a sand colour, a reddish-brown and a dark brown. The blind arches are different to any other house, so people will know it's us." In the villages, the owners of the new house may do some of the unskilled and semi-skilled work, and often carry out much of the decorative external paintwork. The pillars of verandahs, door and window surrounds, and the walls facing the street are favoured places for geometric patterns or stylised flowers and birds; courtyard gates are often elaborately or distinctively patterned. Women have a large say in the layout, details and finishes of their houses. One of the most attractive villas, seen in 'Ar'ar in Saudi Arabia, had much of its design and all of its decoration carried out under the direction of the family's elder daughter.

The study of the vernacular architecture of Jordan and Syria has been undertaken especially in the last two decades. See among others; Aurenche 1984; Biewers 1992; Dakar 1984; d'Hont 1994; Seeden and Kaddour 1984; Kana'an 1993; Kana'an and McQuitty 1994; Khammash 1986; Layne 1994; McQuitty 1986; Mershen 1992; Noca 1985; Seeden and Wilson 1989; and Shami 1989, while earlier studies by Canaan (1932; 1933) and Thoumin (1936) should not be overlooked.

The earliest descriptions of houses in the rural areas of the Bilad ash-Sham, to our knowledge, are those of Burckhardt (1822) and Buckingham (1825). Burckhardt (1822: 292–3) describes the peasants of the Hauran living in the ruins of ancient buildings and in newly built villages, with every house having a *mudhâfa* or guest room; this was where the men of the household slept, and where coffee was made. The houses of Karak (1822: 388–9) "have only one floor, and three or four are generally built in the same courtyard. The roof of the building is supported by two arches, much in the same way as in the ancient buildings of the Hauran, which latter have but generally one arch. Over the arches thick branches of trees are laid, and over the latter a thin layer of rushes. Along the wall at the extremity of the room, opposite the entrance, are large earthen reservoirs of wheat (*kowâri*). There is generally no other aperture in these rooms other than the door.." Buckingham (1825: 33–4) says "The houses of as-Salt are very small; each dwelling, with few exceptions, consisting of only one floor, and this having only one room, subdivided into recesses..... They are mostly built of stone; and, where necessary, a few pointed arches are thrown up on the inside, to support a flat roof of trees and reeds plastered over with clay. The interior of the dwelling is generally divided into a lower portion for the cattle and poultry, and an upper part raised as a terrace, about two feet above the ground floor, for the use of the family. In this raised part the fireplace for cooking is generally placed.........In the upper division are the beds, clothes-chests and provisions; and....there are again other subdivisions made in the upper part of the house by walls, shelves, and recesses, all formed of dry mud or sun-baked clay, without being white-washed or ornamented in any manner. There is seldom any aperture for light, except the door...." A Beni 'Amr described the use of his parents' house in the fifties similarly; "there were stores at the sides, we used the back which was raised up a bit, and the animals had the front." Buckingham also comments (1825: 62–3) on "inhabitants of caves" (between Amman and as-Salt) who were herders and cultivators, who "deem them far superior to buildings of masonry, and consider themselves better off than those who live in tents or houses." Caves are and were used in the winters by herders for shelter for themselves and the animals, with sleeping platforms, niches in the walls for storage, fire-places, walls to contain animals or stores, and holes in the roof for ventilation and light as typical features.

Information from nineteenth century travellers and local people indicates that the existing old buildings in the present villages of the north Karak plateau were built as stores at least from the 1870s. People lived in tents when looking after their flocks and land, and had houses in the town of Karak, used by a variety of family members at certain times of the year. al-Qasr buildings had *râwiyya* (built in sub-divisions often rising to the roof) rather than the *kowâri* for grain and other household provisions as in as-Salt to the north and Dana in the south (Kana'an and McQuitty 1994). In the Christian village of Hmud, some four kilometres southeast of al-Qasr, Khammash (1986: 23–4) noted both *râwiyya* for hay and grain storage and fixed or portable mud *kowâri* for flour. Karak buildings, like the older Hauran houses, used stone arches from shallow springers for supporting the roof. Excavations at Khirbet Faris on the plateau reveal that house structures at earlier dates are not the same as those traditional buildings from the nineteenth century, but re-used Byzantine buildings or barrel-vaulted buildings round a courtyard (Kana'an and McQuitty 1994: 149). Kana'an and McQuitty discuss construction phases, building methods and uses of two courtyard complexes in al-Qasr dating from the beginning of this century, built as storage areas, and comparable to fourteenth and eighteenth century examples from northern Syria.

The building of stores has been seen as part of the sedentarisation process by Aurenche (1992: 46) and Noca (1985), or to reflect a greater interest in grain production for commercial sale (Lancaster 1995: 115, 110). Kana'an and McQuitty (1994: 149) emphasise that "the early houses are definitely not skeuomorphs of tents; as the products of a semi-nomadic population's priorities they were a complement to tents and performed a different function." Various members of al-Faris branch of the Majali remarked that grain and produce had always been stored in caves, "but then people began travelling and saw others building storehouses so we did." Family storehouses, like those in al-Qasr, Khirbet Faris, and other focal points of family agricultural lands, were built close to threshing floors or *baidar*, where limestone bedrock is on the surface. Threshing floors were associated with storage in caves, cisterns or buildings, and had the capacity to develop into village sites; many are at the sites of earlier villages or *khirba*. Storage is inherent in grain cultivation and dairy processing, the basis of household food consumption in much of the region, and documented from many sources; goods, produce and tools were and are left in caves, in skins and bags hidden under rocks or in the sands where people expected to return, left at shrines, or carried with its owners. A Fuqara talked of his father and uncles leaving property at the shrine of as-Sa'idat in the Wadi ibn Hammad, and he himself regularly stores *tibben* and tools in his family caves in the wadi. A Beni Hamida pointed out that the shrine of Sulaiman ibn Da'ud continues to be used as a safe depository, while Musil (1908: i, 87) saw Beni Hamida ploughs at a now vanished holy tree. A Beni 'Amr recalled that Hammam Umm Sidre, which belonged to all the people of the area, was used in the same way. Household provisions or tools were stored, wealth accompanied people as animals or weapons. A change in political and economic circumstances where land not only has a value for agriculture, but to be owned has to be registered by and pay taxes to the state, and land unregistered belongs to the state, transforms ideas of wealth and political representation. Some families are quick to take advantage of the scenario, others are not.

This is evident in the history of land registration and the building of houses/stores in al-Karak (Lancaster 1995: 114–5), with Majali, Christian tribes and Beni Hamida being quicker to register land and build stores than Beni 'Amr, with the exception of al-Jaraydat section of Dimneh, and Fuqara. Sulaiman Majali built the first store by 1875 at al-Qasr; by this date, agricultural development, linked to land registration under the Ottoman reforms after the Crimean War, had started in al-Balqa by former merchant families like the Abujabers, and tribal leaders like the al-Fa'iz and others of the Beni Sakhr (Abujaber 1989: 136ff). An elderly and respected Majali said that an bad outbreak of malaria forced people to use the plateaux rather than the *wudiyân* around the 1850s.[26] People were unclear as to the exact processes of land registration but thought that land was registered as tribal or family land under the Turks, although there was individual (*mulk*) land ownership within tribal and family land. Under the Mandate, registration "had to be in the names of individuals who had to have a house, and that was when we registered the houses as houses even though we didn't use them as houses. We always lived in tents, the houses were stores for grain, seed, equipment, the animals used them sometimes and workmen slept there."

---

[26] A member of a northern family gave a similar reason for his ancestor's move to Irbid from the valley.

An account of the early development of al-Qasr (so called because of a Nabataean 'castle' there) by one of the al-Faris Majali indicated that Faris ibn Salamah never built at al-Qasr. The sons of Salamah had their stores at Khirbet Faris or Tadun, where they had their threshing floors, while the sons of Khalil (the cousins of Salamah) had stores at al-Qasr, in and around the *qasr* itself. "All around the *qasr* were the threshing floors, especially to the south and east, and that was where they built their stores, which eventually turned into houses. Beit Ibrahim (ibn Khalil) has five arches, and is dated 1909. It was built by the *ashîra* (here extended family) for them to store their produce. The grain they couldn't store, they sold. Beit Ibrahim now belongs to Sulaiman ibn Muhammad ibn Ibrahim and his brothers, in whose names it is registered. The buildings to the south of the courtyard of Beit Ibrahim were built by Ibrahim's sons, and the stone houses were built in the 1960s by other *ibn 'amm*. The new houses of reinforced concrete or concrete blocks were built about fifteen or twenty years ago (early to mid-seventies) in the courtyards or *haush* of the old houses, which became animal pens and stores. But when we lived in tents, we spent the early part of the winters or the spring at Azzur, depending when the cisterns filled, or at Ain Jubaiba where there was running water. We came into Khirbet Faris or al-Qasr to get stores when we needed to. Faris was as big as Qasr then, but Qasr grew because it had the Turkish and then Mandate government road, with the Police Post and watering point, and Faris had only the old Roman road."

Jaussen records the building of houses/stores by the Majali (and Zraykat) at Rabba, Yarut, Qasr and Faris; by Beni Hamida at Sirfa and Fagu', by Christian tribes at Hmud and Smakiyya, and by Beni 'Amr at Dimneh by 1902 or 05 (1948: 245). All these places are mentioned as watering places with cisterns owned by the same groups by Musil (1908: iii, 97–105) at about the same time. Glueck, quoted by Miller (1991: 33, 38–9, 60, 62–3, 66), visited the villages of Rabba, Qasr, Fagu', Imra (Beni Hamida), Smakiyya, Hmud and Jada (Mesarwa and 'Amr) in the thirties. In 1968, from Gubser (1973: map3) the villages in our area were; Rabba, Qasr, Yarut; Dimneh; Majdalain (mixed Palestinian, Christian and Majali), Imra, Sirfa, Fagu'; Hmud, Smakiyya (Christian); Judaida (Thanaybat Mu'aita); Shihan (Palestinian), Jiddat al-Jbur (Jbur Majali); Jiddat al-Sajayda, Mis'ar, Mughaira, Abu Traba and Ar-Riha ('Amr). By the 1979 census, additional villages had come into existence, such as al-Mujib in the Wadi al-Mujib, and Aliya. By 1992, there was another Beni Hamida village west of Imra, and a Rashaiyida village east of Smakiyya. While all these villages, and in the case of Qasr and Rabba now small towns, are associated with specific tribal groups, there are in nearly every village other groups who came for protection or employment or as guests. Al-Qasr has Palestinian families who came "eight generations ago", Mbadiyin families who came in the twenties for protection, and Beni Hamida families who used to work as share-partners. Dimneh and Sirfa have Fuqara families; Yarut, Beni 'Amr and Fuqara families; and so on.

The changing priorities of local people is seen by Kana'an and McQuitty (1994: 146–7) to account for the change from stores to houses, especially a decline in the need to store grain from the 1940s, with the change from a subsistence economy to integration in a wage-based economy. Local users also connect the change to the droughts of the mid- to late forties (Lancaster and Lancaster 1993), when households came in from outlying encampments like Faris to family stores and shelters at Qasr where the Police Post had a watering point and a recruitment centre. "In 1949, I was a young boy and we stayed in Beit Ibrahim because of the drought. Beit Ibrahim was very full of people at that time. We were there because of getting water from the Police Post – that was the old Police Post from Emir Abdullah's time, and it was in the Qasr (the Nabataean temple ruin). They were recruiting for the army there, and people joined because of the drought; they got five JD a month. When the drought was over, they didn't want to come back, and there were other jobs as well, in the police and government. The Palestinians came and they worked in agriculture. Gradually people moved into Qasr, there was no one reason but there was the primary school, and the road. The line of the present main road was done in 1952, they moved it a block to the west, and my wife's father built a row of shops along the road. That was a year after I went to secondary school in Karak. My father built a stone and mud house south of the Qasr in 1956. I remember the first time I saw a house of reinforced cement, I was so impressed with so much space, the big windows and the amount of light. I couldn't work out how the roof stayed up without arches and beams. When more Palestinians came across after 1967 they were used to having modern houses with water and electricity, and big windows, and they really were a big influence. Some of them were builders, and they needed money, so they worked for us building new houses." Gubser

(1973: 39) gives a table of new buildings in Karak city and Karak district between 1952 and 1961; the al-Qasr area is not isolable, but at both dates the great majority of people lived in rough stone houses or tents, with houses of reinforced concrete appearing after 1952. Most skilled and semi-skilled workers in the construction industry in al-Karak were, according to Gubser, Palestinians from outside the district, while the unskilled labourers were local. This agrees with local information concerning reinforced concrete buildings, but not for stone buildings, where skilled workmen were either Circassians or local people, especially Christians; it was generally held that anyone could build a rough stone house. Whether workmen were used or not seems to have been personal preference and expense rather than a chronological development as suggested by Kana'an and McQuitty (1994: 142).

The gradual change from stores to houses coincides with growing participation in the increased administrative and security arms of the Jordanian state, with a perceived need for education, Palestinian influence, and the government's decision in 1974 (Prof. Kamel Abujabr, personal communication) to take over the purchase and sale of grain. From that date, herders and farmers only stored feed for animals, mostly as *tibben* (chopped straw), and kept in old buildings or caves.

Summer cisterns for animals, threshing floors, and stores for grain as a focus of family groups linked by descent and marriage are clearly linked with the development of villages as recipients of governmental infrastructure. The physical growth of village structures over the years is a function of the growth or decline of its constituent groups. In Karak, these are the *khâna*s, the *khâna* being the lands, arable and irrigated, buildings, cisterns and descendants in the male line of a common ancestor or grandfather. This ancestor was the one who "developed the land and so owned it."

Dimneh has eight *khâna*, all registered in the thirties. The *khâna* lands radiate out in rough segments from the cisterns and caves that were the original focus. H K's original house still stands in the centre of the present village, a small one-roomed stone house adjoining his brother's identical house, built in the mid to late thirties by a *muhandis* or builder from Karak, who brought his own workmen (figure 13). The stones were local, and the wood for the beams came from the wadi. Both brothers used the houses as stores, like their caves. The cisterns and caves associated with this pair of houses are now under the road or buildings erected later by other

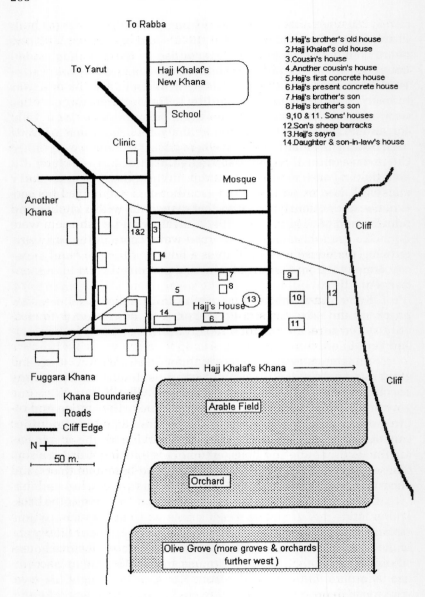

Figure 13 Diagram of a Khâna

khâna members. His second house, on land to the west, was built in reinforced concrete in, he said, 1943 – and his brother built one nearby; both had two rooms. This westward move both extended the existing *khâna* and established a new. The existing *khâna* contained the houses of the grandfather's descendants, the new was himself, his brother and their children. Both brothers dug cisterns, now under a later road, at the north of the site; "the cisterns were for summer water for us and our animals. But sometimes we used them for storing grain, it kept well. You lose the topmost layer, but the rest was fine. If we were storing grain in it, it was a store; if it was holding water, it was a cistern." When two of his brother's sons built houses on land to the southeast, H K enlarged his second house by a courtyard extending eastwards, with a kitchen and additional store against its walls. A lavatory and wash-room were added to its eastern end when piped water came in the early seventies. This house was lived in as a house, while the family continued to live from herding and in winter and spring lived in a tent in the wadi.

In the mid-seventies, he built his third house, a villa, a little to the southwest of the second. This is a cement house with electricity, three reception rooms, inside lavatory, washroom and kitchen. The second house was a shop for a few years but is now a store; its kitchen is used for cooking for feasts. With cooking and washing moving inside, the third house is double the size of the second. During the eighties, he and his sons built four villas in a row on the western edge of the bedrock, above the garden and olive grove started in 1968. Most of the money came from gratuities and earnings by the older sons in army service. Each son expects in time to build another floor on top of his existing villa for a son. These villas have storage in basement or semi-basement floors, and the rooms have high ceilings; floor plans are variable, but each has two entrances, men and guests at the front, and women at the back. The public entrance leads directly into a reception room, with a second winter reception room behind. The women's or family entrance leads into a wide passage with a room used for sitting and sleeping, and the kitchen. Washrooms and further sleeping accommodation are laid out either side of the passage. The villas have gardens with flowers, herbs, vegetables and fruit-trees around them, and a shaded place for summer sitting. For the remaining sons' houses and for the eldest grandsons, land has been bought at the eastern end of the village. While skilled workmen, led by a son-in-law, organised the work, much of the unskilled work was done by the family. H K considered using the site of his first house, in the centre of the village, but concluded it was cheaper to buy land and build on an open site than to demolish and build on what is now a site with restricted access. This new land, detached from that of the original *khâna* and its extension, becomes part of the *khâna*; the lands need not be, and often are not, contiguous. Over time, the *khâna* land changes shape; within it, members realign and shift shares.

House construction methods and house contents for three villages in the Shera mountains, including Dana, are described by Biewers (1993: 20–31). Burckhardt visited Dana in 1812 (1822: 409–10), as did Musil in 1900 (1907: ii, 277). The village clusters along a long rocky outcrop, although isolated houses have been built in various gardens. The village buildings were used as stores and as shelters when working in the gardens. In winter, people moved west down the Wadi Dana to as-Safra with the flocks and lived in tents, in the summer they were harvesting on their arable land to east and living in tents, while they had more stores at Khirbet at-Talaya. "We used the buildings for shelter when working in the gardens, and for odd nights in bad weather, but really they were stores, and we lived in tents. Then most people moved to our summer cisterns at Jabal 'Ata'ata up by the main road when we built our new village, and we called it Qadisiyya. We live in houses there, with electricity and piped water, we have television, refrigerators and washing machines, and we cook on gas. There are proper schools, a health centre, the *baladiyya*, shops, and the bus and taxis. We use the tents for shelter in the fields when harvesting, in the spring with the sheep, or for weddings and funerals. Some people move to Dana for the summer for the gardens, and a few of the old people stayed there all the year, but there wasn't any electricity or piped water. The Friends of Dana society help people renovate their old houses, which is good because as people get to retiring they often want to move back to Dana for the gardens and the quiet. But the old houses are difficult to live in, not just because of not having electricity and water, because the government is bringing them down, but the houses are dark and the Society only allows small windows, and of course some houses only have one wall that's free for windows, all the other walls join onto other houses. And then how can we fit in a kitchen for baking bread, or a lavatory or a washroom? Some houses are all on bedrock, so there's no place to dig a cesspit. Really, a household would need two or three structures to have all the rooms of a villa. People who want to live here will have to sort out the problems. Every household has claims on different buildings so people could swap round, or buy and sell."

The 'Ata'ata made their buildings from local stone, with roofs from local juniper wood, canes or reeds from the springs, brushwood and earth. Stone arches, with low springers in shallow foundation trenches or on bedrock, supported the roofs. Inside was the central space, with 'rooms' between the solid supports of the arches, usually with built up floors. These 'rooms' were for storage and often included fixed or movable silos for grain, flour and legumes. These silos were made by the woman of the household from local clays and earths, reinforced with chaff and short lengths of brushwood. Adjacent to the house was a decayed building, also owned by the family, and divided by rough walls for use partly as kitchen, partly as animal shelter and lavatory. In time, roofs weaken beyond the care of regular maintenance as the brushwood packing decomposes and rainwater starts to seep inwards. Roofs begin to disintegrate at the corners of walls, along the length of the arches, or where the external surface of the roof is less than uniformly smooth so that hollows are formed, allowing rainwater to collect and permeate the earth and brushwood packing. Rot can then occur in the roof-beams along the line of the wall or arch until the roof becomes detached and sags along the whole length of the affected wall. During rainfall, earth from the roof slides towards the gap, and falls into the building. Once the roof has given way, the building needs major repair. Parts are usually walled off to form a kitchen, *tabûn* shelter or animal pen. *Tabûn*s are the bread ovens made by the woman from local clays; the oven is built up in a series of coils over a few days, and sun-dried, in the same way as the silos were made (Biewers 1993: 35–6; McQuitty 1984).

This use of two or more structures to encompass the necessary functions of household living parallels village houses with courtyards, or the tent with *haush*. A decayed building starts as a new or renovated store and seasonal shelter. The following description of the process comes from Dana, and recalls that of Schumacher in the Jaulan (1889: 42–3), where the inhabitants built using old basalt building stone, oak beams for the roof, then oleander brushwood, damp earth and finished with clay and *tibben*; "The family inhabit the same hut till the roof commences to fall in under the weight of the yearly increasing layer of clay. Then with the co-operation of the relatives, another hut on another part of the ruins is built." Roofs should last for about seventy years, so the new or renovated structure would have belonged to the present owner's grandfather. "The building we're using now for the kitchen and for the goats was the house I grew up in, and at that date, the kitchen and the animals were where this renovated house is. Some of the walls aren't exactly the same, but the arches are in the same places. When we came to redo the roof, most of the juniper beams in the old part were fine so we used them again and got just a few new ones. I know that they were first put in by my great-grandfather, and re-used over and over again. Next year my wife wants us to redo part of the ruined house for a bathroom with a shower; she has her kitchen in a bay of the new house, as you know, and she uses out here for the *tabûn*, and if we're entertaining. We like living down here in the renovated house, we have quiet, and clean air, and marvellous views from where we sit out in the evenings. Our house in Qadisiyya is where my parents live, next to my brother, and our eldest daughter lives there during the week to go to the secondary school."

The inheritance of structures down the generations is affected by inheritance in the family estate, where male descendants have equal shares and female descendants half shares, and by the preferred marriage of first cousins. Thus some married couples end up with a greater number of claims on one property rather than another, as each receives shares in a number of properties. It is common in Dana to hear someone say "I have four houses" or "We own three houses"; they do, but as claims on rather than outright disposal over these properties. Reconciling claims over a building or garden provides much of the dynamics of social life. Temporarily irresolvable claims are a stated reason for letting a building decay, until most potential claimants have resolved the problem for themselves by finding an alternative; the building is then taken over by a family member of the younger generation who can reconcile the diminishing claims of other parties. The needs and alternatives open to potential claimants are taken into consideration by the family group in deciding which claimant gets what, while the act of starting renovations is often a deciding factor, although initially arousing opposition.

The layout of buildings in a village is ideally a series of segments radiating from a centre or centres in concentric rings. Visually, Dana presents such an image, although the distribution of inhabitants only partially and inexactly mirrors the image; houses at the centre are not occupied by the most senior members of the family group, as at one stage of development, nor by the most junior, as at another, while a household's neighbours are closely related in several paths. The ideal progression is necessarily broken by the different stages of village family groups over time, the variation in numbers in a generation in each household, their marriages, and individuals' choices in economic activities and marriages – which often follow economic co-operation – as some move out, and newcomers come in. The distribution of houses of *khâna* members in Dimneh illustrates that the pattern of outward moving segments is not always possible, and the acquisition of new but separated land solves the immediate problem. In al-Qasr, with population growth, the development of commercial property, and land purchases for government buildings, family members wanting building land often have to buy low quality agricultural land or building land owned by another family member or by another family. In this way, the ideal layout of a village reflecting descent groups fails, while the pragmatic actions of participating members continue the real village. Individuals and families of villages and tribal groups live in the cities of Jordan, Syria or Saudi Arabia; people buy or rent a flat or house, or buy land on which to build, where they can. A group of brothers may buy or build together, but there are not family or tribal blocks, except for those families who were the original owners of land now part of the city area.

Villages based on tribal sections' cisterns, stores and threshing floors often have old cemeteries, sometimes associated with a shrine. New cemeteries are made by the *baladiyya*, and attached to the mosque or church, but some are buried on their own land by their families; "we buried my father here, because his father is buried here, and he wanted to look out over the valley." The assassinated former Prime Minister, Hazza' al-Majali, is buried in a mausoleum on family land in Rabba. The graves have head and foot stones and follow the convention of the local Christian church or Islamic school. In the villages of northern Saudi Arabia, which follow Wahhabi practice, graves have no distinguishing marks. The House of Mourning, the Bait al-Azha', is an important part of urban and village ceremonies after a death; the House of Mourning may be a tent erected outside the house of the deceased, or its formal reception room, where male relations, neighbours, and associates pay their respects to the memory of the dead man. Women have a separate meeting room inside the house, since the number of women paying their respects is smaller, restricted to relations and neighbours. A daily morning radio programme announces deaths and the place and time of the funeral, which takes place before noon if the person died after the previous sunset, and before sunset if the death was after dawn. The House of Mourning allows for those unable to attend the funeral to pay their respects since it continues for a period varying from three to seven days and in some instances, for forty days.

Some present cemeteries have a long history of association with shrines. Palestinian shrines are discussed by Canaan (1927) and Jaussen (1927). al-Harawi (1957) mentions many of the shrines of the wider region in his guide to the places of pilgrimage. Two well-known Jordanian examples are Yajuz, north of Amman, and Nabi Usha near as-Salt. Maqam Nabi Yusha and Arasat Sulaiman ibn Da'ud are both near Sirfa, on the western edge of the Karak plateau, where Jabal Shihan is said by Yakut, quoted by LeStrange (1890: 533), to have been the place from where Moses saw the Promised Land. Yakut mentions Nabi Yusha, while az-Zahari, quoted by Hartmann (1907: 46, n.1) mentions both Nabi Yusha and Sulaiman ibn Da'ud, and that Sirfa was on the route from Damascus to Karak (1907: 82). These shrines were important places of local pilgrimage and as cemeteries for Beni Hamida at the turn of the century, and Arasat Sulaiman ibn Da'ud remains so. Musil (1908: i, 91) photographed the building in 1896; the west wall has now gone, together with most of the cut stones from the grave of Sulaiman in the courtyard. Sa'idat was a shrine of Sa'id and his sisters, the ancestors of the Fuqara, and is a low cairn orientated east-west above a spring on a spur overlooking the western end of the Wadi ibn Hammad. When Musil (1908: i, 375) was shown the site in 1902, there was a holy tree, now gone, although our sponsor and guide said he recalled his grandfather talking about the tree. In the cairn were poles with clothing and rags tied to them; stones had marks of libations, and an iron ploughshare and a basalt olive mortar had been left there. Several graves in separate groups were around the shrine, but "although there is nothing to stop people being buried there now, they just aren't". As well as being sites for burial, people also came to shrines for healing, physical, mental and social, to place goods for safe-keeping, and some served as places of worship.

Nabi Usha near as-Salt was visited by Burckhardt (1822: 353–355) who describes the building, adding that the "tomb is much resorted to for commercial purposes, and like Mekka and Jerusalem, is transformed into a fair at the time of the visit of the pilgrims." The main article at the fair was *kilw*, the plant ashes used in soap manufacture, brought by the Beni Sakhr and other tribes of al-Balqa, and purchased by merchants from Nablus, where there were soap factories. The purchaser paid duties on every camel-load of soap ashes to the chief of the Adwan tribe, and a lesser amount to the town of as-Salt, which divided the revenues among the public guest-houses of the town. Canaan (1927: 195–208) portrays the events at a pilgrimage site with market in Palestine in the 1920s.

Present day commerce takes place both at wholesale and retail markets and shops, and outside such defined places of commerce. Jordanian rural areas supply wholesale markets at local centres for the sale of vegetables and fruit, or live animals for meat. These buildings are concrete block structures, with an office, scales and weights, display areas for goods, and shelter from the weather for participants, within a walled yard. The fruit and vegetable market in Sakaka is similar, though unwalled. Local administrative authorities often own the wholesale markets, and manage them through an appointed officer, or rent out the running of the market to a third party. Wholesale markets are also owned and run privately. Local councils also own shops, workshops and stores, for rent, and private landlords build shops and stores as investment opportunities. These are all concrete block structures with electricity and water, metal shuttered doors, and road access; sometimes flats are built as a second floor. Using an old building for a shop is common in town and village, although many, especially in the main shopping areas, are purpose built. All commercial buildings are registered and pay a fee to the local council. Commercial activities also took and take place without formal markets, sometimes alongside existing markets as in al-Jauf, where bedouin bought dates in the market or at the grower's house. Doughty ([1888] 1936: i, 184, 339) says in al-Ula and Taima that people sold their produce of corn and dates in their own houses, while traders coming from outside hired rooms or a house in which to sell their goods (i, 338).

The processing of grain and seeds may need constructions, or leave only indications of use. Pounding holes are common in archaeological sites and around contemporary villages and camp-sites. Recent use was for pounding up bulky plant parts before the finer work of grinding for medicines, and for crushing *butm* and olives. "In the past, we waited until the olives were really ripe, when they were black. We put them in the sun to dry, and then we broke them by pounding with a basalt mortar in these holes (in local outcrops of basalt). Then we put them in these square-cut trays in the rock, put really big stones on top of them, and the oil would come seeping out. It collected at this lip at the front of the rock and dripped down into the pot or dish underneath. Or we ground the olives in the household grinding mill, and then put them in warm water and heated it a bit more; then we let it cool, and we skimmed off the oil on our hands and scraped our hands over the rim of a bowl so the oil trickled into it. We did *butm* fruit in the same way if we wanted them for oil, they must be left until they are blue-black. Some people used them for flour, and then they were ground in the handmill. People don't use *butm* now. Then there were big olive crushers, and now we men take the olives to the presseries, we use the government one at Zallum." Most presseries are built by local authorities, but a few are privately owned. Thoumin (1936: 167) describes horse or mule driven olive presses with basalt or limestone stones in the Syrian interior, owned by a village or by a private owner. Every apricot orchard, or sometimes every section of an orchard, had its press made from baked earth (Thoumin 1936: 168); the fruit was placed in a basin eighty cms high, and pulped with a pestle. The juice and pulp trickled down into a lower basin through a hole. This pulp was taken out and put on wooden planks to dry in the sun.

Grain was ground in the tent or house by the women in stone hand-mills, sometimes bought, sometimes made by the women if there was suitable stone around. Ahl al-Jabal women recalled making handmills; "just look for suitable basalt, the right sort of size, and with one flat surface of the right sort of texture, and then shape it by tapping away with a smaller harder stone until it's right. And you make a hole in the centre, and then you shape a stick for the handle. You need a bit of iron or really hard wood for the axle. Anyone can do it." A similar description is given by Musil (1928a: 91) for Rwala women, who also used pestles and wooden mortars. In the Karak area, where watermills were in use, the women ground grain in handmills only for *burghul* or *farîkha*.

Watermills were the usual means of grinding corn in the Bilad ash-Sham (McQuitty 1995) at least since the tenth century AD; taxation on watermills in the District of Karak is mentioned in the sixteenth century Ottoman registers (Hutteroth and Abdulfattah

1977: 72; Bakhit and Hmud 1991: 87). Watermills were common in the Arab Middle East. In Oman, mills were associated with *falaj* irrigation (Costa and Wilkinson 1987: 56–7), and some Syrian mills were powered by *foggara* water (Thoumin 1936: 164ff; Haj Ibrahim 1990). In Jordan mills used seasonal flows of mountain streams, and were widely distributed along the *wudiyân* running into the Jordan Valley and the Wadi Araba. The surviving mills are horizontal wheeled mills, and either single or double penstock types; both exist in Wadi al-'Arab and Wadi Hisban, while Wadi ibn Hammad had only single penstock mills (McQuitty 1995: 746–9). Thoumin (1936: 165–6) describes the working parts of horizontal wheeled mills in Qalamoun in the early part of this century, and says that each pair of wheels could grind five hundred kilos of grain a day. Rebuildings from the eighteenth century are known; Conder (1889b: 129) refers to the mills at Sumiya and below Hisban erected by Dhiyab of the Adwan in 1191 AH (1777 AD), and Gardiner and McQuitty (1987) have tentatively dated a mill in the Wadi al-'Arab to the same date. This dating would be consistent with Seetzen's 1806 meeting (1810: 30) with peasants near al-Husn carrying their corn to a mill for grinding. Presumably other construction and re-buildings took place but were unrecorded.

Rogan (1995), working from Late Ottoman land registers for as-Salt and legal documents, finds an increased investment in mills from the 1870s on by wealthy merchants and tribal shaikhs, and considers that, due to the expense of even restoring a mill which took three years' rent to recoup the outlay, the initial investment in mills and the associated infrastructure of canals was almost certainly from the Burji Mamluk state. This is not wholly convincing. Local people see no great costs in constructing water channels. Travellers refer to the millstone industry from the Ladja northeast of the Hauran (Buckingham 1825: 166–7, 283–4; Merrill 1881: 24), with the stones exported by camel to the south, Jerusalem, the Palestinian coast, Egypt (Hasselquist 1766: 275) and Qalamoun, while Merrill (1881: 190) saw millstones being cut in a quarry south of Wadi Yabis. This demand for millstones implies regular construction and/or renovation of mills, since worn millstones can be redressed in situ. The 1596 fiscal register records four mills in Wadi Karak, as does Seetzen in 1810 (1854: 417); Tristram (1873: 78) saw the water from Ain Sara below Karak turning a mill. A recoupment period of three years does not seem excessive when shares in these investments or outright ownership could be bought and sold;

a mill next to Karak was co-owned in shares by four people in 1596 (Bakhit and Hmud 1991: 33).

Jaussen (1948: 31, n. 1) records the use of watermills in Wadi ibn Hammad and a steam-operated mill in Madaba in 1905. The watermills in the Wadi ibn Hammad stopped working in the thirties, following the installation of imported diesel mills in buildings near the stores and threshing grounds on the plateau. Small motor mills were being installed from 1924 in Qalamoun and the Ghuta (Thoumin 1936: 166). The Wadi ibn Hammad watermills were owned mostly by Majali, and shares in mills were bought and sold "all the time". A few people remembered mills working in the late winter and early spring when the water coming down the wadis was at its highest. Water was channeled off the main stream to drive the mill. "When you wanted your grain milled, you booked a date with the miller. The millers were hired from Palestine. If you just turned up with your grain, you might have to wait three or four days. Your camels or mules, horses or donkeys carried it down, and you paid the miller in grain or cash, but never in flour." ar-Rabba, al-Qasr and Jiddat as-Sajayda had diesel mills. The present flour mill in Jiddat as-Sajayda was installed in 1942, and made by Ransome, Jeffries and Sims of Ipswich; a Ruston belt driven diesel engine powered it. Both machines sit in an old one-arched stone building, next to a much smaller diesel mill, which was the first in the village, in the 1930s.

Lime was needed for plastering cisterns; house interiors were also plastered, although this was often with mud and *tibben*. A lime kiln near Ain Jubaiha west of al-Qasr was built in the late 1950s; "my father built it because there was good stone right there, and plenty of wood for firing as people were clearing trees in the slopes and the *wudiyân* for agriculture." A nephew of the owner described how lime was made; "the stone, white stone, has to be without nodules of flint. The soft limestone is cut up, and you build it into a pile until it is the shape and size of my shop (about four metres by two and a half). The top is corbelled, and at last the final stone goes in. A little doorway is left at the bottom on the west side; you need the west for the wind to provide a good draught. You build a fire inside the doorway, and first you use chaff, then *tibben*, then wood and then you feed the fire with wood and *tibben*; it takes several loads of *tibben*, because the fire burns for three days and nights. Then you seal the entrance, and leave it to cool as slowly as possible. Then, it all collapses in on itself and that's quicklime.

Quicklime was dangerous, and when the water to slake it was poured on, it exploded. For one bucket of quicklime you got ten of slaked lime. We used the lime for cisterns and for general building. Making lime stopped in about 1975. Everybody made lime, but usually people did it on their threshing floors because the *tibben* for firing was right there – at that date we didn't use it for feeding animals. It stopped because people started using cement, and then *tibben* became valuable."

Industrial pottery kilns appear to be largely lacking in the rural Bilad ash-Sham in the recent past. In the present, some pottery, especially water jars and also plant pots and cheap decorative pots, are made and sold by Egyptian men at various places adjacent to main roads. Formerly, women in the villages made domestic wares after the harvest. An elderly Christian woman in Hmud described how she collected clay from a place in the wadi; "I looked for grey earth, that's the right stuff, and brought it up on the donkey. I used *tibben* or finely pounded sherds I found on the old *khirbas* for temper. I made *jarra* (jars) and *tanâjir* (casseroles), and *tabûn* (bread ovens). I coiled the clay, building it up until I had the size and shape I wanted. For the firing we used dung, and some wood, oak and pomegranate, what we had. I made them for the family, and I might give a few as presents, or make them for people if they asked." This was the common pattern in Karak and the Shera. The limited pottery repertoire was extended by the use of wooden vessels and containers, and copper vessels and trays, later replaced by imported enamelware, aluminium, plastic and glass. Musil (1908; iii, 138–40) lists the kitchen utensils in Karak, mostly locally made of wood, copper vessels and jugs bought from towns, and earthenware storage jars made by the women (also Crowfoot 1932). Women in the north Jordan Valley made a greater variety of pottery, including cooking pots, basins and storage jars (Mershen 1985), as did those of Busra in the Hauran, described by Bresenham (1985), where in the late seventies a woman potter was making water jars and bread moulds, and had earlier also made grain silos and ovens. None of this was wheel-made. Wheel made pottery comes from industrial potteries; at some sites, sherds of black pottery are found which people remembered as coming from Gaza by train as water-jars. Mershen (1985: 76) says these were found in north Jordan villages using basalt temper for local wares, which were therefore not waterproof. Very little pottery is made in the villages now, except for *tabûn*s. Pottery was also dug up out of old pre-Islamic burial sites, in the Karak region at least; "if we wanted pottery, we went and dug it up. There are lots of places with useful pots". Kaddour and Seeden (1984: fig. 22) show a Bronze Age pot in use as a salt container in a village on the Euphrates. In the *harra* and the *hamad*, there was virtually no pottery in use except for coffee cups, which were imported Chinese ware, bought in urban centres or from travelling merchants. Large storage jars seen in the oases of al-Jauf were brought down from the Hauran by men from the town when they went to the Hauran for harvest work, or from Basra. d'Hont (1994: plates 49–54) has pictures of industrial potteries on the Khabur river, producing mostly water jars.

Villagers in the mountains of the Bilad ash-Sham made gunpowder (Burckhardt 1822: 250) in the proportions of one part of sulphur, five and a half of saltpetre, and one part poplar charcoal. Saltpetre was manufactured in the villages bordering on the Ladja (Burckhardt 1822: 214), by boiling up salt-impregnated earth; "the boilers of these manufactories are heated by brushwood brought from the desert." Doughty ([1888] 1936: i, 410), travelling some seventy years later, mentions several times bedouin collecting salt-impregnated earth from ruins, caves and watering places where animals regularly stood and urinated; "They gather tempered earth, when they have tried it with their tongues, under any shadowing rocks that since ages have been places of lying down at noon, of the bedouin flocks. This salt-mould they boil at home in their kettles, and let the lye of the second seething stand all night, having cast in it a few straws:- upon these yellow nitre crystals will be found clustered in the morning. With such (impure) nitre they mingle a proportion of sulphur, which is purchased in the Haj market or at Medina. Charcoal they prepare themselves of certain lighter woods, and kneading all together with water, they make a cake of gunpowder, and when dry, they cut it with a knife into gross grains; such powder is foul and weak and they load with heavy charges." A Sulaib recipe for gunpowder from 1909 (Musil 1927: 233) mixed brimstone, brought from the western edge of al-Batin, with butter, and heated it until the sulphur separated from the brimstone; then crushed charcoal was mixed with the sulphur, and the gunpowder was ready.

Salt collection for sale to towns, villages and tribespeople, was carried out at Palmyra (Weuleresse 1946: 307), near Jerud in the Qalamoun, at Azraq, on the shores of the Dead Sea, and in the Wadi Sirhan at Kaf and Ithra (Musil 1927: 326). Tribespeople also collected salt for themselves and for sale from saltflats (*sabkha*) in wadis south of the Dead Sea (Musil 1908: iii, 146–7) and rock salt (Musil 1927: 363) from places in the *badia*. People dug shallow pans at the edge of a *sabkha*, allowed the salt water to flow in and then to evaporate; it was then collected and put in sacks. A Druze recalled the salt business at Azraq from the early twenties; "salt we collected at Azraq was taken to the Hauran and Jabal al-Arab when we moved up for the summer. We carried it on our cows and camels to exchange for wheat. If there was no winter rain, we couldn't grow crops at Azraq, so we would work on the salt and then move to work for wages on farms in the Hauran, wherever it had rained, or Palestine or Lebanon, and the women stayed behind living off stores. This was the way of life until the fifties. We paid the government a yearly tax for collecting and selling salt. The Ahl al-Jabal did it too. 1968 was the last salt caravan using camels, and then in the seventies we used cars and trucks, but only as far as the border, because there were state borders and customs then. Then the government developed the salt into an industry with a factory that cleaned and refined it and packed it. Then I worked as a mechanic there."

Quarries for stone and gravel are owned by those who develop them, usually those associated with the land on which the quarry is situated. Some people put in gravel-crushers and buy fleets of lorries themselves, others contract out such work or arrange share-partnerships. Most supply gravel to road building and to cement factories; others supply stone for facing buildings.

Other diggings were excavations to find antiquities for sale. Dealing in antiquities appears to have arrived with western tourists (Issawi 1988: 388 quoting al-Qasimi), and excavating and selling antiques was quite important for some rural individuals in Syria in the thirties (de Boucheman 1936: 82–4; Weuleresse 1946: 307), and later from reminiscences. Excavating for antiquities for sale is still common in Jordan, Syria and Saudi Arabia.

Processing of materials, whether of foods or stone and minerals, in the past often required no obvious structures, or few that left lasting and unambiguous remains. Nor were specialists necessary; in local memory, everybody was capable of carrying out these processes, they were part of people's knowledge. Processing was often, but not invariably, for household consumption but most of these products were also available for purchase; other processed products were for sale. Although products varied over time, amounts processed and sold throughout the Bilad ash-Sham added up to a considerable total production.

Roads, as routes, were important in the past and in the present for the passage of people, animals and goods. Many routes have their origins in antiquity, and follow natural features of the landscape that facilitate travelling. Inscriptions and documents enable the use of some routes to be established in detail for the Roman to early Islamic, and then the Mamluk administrations (Bisheh 1989; Bowersock 1983; Dussaud 1927; King 1987; MacAdam 1986; Musil 1927: appendices V and VI; Musil1928b: appendices II–IV). The Ottoman Sultans reorganised the Pilgrim Route from Damascus (Petersen 1995), and ensured that this was normally kept open, with groups of local tribesmen being responsible for security of people and goods. Alternative routes were always accessible to travellers who followed local customs, travelling with a series of sponsors (*kafila*) or companions (*rafiq*) who provided restitution of goods and personal security for payment, and who handed their protected onto other responsible men from other tribes when their own influence ended. Grant (1937: 176–9) refers to the accounts of western merchants and travellers from the sixteenth to the eighteenth centuries crossing the desert between Aleppo and the Gulf where this system of contracts and guarantees was the recognised and effective way of travelling. It also worked on the less wellknown routes crossing the length and breadth of the Bilad ash-Sham and the Arabian peninsula. Some routes were the 'possession' of a particular group, whose men had the right to provide protection for a stage of the route or the whole route; one was the section from Rakhama to Shobak of the Hebron-Hijaz road, for which the Amarin provided sponsorship. In all travel, arrangements for sponsorship needed to be made with members of each local group in turn, so that the traveller was handed on to known sponsors for his safety; the sponsor need not be the shaikh, but any respectable man. Palmer and Tyrwhitt Drake (1871) give a string of examples. There were always alternative routes to a particular destination, and alternative markets. Musil (1908: i, 20–2) discusses the trade routes in the Karak region; he mentions five north to south, with the Ottoman pilgrim route (Peterson 1995: 299) making a sixth, and seven west-east routes with the further addition of a very difficult route. These do not include routes further east between the Hauran, Hijaz and Nejd using the Wadi Sirhan, listed by Burckhardt (1822: 665–6). There were other routes not listed by Musil, but on one of which he travelled (1908: i, 32), which was yet another alternative route to Damascus or Jerusalem, used when the preferred Ghor as-Safi road was impassable because of feuding between various tribes. Musil's guide pointed out on the road many small towers, built for protection of travellers from raiders.

These routes were rarely surfaced or graded tracks, and passable only on camel or horseback, or on foot. Graded tracks for motor vehicles following the main routes existed from the Mandate; the King's Highway was made suitable for motor vehicles between 1933 and 1939 (Amadouny 1994: 143). In the eastern *badia*, the Royal Air Force route between Jerusalem, Amman and Baghdad was marked on the ground by a furrow ploughed by a tractor and plough, and later by black stone cairns. This furrow was followed by the Nairn bus service for six months in the winter of 1925–26. The Nairn transport company (Grant 1937: 270–89) had been running services from Damascus to Baghdad but the Druze Revolt of 1925 – 1927 caused the company to re-route their service. The Jordanian route lasted only six months because of the slowness and difficulty of the *harra* stretches. The pipelines laid by the Iraq Petroleum Company and opened in 1935 between Kirkuk, Haditha, and then to either Palmyra, Homs and Tripoli, or Rutba, Mafraq and Haifa, have been the foundations for main highways. All weather surfaces started in the 1950s (Gubser 1973: 112). In the Karak area, the main roads from the Wadi Mujib, the Ghor and the south to Karak city were surfaced in 1956. The road from Karak to Qatrana, which linked the area to the Desert Highway between Amman and Aqaba, was not surfaced until 1965, delayed by Majali "because their major villages, ar-Rabba and al-Qasr, lay on the road to the Wadi al-Mujib and they feared the loss of revenues and influence." By this date some Majali had built shops in al-Qasr; at about this date, the main road through al-Qasr was moved a block to the west, and a new axis of commercial buildings developed. Later, when al-Qasr became the seat of a *mutasarrif*, government offices were built in this street. Local road building and maintenance is the responsibility of the local administrations, and by the 1990s all villages in the Governate of Karak were connected by a network of surfaced roads.

Routes in the eastern *badia* are made by people of the region going between grazing, water, markets and service centres, by agents of state governments policing borders and smuggling routes and keeping the peace, and by companies exploring for oil, and gas, and extracting gravel and stone. Road construction not only marks the landscape with new roads, but also with the quarries and equipment needed for their building. The tracks made by local users have little construction; boulders might be moved, notorious holes filled in, but that is about all. Sometimes people mark the turn to their tent by an inscription, paint mark or cairn, while the ar-Rishas tracks are marked by piles of earth or marked barrels.

Field systems are the last structures considered. It might be thought fields in the different areas of the Bilad ash-Sham would vary with the different agricultural systems, but their users talk about field systems in surprisingly similar ways, whether the fields are in the Ghor, the plateau or the *badia*. Most of the visible structures are to do with the direction or conservation of water and made with this purpose in mind. The materials come from managing water, or from stone clearance which in itself can create field structures. Boundary markers are often natural features, trees or large boulders, now added to by survey markers.

If water is being channelled from springs on steep hillsides at Dana and other places in Ajlun or around Jerash for example, terraces are created to hold the earth. Here terracing uses stones that need to be cleared, creates stone-free level areas of deeper soils for tree cultivation, and these use water efficiently. A series of terraces are built from the bottom of the slope up. Their exact form appears to be a function of the degree of slope and the amount and nature of rock, but in general a terrace has a perpendicular retaining wall of drystone boulders about a metre or so high, and the terrace follows the natural contour. In some cases, the wall is made of two parallel walls filled with rubble. If the wall is a boundary, it is made of two parallel perpendicular walls filled with rubble, while if the wall is to have metal poles installed for further fencing or for grapes, the outer wall leans in at an angle of maybe fifteen degrees. Terracing is also a feature of a decision to use land on a hill slope for fruit tree or vine cultivation, using seasonal irrigation; such a decision often occurs when a man inherits or purchases an area of land too small for arable cultivation, does not live and work in the area, but has the financial resources to build terraces, fence the property, install irrigation and hire a workman; often a deep litter chicken-house is part of the enterprise. Terrace building, like water-channelling, can be done by all in the countryside, although gangs of young men who specialise in terracing can be employed. Channelling spring water on more gently sloping hillsides, as for example in the Wadi ibn Hammad, may mean the ground is built up in places for the channels, but the gardens are on the downward slope from the channel openings. Here there is little use of terracing, only stone clearance from the gardens to make small walls at the lower boundaries of the gardens, which help to retain earth and water.

Similar water retaining walls are a feature of olive groves in the Aluk area, famous for its oak woods. These walls are built across the slope, to retain water and so increase soil moisture levels for the olive trees. In the fields of the Kerak plateau, or the slopes west of Sirfa, walls built with field stones along the contours are made to retain water. Such walls also delineate the current owner's boundaries, together with older walls of the larger boulders against the contour. These walls often indicate the larger divisions between tribal or family sections from a division of village lands, whereas the walls along the contour show more recent shares through inheritance or swaps.

Walls that divide land into fields may be a function of water management rather than of separating blocks of land. A farm in the *harra* developed by a Sharafat in the 1930s had a multiplicity of walls which "were all to do with channeling water. Keeping the animals off growing crops was not really important because the animals were being herded or in their *haush*. When you make a channel, you uncover stones and take them out, and then they have to put somewhere. The easiest thing is to make a wall as you clear the channel. That is how these walls were made. Some of the channels have a wall each side, so they look like pathways. Some channels took water to fields, some to cisterns, and they can be very long (many were over a kilometre long, some nearly two kilometres)....... The water can flow uphill. If the wadi really runs, the force of the water does push it uphill between the walls. And you went along with a mattock, banking the channel up or digging it deeper so the water didn't escape. And you opened or closed the channels to get the water where you wanted it. The fields over there to the west, they were the last ones to be cleared, in 1968 or 69 perhaps. They only got rainfall, and they never worked. Their walls are only stone-clearing, and that was just rough. If you are ploughing with animals, donkeys or a camel, stones aren't important. You go round the big ones. We weren't the only family making a farm and channeling water, there were others from the tribe. There was one further down the wadi, and one further west on a side wadi."

Fields were smaller when most ploughing was by animal, they "were strips, the size a pair of oxen could plough in a day. Now the fields are much larger, although of course fields are owned by people in strips, some of them very small." Stone clearing is more necessary with the use of tractors for ploughing, and in the expansion of irrigated agriculture with the laying of plastic pipes for watering, and plants being sown through polythene. In the Hauran and east of Mafraq, fields are sometimes bulldozed clear of stones. More often, large boulders are moved by bulldozer or left in situ, while smaller stones are moved by the work of the entire family. How the stone clearing is done is eclectic. In many cases, stones are thrown into wide, low heaps along the boundaries of the field, and then built into walls. If there are still more stones to be cleared, sometimes short wide walls are built as windbreaks, especially east of Mafraq, and where the field will be planted with vines, figs and olives; in the same area, extra stones may be left in regularly spaced heaps, and vines grown against them. On the Jabal al-'Arab such clearance piles form mounds with a small, roofed shelter built on top. An alternative seen in the Karak area is to build circular solid 'towers', up to two metres high, in the field.

The current position of living in houses in villages and small towns in the rural areas of the Bilad ash-Sham has earlier precedents, even in the badia regions. Local people connect the present development with a transformation in the method of access to the important service sector of multiresource livelihood or, in their words, the need for education. Education sums up the incorporation of local political structure and practice into those of the regional nation states, together with regional and global economics. This participation provides, directly and indirectly, the increase in money supply that permits the funding of imported materials, technologies and specialised labour rather than earlier self-sufficient, unspecialised practice. While such participation in the wider economics is widespread throughout the region, not all manifestations of settlement are equally viable. Some grow, some decline, some remain static, a few see-saw between growth and decline; the judgement as to the viability of any village or small town compared to any other is inevitably short-term, over perhaps twenty or forty years, apart from a few regional centres like Karak city, as-Salt, Ajlun, Tadmur, or al-Jauf. The reasons for growth or decline vary, and are related at least partially to the reasons for the establishment of the village.

Although people say constantly that their reason for settling is for their children's education, the causes for the development of villages are concerned with the perceived need to register land for agriculture. Registering and cultivating land was and is the only way of holding land in state terms, while tribal agricultural lands and individually owned land were recognised. Many villages, especially in agriculturally marginal areas, began shortly before or after the land was registered by the inhabitants, based around cisterns, stores and/or threshing floors. Others had longer histories if based around mountain springs and orchard cultivation, like Dana, the three longterm villages southwest of Karak, and some Ajlun villages, which all produced raisins and dried figs, purchased by merchants. Some have shorter histories or remain static, like those in the *wudiyân* of northern Karak (Lancaster 1995). In the *badia*, the ar-Rishas declined.

The ar-Rishas started in the late sixties as bases for political and economic action by a part of the shaikhly family of the Rwala, after the Ba'ath came to power in Syria, to wage economic retaliation against the Syrian government for expropriation of tribal assets. This attracted much support among tribesmen still suffering the economic results of the drought of the late fifties and early sixties. In 1971, ar-Risha an-Nuri had a number of stone or cement houses of family members and employees, groups of tents from various Rwala sections and from protected Fuwa'are families and an Ahl al-Jabal protected family, a school and a *sûq*. By 1979, some of the personnel had changed; some families had moved to Ruwaishid or Saudi Arabia, and few families had joined. Negotiations in Syria for the return of assets or compensation were underway. Profits from smuggling were lower, and many took their profits and invested in sheep-herds and/or small businesses. The suq was closed by the Jordanian authorities after complaints from the Ramthawi traders in ar-Ruwaishid that the greater variety and lower prices of Syrian goods at ar-Risha undercut them. By 1985, a few 'villas' replaced earlier concrete houses, the suq had re-opened on a small scale, shaikhly organised crossborder trade had ended with the agreement on compensation, and most Rwala had moved to Saudi Arabia. The inhabitants who remained were some of the Fuwa'are, a few Rwala, and an Umur. These 'villas' are walled courtyards with living quarters built on the inside of the west wall; the kitchen is a separate building, forming an L along the north wall with the women's entrance. The south wall has the men's entrance, with sometimes a men's guestroom built on the exterior of the wall adjoining the main building. Most have their tent pitched adjacent to the house. In 1996, there are about fifteen villas altogether, well dispersed, and most have changed hands several times. The school has expanded. Current residents are some Fuware, a few Umur, and a Rwala, all sheep traders and herders, with regular visits by the shaikh's agent, and rare visits by male members of the family. The suq is intermittent, active in late summer when tentcloth and poles arrive with Syrian traders, or if there is a spring. Ar-Risha's decline was caused by a lack of economic and political activity; "nothing happens here".

Ar-Ruwaishid started as a pumping station in the thirties. Since the closure of the pipeline after 1948, it has been a customs post, administrative centre, and local service centre for travellers and long-distance lorry drivers, and local herders. Some IPC employees opened shops or restaurants. Police and other security forces have always been present. In the early seventies, the town catered for the transit trade between Syria and Saudi Arabia, and had a mixed population of Palestinians, Ramthawis, Syrians and bedouin. Some bedouin had houses there for their children's primary and secondary education. By the late seventies, the Syrian – Saudi Arabia route was closed, and the town depended mostly on transit trade between Jordan and Iraq. By the mid-eighties this ceased with the closure of the border between the two countries, and all Iraq's imports from the Mediterranean came through Syria. Ruwaishid's garages, mechanics' workshops, restaurants and shops mostly closed, and population fell. Most Rwala moved to Saudi Arabia, a few went to Amman. After Syria closed its borders to Iraq during the Iran-Iraq war, Iraq imported through Aqaba and Ruwaishid expanded fast. This has been maintained after the Gulf War, and manifested itself in the re-opening and rebuilding of garages, repair shops, restaurants, hotels, *sûq* and shops. Administrative services increased, with an outpost of the Ministry of Agriculture and an *alaf* depot, a new well and pumping station, secondary schools and a health centre. The population remains mixed but the personnel have changed. Most businesses are owned by Ramthawis or a few of the Sha'alan, and staffed by itinerant Egyptians, Syrians and Palestinians. There is little sense of community in Ruwaishid, unlike its crossborder neighbour, Turaif, in Saudi Arabia.

CHAPTER 7

PRODUCTIVITY, DISTRIBUTION AND CONSUMPTION

Productivity is founded in the wider domestic group on the resources and labour available to group members through accepted social practices, and is multi-resource. Labour and resources are flexible in personnel and practice. Labour may be the individual's own or from his family or wider domestic group, that of a partner, or hired by the season, day or task. Use of resources can come from ownership, renting, partnership or indirectly from wages as a labourer. Members of wider domestic groups may be simultaneously owners, partners, tenants and labourers in a variety of enterprises. Ownership in customary law is rather preferential access to resources and control over surplus than outright control of disposal. What is owned are the means to production; so, collection, storage and distribution of water that make land productive are owned, together with the land; tools, draught animals and their modern equivalents are owned; domestic animals; mills and other processing plant, and so on. The owner, as the person or body with rights to dispose of property, may be an individual, man or woman, or a three generation group of inheritors from a common grandfather. Such a unit also has within itself members with property acquired in their own right by inheritance from maternal relations, gifts and purchase. Ownership is strongly connected with use, but contains longterm rights to access that can be held in abeyance for years. Rainfed arable land is, in customary law, associated with identified descent groups; those using the land have right of usufruct and have preferential claims over the crop, so cultivation confers preferential access through use by identified and participating social persons. Irrigated land, because it is developed, is owned by those who developed it. The provision of administrative, jural and protective services — ruling — seen as necessary for the maintenance of peace so that people may obtain their livelihood, may be from within the groups using the land, from families of other groups, or by agents of a centralised state. Processing, manufacturing and distribution facilities are owned but their effective functioning depends on government, keeping the peace, by local or state agents. Credit and finance services similarly require jural support.

The agents of 'government', whether from within the producing group, a local family, or a state, can be considered as 'enabling partners' of production and entitled to a proportion of surplus. Which entity provides 'government' or 'rule' is seen by participants in the countryside to depend on the amount of surplus created by or passing through the region. If there is surplus, 'the state' moves in and takes this wealth for itself, if there is little, then 'ruling' comes from local groups including those who cultivate. A state will, in some fashion, own rainfed arable land and pasture lands and derive wealth from these, either through taxation on agricultural and pastoral production or, for rainfed arable land, may hand such land out to its functionaries in lieu of cash for their support, rent out lands as tax farms, allow lands to be registered in the names of current users in return for taxes and registration fees, or rent out lands to citizens. These options have all been used at one time or another; at present, Jordan and Syria, as inheritors of the late Ottoman systems, use the last two.

Livelihood in local usage implies 'subsistence plus', 'subsistence with some profit' so that a family actively participates in economic, political and social life. As well as food, clothing and shelter, subsistence includes social expenditures on formal and informal hospitality, weddings and funerals, jural and health costs, and some surplus for paying old debts, extending credit or investing in some enterprise. 'Subsistence' is thus largely equated with consumption, where consumption is sufficient to allow the living of 'a good life'. The ideology of individual autonomy and jural equality in no way envisages equality of economic reward, although it expects relatively open economic opportunities to be available through customary social practice; after that, economic activity is on the head of each participant and his wider domestic group. Production strategies divide into those from which a family lives and those which bring profits: a Beni Sakhr family summering on the Karak plateau said "we live from our camels; most years the sheep are profits. This year we are keeping them" (from stored profits). Official recognition of a difference between productive resources yielding basic livelihood and those from which profit was possible can be seen in the Negev under the Mandate, with its thriving grain market; here, plough camels were liable for double the tax on milk camels (Ben David and Kressel fc). In the early Ottoman tax registers (Hutteroth and Abdulfattah 1977: 82) cattle, like horses and camels, as draught animals were not taxed; sheep, goats and bees were taxed for they provided commercial products, and grain was taxed as a basic product.

A notable feature of a customary system where access to resources and a right to usufruct is more relevant for livelihood than ownership is the 'handing-on', either by consignment or sale, of processing and distribution to others. 'Ownership' in the sense of control of productive enterprises is self-limiting, since at its most basic it is difficult for one party to command or buy the labour of another, where both are free autonomous individuals. This is why partnerships, contracts, and handing-on are popular and successful, widespread and of long standing (Firestone 1975: 185–209; Goitein 1967: 164ff).

An individual, as a family member, has a right to livelihood from access to and his share in family resources, a responsibility to contribute labour, skills and capital to the management of these resources, and/or to develop other resources. We have seen, in the section on agriculture, how one of the inheriting sons may manage the family land and divide the profits among his co-heirs, with an additional share for his time and labour; and comparable situations are common in families where the main family resource is animal flocks or trade. The family resources, the estate of the inheritors of a common grandfather in the male line, are not an immovable block owned conjointly by *a priori* defined family members but rather shift in exact membership and assets over time as each co-inheritor manages his share in terms of his own needs and abilities, while taking account of the wishes and inclinations of close family members. He may have to realise his share to pay some pressing obligation; as part of a marriage transaction; or to develop another resource that he thinks more profitable. In many instances, the asset will be transferred to a member of his close family, while sometimes assets are lost to creditors, to the men of the family of a son's or brother's bride, or in compensation for blood spilt by a family member to the victim or his or her family.

In addition to his (or her half) share in the family assets, each individual pursues economic opportunities that arise within the locality and the region through the processes embedded in social practice. Women's economic activities may be carried out through an agent, or are with other women, so they are less visible, and usually on a smaller scale. As a man develops his own economic interests, he may offer shares in a projected enterprise, often a development based on a family interest, to brothers and close cousins in return for investment. The decision is up to each individual as to whether he becomes a member of this new share partnership or not, but if there is no sharing of the risks by putting in capital of one sort or another (money, materials, land, labour), then there is no sharing in the profits, which are shared among the partners in the proportions of their investment. If non-investing family members fall on hard times, other wealthier members support them and their dependents, and in any case, they may well be given gifts from the profits; but these gifts and support are either prompted by affection or because of real need: there is no general right of access.

Share partnerships were and are in use in trade, industry, agriculture and livestock rearing. For a share partnership to be lawful, both partners must contribute to the enterprise, and the nature of the contribution of each affects the share of each in the proceeds. A contribution of labour only is unlawful, because there is no input of any sort of capital and therefore little risk; labourers would either be fed, clothed and housed, or paid daily or weekly wages. Abdel Nour (1984: 79–81) discusses types of share contracts in agriculture. Examples of share partnerships in different fields of enterprise are given later in this chapter, all of which were observed and discussed in informal settings. The contracts for agricultural share partnerships in the Karak region are often but not always witnessed by local government officials. Herding and milking share contracts are invariably witnessed by people present at the tent where the contract is established, all of whom have the ability to be witnesses by virtue of their known identity and reputation. Share contracts for business and trading enterprises can be formally drawn up and registered at a court, or informally between partners with neighbours as witnesses, depending on the closeness of the relations between the partners and the value of their investment. When a partnership ends, which may be after an agricultural season or herding year or after several years, the assets are divided according to a formula agreed at the beginning of the partnership among the partners, after all outstanding payments have been made. Registered partnerships must be wound up by the presentation of accounts by partners to the registering authority. Disputes between partners may be settled formally at a government agency or court, or informally by a mediator acceptable to all partners.

'Handing-on' by sale commonly occurs at the points between production, processing, manufacturing, transporting, and marketing. Shunnaq (fc) describes the handing on of milk by sale to processors of dairy products in north Jordan, where the processors are incoming families of Syrian or Armenian descent. In central Jordan, commercial dairies are usually Palestinian owned, who buy milk from local producers. In the eastern *badia* professional cheesemakers, always said to be from Ramtha, come out and contract to buy sheep's milk delivered by the herders of the sheep and make cheese for urban markets. In these examples, 'handing on' depends on the supply of the raw product and the market demand, for there must be sufficient to give profit to a middleman. The curing and sewing of sheepskins for cloaks are two other industries employing middlemen processors. In Mafraq, sheepskin cloak makers are Syrians who rent shops and buy skins from merchants in the Sukhne area of Zarqa; they bring the cloth backing and trimmings with them. Their shops also sell Syrian goods for herders — bells for sheep and donkeys, shears, coffee roasters, spindles, tentcloth and clothing — and tent pegs and pins, and coffee pot stands, locally made from iron building rods. The businesses are profitable because the men leave their families in Syria, where living is cheaper.

Animal trading is by a series of sales, of varying kinds, from producer to trader to end user. Breeding animals are sold by a herder to a trader who sells to another herder, or to another trader who then sells to a herder. Animals for meat are bought by small traders who go round encampments, and then sell animals to bigger traders at markets outside small or large towns. Both breeding animals and those for meat may be consigned by their owner to an individual for sale. Here, the owner states the price he is willing to accept, and the consignee agrees to deliver this price; if he can sell the animals for more, that is his profit, if for less, he accepts the loss. Small traders usually buy for cash, and the price is agreed between the owner or his agent and the trader; both know the current range of prices. Livestock markets, as at Hafr al-Batin or Sakaka, provide a known location for sales. The sales themselves can be made in a variety of ways, by auction, by closed bids, bids through brokers who move between sellers and buyers, or by agreements between a seller and a buyer. An auctioneer or broker takes a percentage of the price. Sometimes cash payments are demanded, in other cases credit is extended for a certain period, sometimes other commodities are part of the price. Really big traders buy from big herders in the *badia*, or from traders at markets where there are slaughterhouses, as at Zarqa. The big traders buy a hundred to a thousand animals at a time, and arrange for their slaughter at the slaughterhouse, paying a fee to the slaughterers. Just before the animals are killed, the trader sells them to the big meat buyers who then sell the carcasses to butchers' buyers. The skins are bought by merchants, and most are sold to Turkey.

The vegetable and fruit market of Damascus is discussed by Bianquis (1978). Peasants sell their produce to a merchant, sometimes directly but often through a middleman. The association between a producer and the merchant is based on a forward buying contract, which ensures that the peasant sells all his produce to the merchant, not only the amount that would repay any credit he has been given. The loading and unloading of lorries is under the control of a group of porters, who carry cases between the door of the lorry and the pavement of a shop; from the pavement into the shop cases are handled by workers employed by the shop-owner. Cases are valuable; some are owned by the wholesale merchants, others belong to producers, while others are sold with the produce, and then bought by dealers who have workers who repair cases for recirculation. The official classification of traders is divided into wholesale merchants, who receive merchandise directly from the producer or from their agents, and retail-wholesalers, who do not have direct relations with producers, forwarders or agents, but buy from wholesalers. Many merchants are both wholesalers and retail-wholesalers, depending on the product and the season. Potatoes and onions are exclusive specialisations, but in general traders divide activities between summer and winter produce. A trader specialising in summer produce is a wholesaler in summer but a retail-wholesaler in winter. In the summer he supplies his winter wholesaler neighbours and his agents in neighbouring towns; in the winter, his business slows down, he only keeps on his monthly paid workers and lets the others go, and supplies his customers by buying from his neighbours. Merchants from Homs, Aleppo or Lattaquie stay for long periods in hotels around the market, and forward their merchandises by brothers or nephews at home.

The Sakaka fruit and vegetable market operates like the central market in 'Unayzah, described by Altorki and Cole (1989;181–2), where the produce is sold by auctioneers. In Sakaka, the market is divided into an area selling local produce, vegetables and melons, and that for outside produce, mostly imported fruit and a few vegetables. Local produce is sold early in the morning, by the case or small truck load, while imported fruit and goods are sold at

9pm, because it is driven up during the day. For both categories, auctioneers auction goods to big buyers, who then sell it on to small purchasers from local shops, large households, shops in surrounding villages and as far as Turaif. Prices are good at the beginning and end of the growing season for each crop, but poor in mid-season when produce can remain unsold.

In the small-scale production of garden and orchard crops around al-Karak, a son or nephew with a pickup or small lorry delivers the produce to market and sells it. "I grow tomatoes, or grapes, or melons; he markets them" is a constant comment. Both producer and transporter have a general idea on what the market price should be. The producer may give his carrier instructions on sale prices, or leave final decisions to him, but the profits (or losses) are shared between producer and carrier as both are active partners in the joint enterprise. A carrier, as well as transporting his father's or uncle's produce, will carry similar crops from members of the wider family cultivating nearby and bring back supplies of dry animal feed, paraffin, petrol, gas cylinders, plastic piping and sheeting, and sacks of flour or sugar on commission for members of the wider domestic unit. He also takes women, children and the elderly from the group to a clinic, local government offices, or town for shopping for a wedding. The carrier may also sell melons, grapes or tomatoes from his family's gardens to tents on the plateau. These additional businesses are the carrier's own, although other family members may have shares in it since they may have contributed to the purchase of the lorry.

This pattern of a transporter for a producer having his own subsidiary business is common. Long distance lorry drivers employed by transport companies taking goods from Aqaba to Iraq often buy on their own account goods like snack foods, pharmaceuticals, local cheeses and dried yoghourt in Mafraq, as-Safawi, or ar-Ruwaishid for sale in Iraq. A merchant in al-Jauf supplying herding families said "I order my goods (clothes, quilts, coffee pots, roasting spoons, coolers, tongs and bellows, dishes, trays, kneading trays, pressure lamps, tools, tentcloth, tent pegs, rope, water barrels, garden tools, wheelbarrows etc.) from travelling merchants from Syria who come with their vans or small lorries. I use the same traders regularly, I say I want five of these and three of those, and they know the sort of things I sell. The Jauf dates go up to Syria from time to time; they're bought by the traders who supply me and people like me, and other passing lorry drivers who have space or think they can sell them, or have orders. It's small scale, but constant. al-Jauf isn't far from the turn to Medina and Jiddah, and our dates are famous." These small-scale traders from Syria have regular customers among the merchants in Jordan and northern Saudi Arabia who supply herding and gardening needs; some fulfil orders for goods, and some sell speculatively.

De Boucheman (1939: 85–93) describes different types of trading and transporting by the inhabitants of Sukhne during the early years of the Mandate. Traders on foot sold clothes, town foods and cheap cosmetics and jewellery to Sba'a and 'Umur tribespeople in the *badia*, as Burckhardt (1831: i, 191) described Damascene pedlars selling to tribesmen, and Musil (1928a: 124–5, 269–70) the small traders from the Euphrates towns of Kubaisa or Rahba with the Rwala. These traders extended credit, being paid later in *samn*, wool, camelhair, and sheep, which they sold in the towns. Bigger merchants carried *kilw* (soap ashes) from the *badia* to Aleppo and Antioch. Here they sold the *kilw* to soap merchants who then entrusted soap to them to be sold in southern Turkey, Kurdistan, and the Jazira where they picked up provisions for Sukhne and the tribes. In de Boucheman's time, two-thirds of the male population of Sukhne were transporters who also traded on their own account. He might be employed by a forwarding agent, especially for cross-desert trade, but he would also be trading on his own behalf in a wide range of goods, legal and illegal. He might buy up *kilw* ash, and if he did not own his own camel/s, rented animals and either paid a flat charge or shared the price he received for the *kilw*. Sukhne trade was compared by de Boucheman to Doughty's description of the towns of Qasim ([1888] 1936: ii, 312, 457), and the similarities are clear in the accounts of former 'Unayza cameleers (Altorki and Cole 1989: 70–2), where one transporter was commissioned by merchants to buy goods for them, others instructed their agents to deliver goods to him and others gave him money to make purchases for them. Métral (1993: 199–205) brings the Sukhne trade up to date, reporting how trade has diversified but remains dependent on the presence of the tribes and the activities and production of the *badia*, while the small trading is now carried out by the young as an apprenticeship. The use of motor vehicles has led to the repair of lorries and tractors, the sale of water pumps, and contracting of agricultural equipment as additional enterprises. Small artisanal and industrial workshops in Syria (Perthus 1992: 222) depend on merchant-mediators for raw materials and spare parts, as do those of Jordan and Saudi Arabia. Some of these merchants may be tribesmen (Lancaster 1981: 109–111), as may be the owners of small workshops in the small towns on the highways in eastern Jordan and northern Saudi Arabia.

The development of *mahatta* (garage complexes) is common just outside rural small towns. The site-owner acquires a licence for a filling station and builds this; he also builds a restaurant or rest-house, often a mosque, and sometimes a supermarket, selling travellers' supplies and bulk goods for small and opportunist transporters/drivers. He also builds outbuildings for a mechanic's workshop, panel-beating, electrics, and puncture-repair. These enterprises may be run by employed hands or a series of self-employed men, often Syrians or Egyptians, who make individual partnerships with the owner. Some *mahatta* have side-lines of agricultural contracting businesses, selling seed and fertiliser; or depots for the marketing or transporting of sheep. It was common in the late eighties to see signs for 'sheep rest-stations' in northern Saudi Arabia, but these had disappeared by 1995.

Mills provided an example of 'handing on' as a joint venture by their owners, who owned the land on which the mill was built, organised or did the digging of water channels, the provision of building materials and the millstones, and organised and paid the millwrights who built the mills and dressed the stones. The owners employed millers, often from Palestine, who saw to the grinding of grain for a fee, which they handed over to the owners.

'Handing-on' is also evident in the expansion of large-scale commercial grain and other agro-industrial crops in the dry-farming areas of late Ottoman Syria, where the state distributed state land to agents or middlemen for cultivation in return for taxes and produce. While in some ways this could be likened to a joint venture, the state received crops and or cash in lieu at pre-agreed amounts (although these were negotiable in bad years) as rents for the use of land rather than a share of profits. Schilcher (1991b: 185ff) compares this to the European 'putting-out' system, and sees the two ideas of 'use' and 'exploitation for profit' as the key. These twin concepts permeate the whole of productive enterprises and the acquisition of livelihood in the wider region, whether the protagonists are the state and its agents, urban merchants, rural notables and inhabitants of the countryside. They existed in the past (e.g. Abujaber 1989; Doumani 1994) and the present, where recent examples of state development of agricultural projects and the reactions of their participants have been described by, for example, Bocco (1989a) in Jordan, Métral (1984) in Syria, and Hamza (1982) for Wadi Sirhan. These state organised agricultural developments were set up to engineer social change, by settling former nomadic tribal pastoralists or to break the power of certain landed groups. Other state organised agricultural developments have been 'handed on' to certain families of political importance, ostensibly as individuals among them have the financial resources and the interest in further developing agricultural crops for processing and so aiding general economic progress. Such enterprises are often seen locally as the state rewarding certain individuals for political support.

Modern examples of 'handing on' are agencies for import, distribution and services. The state is often a player here. In Syria, the state or its agents is a partner in a joint enterprise with a foreign national, where the state provides land, water and electricity and roads, and the foreigner the financial resources; he also is the one who initiates the whole enterprise and its direction, although the state must approve. In Saudi Arabia, the legal right to sponsor business with foreign companies is the gift of leading members of the ibn Sa'ud. At a more local level, an agent importing, say, car tyres rarely organises the distribution to sale outlets; that is a separate enterprise belonging to another person. Cross-border trading by tribesmen illustrates the processes. During the seventies, goods imported into eastern Jordan from Saudi Arabia by tribesmen were for local use, like tyres, spare parts and fuel, or for export to Syria, like cigarettes, cloth and electrical goods. Tribesmen held that they were exercising traditional trade across government imposed borders, while the different nation states' currencies, customs duties and economic activities made this trade more dangerous and profitable. The various governments saw cross-border trade from administrative and political perspectives. Tribal leaders obtained a general waiver from ibn Sa'ud and from the Palace in Amman for importing goods, and organised large scale imports of goods for local needs and for export on into Syria. Men from tribes unable to enter Saudi Arabia bought supplies from Rwala or Sba'a and then took the goods into Syria for sale. Rwala often bought their own goods from merchants in Turaif in Saudi Arabia, brought them to Jordan, and either took them into Syria or sold them to others who took them on up. Others commissioned people to buy them two or three cartons of cigarettes, a roll of cloth, a television, and to take it up and sell it for them. Tribesmen bought the goods and transported them, then sold them to retail outlets; no-one was a wholesaler in Saudi Arabia, a transporter, and a retailer in Syria.

The flexible division of labour within a wider domestic group permits both wider family livelihoods and the establishment of new enterprises by the young or retired. This spread of enterprises and flexibility of labour exists in all modes of livelihood, whether based in arable or irrigated farming, livestock rearing, trade, transporting, or small-scale industry. Multi-resource production and labour fluidity is seen to require a wide range of means of access to production. A date farmer in al-Jauf commented that share farming and renting gardens, trees or date crops, were common because the number of men varied considerably, as some decided to work in the Hauran, Hijaz, or Qalamoun, or joined the *ageyl* or armed forces. While there are families whose livelihood strategies are concentrated on employment, few are without any owned resource of land, flocks or business enterprise. These owned and developed enterprises are analogous to the *khâna* of land, buildings, cisterns and people described by Beni Hamida and Beni 'Amr of al-Karak; over and within the generations there is a constant regeneration and redefinition of family assets and resources. This appears to have been the pattern in the past from available documentary evidence as, for example, indicated by Doumani (fc) documents concerning family *waqf*s in Nablus, and inherent in general social practice.

Relations of regions and people in differing economic circumstances are augmented by a multiplicity of cross-cutting contexts of distribution between surplus and deficit. Doumani (1995: 203) sees "the interdependency of economic activities to be the organising framework" as general, focusing on the Jabal Nablus area of Palestine where he finds horizontal layers of social classes cut by vertical ties of patron-client relations. Gilsenan (1977) analyses the system of the Akkar plain in North Lebanon where land is owned by lords of Kurdish descent with power and Arab religious shaikhs with authority, and worked by landless labourers. He sees the horizontal strata of landowners versus labourers as the crucial facts, maintained by landowners who are favour-givers keeping labourers as favour-seekers dependent by ensuring everyone thinks in terms of 'gaps' to be filled through the favour of the lords and shaikhs; that is, vertical face-to-face relations are localised, while the ownership-dependent labourer relations are generalised. He argues against seeing these face-to-face contacts of peasant-lord as patron-client relations which satisfactorily explain social relations, since it is the "structures of domination from which they were generated" (1977: 182) that are crucial.

Patron-client or alternative lord-peasant relations are a channel for ameliorating a deficit situation from surplus. Doumani and Gilsenan indicate that such relations were directed as much by those in surplus towards those in deficit as the reverse. Doumani shows how merchants used *salam* contracts (advance payments) with peasants to ensure regular supplies of agricultural produce for trade (as do Damascene merchants in the fruit and vegetable market) or industrial supply. Gilsenan describes lords and shaikhs using the giving of favours to maintain the power system. In Jabal Nablus, at least, patron-client relations between merchants and rural lords existed alongside active networks of peasants lending money to each other, buying and selling lands, and entering into business partnerships long before the 1830s (Doumani 1995: 165). *Khuwa* relationships achieved grain, *samn*, or tentcloth, for example, for tribespeople who supplied protection and restitution to sheep and goat herding tribes and villagers in return. Bedouin tribal shaikhs are assumed to build up client retainers and supporters through gifts, hospitality and protection. Both shaikhs and tribesmen point out such activity is inconsistent with the underlying moral premises of their society, since generalised hospitality and protection are open to all and from all, and are diffuse rather than one-to-one reciprocal relations. One-to-one methods of gaining something, like protection from being raided when using an area outside one's own, were contracts between parties who could be held accountable, involving reparation and restitution on non-fulfilment, and of different types; the *rafīq* and *kafīla* for merchants and travellers, and *khuwa* for use of land with a proportion of produce from such use in return for protection. The employment by shaikhs (and others) of agents and armed guards was also by contract.

The denial of patron-client ties is not an ideological response that covers actual relations. Both bedouin and peasant groups exercise 'right behaviour' of individual autonomy and responsibility, although there are tribal, family and individual interpretations of particular circumstances. During the smuggling, Rwala shaikhs provided a free field of supply while it is said Sba'a shaikhs made a charge on each item for their trouble in getting the goods to transporters. Casual work at rural archaeological excavations is arranged by local tribal leaders, local Dept. of Antiquities officials and archaeologists; some tribesmen are said to pay a percentage to the local leader for his efforts in getting a quota of jobs, men of other tribes do not. Rwala shaikhs go to considerable effort to make complainants resolve disputes themselves, and make no charge for their mediatory services. They insist that those who come asking for help can be under no obligation except, if the beneficiary wishes, to spread the donor's reputation as an exemplar of the efficacy of the tribal system. Azazma share-workers deliberately move landowners each year, and consciously invest in machinery and equipment to be share-partners rather than labourers while maintaining their flocks to ensure they have the basis of livelihood in their own possession. Although a Majali landowner may say "we have the same Ghawarna family every year for our partners" the Ghawarna family ensure they are not in fact available each year.

The lack of patron-client relations comes from the multiplicity of alternative partners, resources, markets, and management strategies in the system, and the high degree of internal cohesion of social practice within and between groups in the countryside of the Bilâd ash-Shâm. Interdependence of economic activity is indeed one operating principle, but the methods of access to resources by responsible persons using multiple roles and a variety of contractual processes result in more complex enmeshments than two or three horizontal strata cut by vertical patron-client relations. The practice of 'handing-on' and so of actively not controlling all aspects of production, of agents who are autonomous partners in contractual arrangements, and of the multi-purpose nature of the livelihood strategies of most individuals and all families, are embedded in moral principles as well as a response to variable economic conditions. That each individual has a right to livelihood is accepted by all; so while competition for access to resources and markets may cause irritation, it is also recognised as an inevitable fact of economic life that must be accommodated. The constant gathering and assessment of information by all at the constant sharing of hospitality not only provides knowledge of potential opportunities but also of possible sources of competition and dispute which can then be avoided, negotiated around, or confronted. The avoidance of or negotiating around possible sources of conflict is an active part of grazing management, for example.

The contexts of surplus and deficit are multiple. For the individual and the wider family, distribution from surplus to deficit takes place through generosity, obligatory and optional transfers of wealth and wealth creation opportunities, and credit and debt.

Credit and debt within wider domestic networks are rarely visible, never ending, and circulate between members; land, money for marriages, houses and compensation, animals, cars, tools and equipment are the usual items. Accounts are rarely brought into the open but generally known in outline among those concerned. Settlement of credit extended should only be demanded when the creditor has no other means of satisfying his own creditors, or if he or she feels the debtor is misusing the loan.

Most people prefer to raise money within the wider family and often succeed in this without using banks. An exception is where an individual or family wants to make an investment that demands a lot of money at once, has a reasonable amount of capital as security so he can use the banking system (or is part of the banking system), and is making the investment for his and his children's livelihood; building housing, office and commercial property by individuals and families in Amman are usually financed at least partially through banks. Other exceptions are where a nexus of families has a member in a favourable position in a government sponsored development bank, as with Beni Hamida of Fagu'a in the early 90s; or when government policies support bank lending for particular enterprises and some individuals see this policy as a resource to be used for their profit. In the towns of northern Saudi Arabia, some make money by building; cheap loans are available from the government, so land is bought and building started. When the repayment date arrives, the investor's lawyer finds a reason why repayment is not possible, and the government does not press for payment because to do so would hurt its image. This tactic can be used three or four times, until the building is completed and sold or let, when the money is repaid from profits. The process then restarts.

Agricultural and livestock management, and the physical infrastructure for irrigation, processing, manufacture and distribution involve large amounts of capital. In part, the existence of capital in the form of tree crops, animals, irrigation, processing equipment and so on testify to wealth in the countryside. Rogan (1995: 753–6) considers the costs of building or renovating water mills, as part of the expansion of grain crops in the late nineteenth century, as expensive. Mills cost between $70–210, while a six room house in central as-Salt or 3,000 dunums (300 hectares) of rainfed grain land both cost $35. Musil (1928a: 61, 122, 125, 133–4, 349) has prices for equipment, tentcloth and clothing bought by Rwala at roughly the same date; goathair tentcloth for the smallest size tent cost $29.70, rifles $25.50–54, cheapest camel calves $9, riding camels $76.50–$180, and six-year old female camels, ready to breed from $54–72. Roughly similar prices for camels are given by Musil (1908: iii, 292–3) from Beersheba tribes and Beni Sakhr. Abujaber (1989: 83–4) mentions two sales of land in the same period, one by 38 Zufifa tribesmen from Khirbat as-Suq south of Amman of 919 dunums for $360, and one of a twelfth share in the whole property of a village, its cultivable lands, caves, built area and cisterns, that works out at 10 US cents a dunum. These figures indicate a wide range of land values, but the land sale quoted by Rogan is much cheaper. Compared with other productive assets of land and camels, mills were not particularly expensive, the more so since many were built or renovated by groups of share-holders.

Digging water channels to mills or gardens was not expensive in labour terms, from the times given for building *foggara* at Deir Attiya. Neither al-Nasif, Haj Ibrahim nor Thoumin provide any monetary values for *qanat* or *foggara* construction, but the enterprise was clearly within the capacity of a section of the village community. Thoumin (1936: 47) gives the only cash value for water in Nebk, where five minutes of one-third of the water flow was 20 Turkish pounds, 2,200 French Francs or £17.10 shillings at 1930 values. What use can be made of the figure is doubtful, as no other figures for land, crop, livestock or industrial goods are given. Thoumin (1936: 331) describes the economic condition of Qalamoun villages as 'traditional', with the potential for irrigation limited by the amount of water in the springs and the dry climate. Emigration was common; though while Christians emigrated for good, Sunni Muslim emigrants returned once they had enough money to buy land. Although expansion of agriculture in the region was limited, ownership of irrigated land was desired by Sunni peasants for livelihood, with water and land having a market value.

Spreading investment and running costs through share-holding groups was and is common in all fields of enterprise. People use resources for livelihood both as working capital that cannot be touched and as disposable assets. Resources are regarded as shares in access by group members, rather than at the disposal of an individual. Because individual group members on occasions need to realise an asset, assets of access to a means of production can be sold, although sales are usually either to another group member or members, or made with the agreement of the group. Group resources thus circulate within a group which may include men who have close links to group members through women. Individually developed resources are at the disposal of their developer.

Returning to the means of production in the countryside, the land, there were three main jural categories of ownership and access. *Mulk*, owned, land is developed land with buildings, orchards and gardens around towns and villages, and around springs on more distant village land. *Miri* land is state land over which the state has control of disposal while the population has rights of usufruct. *Musha'a* land is usually described as land communally owned by peasant villagers, practising arable cultivation and herding, and periodically redistributed among members. In the seventeenth century, the jurist Khair ad-Din ar-Ramli ascribed peasants the right of usufruct, derived from production, and the right to occupy land, derived from what peasants added to the land by their labour (Seikaly 1984: 404). In the district of Jerusalem at this date, land ownership was widespread (Ze'evi 1996: 135) by peasants and local notables and members of the governing elite. The latters' acquisition of land was based on privatising state land. Peasants in Jabal Nablus considered *miri* lands as theirs, and disposed of these lands as if they were private property by mortgaging, renting, or selling usufruct rights as reported by Doumani (1995: 157) from court records of the 18th and early 19th century court registers. *Musha'a* land did not exist in all regions. In Jordan, there was none south of Ajlun. Mandate authorities and some scholars viewed this system as an unmitigated socio-economic disaster; more recently, others have seen these villages to be successful at economic co-operation and in converting to intemsive agriculture. *Musha'a* systems varied from village to village, and villages with *musha'a* lands also had land registered as *mulk* and *waqf* (Antoun 1972;20). In Karak (Gubser 1973: 26), lands were held tribally, and each group divided their land among their families (Musil 1908: iii, 87–8), comparable to *musha'a* systems, where land was divided sometimes in relation to the numbers of mouths in each division of population, sometimes according to the pairs of working hands or plough teams available in each. Access to communal lands was in terms of consumption or labour units of consumption, according to the practice of each communal group. Mundy (1992: 79), in her examination of *musha'a* land in the village of Khanzira in al-Kura district of northern Jordan, finds that *musha'a* "appears as a response to the block imposition of agricultural taxes and as strategies for the minimisation of risk in agricultural production by the equitable distribution and rotation of land".

The legal category of *musha'a* lands ended when the Cadastral Survey and Land Registration started under the Mandate. This began in the north, and by 1947 had reached Karak; it stopped around 1957, and never went further east than Mafraq. Until land was surveyed, it was either tribal or state land. In 1972, tribal law was abrogated altogether, so all unregistered land became state land. As a result, apart from the towns and house lands, virtually no land south of Karak has formal title deeds. The continuation of the survey south is exceedingly slow, as every measurement engenders a dispute not only between users, but with various government departments and NGOs. For example, land at Fainan is claimed by three ministries and three NGOs, in addition to claims under customary law and traditional usufruct of local tribal groups. This confusion has not inhibited development of irrigated land, but sometimes initiated it as, for example, by the Rashaiyida.

Disputes over a resource take that resource out of productive use. A Nusayrat in Husn, on the southern edge of the Hauran, described how he used various pieces of land. He inherited 200 dunums of arable land from his grandfather's estate. His grandfather had had 4,000 dunums, "but some of this land went in brideprice, some inside the Nusayrat, some to the Ziyadna with whom we marry extensively. Now we have only enough land for one sort of crop each year; we have a shareworking arrangement with people from the village. I bought land for a garden, where I grow fruit for my family with a hired labourer. I have another piece worked by poor people from Husn, who don't have land. There's no partnership, they use it to support themselves and I take nothing. I look upon it as *zeka* (alms for the poor). There's more land near Aydin, which we've always had but never used much, and I discovered a Bedu had been growing crops on it for three years. He claimed it was his by use, and I said 'but I hold it by deeds'. I tried to make a *sharaka*, a partnership, with him; at first he refused, then he agreed, but he hasn't done anything. This year, the third since I tried to make the arrangement, I ploughed the land myself. The Bedu, who'd been planning to plough it himself, was angry, and complained that we had made a *sharaka* so I shouldn't have done the ploughing, and anyway, it was his by right of use. I don't want to go to court, I'd rather come to an arrangement or have it sorted out by a third party."

None of the three generation family makes a living from their land; all have been, or are in government service, mostly in the army, air force or teaching, so the retired all receive pensions. One of their senior men said "we don't really have any land left. We lost interest some time ago, and let the people who were working it have the land. What can you do with six dunums, one and a half dunums? That is all you end up with if you hang onto it, because of inheritance and marriages. It's better to cut loose and do something else, and we went into the army."

A Sharafat family of the Ahl al Jabal described a land dispute originating in former generosity to families without land who now refuse to return it. "We haven't sown for four years because of a land dispute. We registered this land in 1948, under Glubb. We used it just like we used other land; when we were there and conditions were right, we'd scratch the surface, sow, go away with the animals, and return to harvest. We didn't register all the land we used, just this bit. We settled here during the fifties and sixties, but as we were in the army the land wasn't important. When Sharafat from Syria arrived after the Ba'ath took over, we let them use the land. Some of them have registered our land in their names with the government. They say that registration under the Mandate wasn't lawful and anyway Glubb registered land as tribal land for the Sharafat and not in individual names; or they say that the land was government land and not tribal land (so non-use meant it was free for re-registration). We say all the land here was Sharafat tribal land. We registered our bit as our family's share of the *hamûla* share of the *ashîra* land. We had the foresight to do this. Other Sharafat registered other bits of tribal land, and those in Syria could have registered Sharafat land there. We see this bit of land as not only for his children and my children, but for our whole *ibn 'amm*. The *ibn 'amm* has more people now and needs this land....... We have pensions, and we live from our flocks, and our sons serve in the army and police, or are teachers and work in the university, but they will need our land."

It is sometimes considered by outside commentators that the Land Registries in each state should provide an effective solution to certain sorts of land disputes. In Chapter Five, local comments should have made clear that files in the Land register record only information at a particular time, it is like a series of snapshots of land ownership rather than a continuing film recording actual practice. As well, the Land Registries demand information presented in a specific way, so local practices are reassembled into an official form. This process has been documented by Mundy (1992) for Ottoman North Jordan; but comments made by people in Karak and Dana would indicate that current Land Registries record local agreements over land parcels rather than imposing a national standard.

Historically, the abandonment of cultivation has been attributed to over-taxation and oppressive measures against peasants by state agents as tax-collectors or for military conscription, or by bedouin, or from 'a lack of security'. Insecurity from disputes outside local control normally meant a local population moving to alternative resources along family networks. Disputes over family assets result in (temporary) disuse, but each family has alternative resources open to it in its own right and by using networks. A succinct appraisal of reported land abandonments by either category is difficult to achieve, since neither peasants, local agents of the state, nor bedouin are simple groups as each can be one or both of the others. In addition all parties actively pursued their own interests as, for example, shown by Singer (1992: 70ff, 120ff, 130) in her analysis of sixteenth century Palestinian peasants and Ottoman officials. Peasants did abandon lands; sometimes they "reverted to a nomadic existence among the tribes with whom they had sought refuge from oppression...... others .... took up alternative occupations, becoming either traders, cotton ginners or ordinary muleteers" (Seikaly 1984: 406) in 17th century southern Palestine, or moved to other villages and landlords as in the Hauran of the early 19th century (Burckhardt 1822: 221–2). The current shift to urban-based employment and part-time agriculture seen in Jordan (Mundy and Smith 1991) reflects an increasing population and the need to maintain family land while participating in and getting livelihood from the wider economy.

Land plays different roles in how people manage livelihood. Abu S., of the Faris Majali of al-Qasr, is a farmer; "We, as the descendants of our grandfather, own a thousand dunums (100 hectares) of land in four lots on the plateau, and a garden at Ain Jubaiba. Most of this land is our share of the blocks of Faris land by inheritance and division, and we have added to it by purchase. My grandfather bought the land at Ain Jubaiba from a Fuqara at the beginning of the century, and registered it under the Turks. We bought a small piece of land on Jabal Shihan from another family of the Faris; they needed the money, they are very generous, always giving feasts and entertaining. Our land is shared between all my brothers, but my father left me the land at Khaima as mine outright because I am the one who manages all our land. We all have equal shares, but that is my extra share because of my responsibilities. I've bought more dunums adjacent to my land at Khaima for myself and my children. My eldest son is in the Ministry of Education in Amman, and I used to work for the Customs – I still do odd periods with them. My brothers live and worked in Amman, and take their shares in cash. I've got a hundred sheep, in with my cousin's flock of four hundred. The land is worked with sharepartners; A-A al-M, who owns the tractor, and Abu S, a Palestinian. A-A is ploughing the Mbayyidin garden that borders our block of land on the west today. We have the same shareworkers year after year, a Ghawarna family, good people, but they couldn't come this year, they were too busy in the Ghor and with their sheep."

This Ghawarna family, members of a three generation *ibn 'amm*, were living in 1990 in three tents at Tadun. Each tent housed a household; tent 1 was Abu Z's, tent 2 was Abu Z's father's sister's husband's brother's household, and the third was Abu Z's sister, husband and children. Abu Z and his uncle had separate *sharaka* agreements, each for 150 dunums. Abu Z's father's sister's husband had his sheep and goat flock present. Abu Z's "animals belong to my brother, myself and our father, and they are in the Ghor, herded by my brother. Our agreements with Abu S al-Majali split the wheat crop fifty-fifty of the gross product, and we provide ploughing, seed, harvesting and threshing; for the other crops, Abu S gets 40 % and we take 60%. The *sharaka* excludes rights to water in the cisterns on the land; these remain with Abu S but we've bought two as a separate transaction. We, the *ibn 'amm*, have always owned land in the Ghor; we've sold some, we've bought some, we rent some, we sharework some, some of our own land is shareworked."

Some individuals and families have a longterm policy of buying agricultural land like certain Amarin families at Hmud. During the twenties, the Haddadin tribe moved to Madaba and sold their land at Alleyan to Amarin. Individual Amarin who saw their futures in the professions sold their shares in land to Amarin who concentrate on farming. Such choices are made by members of all land-owning families; in addition, the division of land from inheritance can result in such small parcels that it cannot support the inheritors. One group of five brothers co-owns a small area of land

and each has its use for a year; only one brother lives locally, and he lives from a clothing shop in Karak, an interest in a factory in Tabuk in Saudi Arabia and other activities, the other four are all employed and live outside the area.

In explaining the uneven nature of land ownership among group members a Majali noted, "land is always spread unevenly among families and members of a tribal group. Land is inherited *per stirpes*, and families have varying numbers of children. Land moves between families because of gifts at marriages, sales and debt. It's always been so, landholding never stands still." Women now inherit farmland among Majali, while in the past they inherited building land and buildings; "the land is hers, she can do what she likes with it. Sometimes on marriage she will make it over to her brothers in front of the Qadi (religious judge), but she need not, and she can leave it to her children even if her husband (and so her children) are of a different tribe." Wahlin (1993a and b; 1994) examines land ownership and inheritance in the northern al-Balqa of Jordan.

The land use of a Beni 'Amr family illustrates changes over time. The former arable land on the western edge of the plateau, owned by the household head and his cousin and inherited from their grandfather in whose name the land was registered, has become building land for their older sons and a large olive grove and garden. This garden began in 1968. The household head acquired title to 127 dunums of government land on the wadi slopes which grows barley, melons or tomatoes. He often shareworks a tomato crop on Majali land in the wadi, and he and his sons rent land for tomatoes grown by shareworkers. He buys a wheat or barley crop for harvesting in some years. He bought poor quality land, now rezoned as building land because of an expanding population, at the far end of the village from a Majali, for houses for his two youngest sons and eldest grandsons. His eldest son bought a dunum of land by the main road, and built a block factory. When the family had built their villas, this was sold. Ten dunums of farm land between two nearby villages is a recent acquisition, possibly as a result of a redistribution of residual shares in family land. This land may become part of a new village, and they are considering building a block of shops with flats above and a commercial orchard behind.

The aim is to provide each son with a house built on owned land, and to provide holdings of agricultural or building land to supplement the diminishing individual shares in existing family land. Finance came largely from the eldest sons' foreign secondments, army gratuities and grants, and credit. All the sons served or serve in the Jordanian armed services; the two eldest retired a few years ago. Initially, they bought a bus and service licence. After a few years they sold the bus enterprise and bought a tractor, plough and pickup for agricultural contracting, and a sheep flock. There was little profit in agricultural contracting owing to competition, so they sold the tractor and plough. The succession of short-term flocks were profitable when the Saudi market was open to Jordanian sheep dairy products and breeding sheep. The eldest brother traded small quantities of *samn* and *jamîdh* in Tabuk, maintaining family connections and making good prices. After the Gulf War this was impossible, and the brothers gave up sheep-herding. One bought a small goat flock and share-farms tomato and melon crops. The other bought a tobacco distribution franchise from the state and has a shop in al-Qasr. Each brother's employment, enterprise or pension supplies his own household, while any one in deficit is helped by those in surplus; all contribute to their parents' support and receive a share of the proceeds of arable crops, olives and grapes grown on family land.

Olive and fruit-tree cultivation fits well with part-time farming. Parcels of land are smaller as a result of population growth and inheritance through shares. For many, land is no longer the main producer of livelihood; people have employment or pensions, and businesses. But land maintains identity and standing, and is a source of value, so people are reluctant to sell unless necessary, or until the amount gets so small it is virtually useless. With the provision of mains water, contractors to dig cisterns, and manufactured fencing, a small parcel of arable land can easily be made into an olive grove and garden. The garden supplies the household, while olives provide an income from the oil, sold to government private presseries springing up everywhere; a tank of oil fetched 60 JD in 1992, and 40 JD in 1993. The work of ploughing three times a year and pruning may be contracted out to a cousin or done at weekends by the owner; harvesting is a family affair in late October and November, or Egyptian labourers are hired. Such changes in land use are common (Mundy and Smith 1991), but a rapid concentration was seen on the plateau northeast of Fagu'a, triggered by the appointment of a local Beni Hamida as the head of the regional Agricultural Development Bank.

Lancaster (1981) described how many Rwala shifted from camel-herding and service provision to sheep-herding, service in state armed and security bodies, employment in oil companies, and used entrepreneurial opportunities afforded by the differences between states' fiscal and trading policies. Since the seventies, Saudi Arabia's emphasis on agricultural development has meant that Rwala, like others, have participated to ensure their ownership of tribal lands in areas with underground water, like al-Juba and the Wadi Sirhan. At the same time, educational opportunities mean that many tribesmen enter the professions, service industries and business. Teaching and medicine are the only opportunities open to the new generation of educated tribeswomen.

The core of a three generation *ibn 'amm* lives in a small village, where they own their share of tribal land and have gardens, "but these gardens don't keep us. The soil isn't good, and we would never have developed them if the government hadn't given grants. We live from our enterprises or from employment. I have a business supplying immigrant labour, and my garden is for pleasure. One of my sons is a doctor and studying to be an eye surgeon, the second is finishing studies in telecommunications and will be in charge of the telephone company here. My elder brother has a pension from the National Guard and has sheep and a garden. His eldest son looks after the gardens. The second has companies in Sakaka, he had a convalescent home, and now he has a centre where people learn computing and he provides office services like typing, translations, and draughting plans and drawings on computer. The third son is in the Ministry of the Interior, and the youngest is a pilot. My third brother lives in Turaif, he's an accountant for the Police and deals in building land; he retires soon, and he's thinking of going into sheep – his wife's family have sheep. If he does, my father's garden and mine here might grow barley and forage for the sheep, as we have centre pivot sprinklers. My youngest brother is in the Police and manages my father's garden; they get fruit, olives, dates and vegetables, and send a bit to the market in Sakaka, and they grow barley for the sheep. My father has a government pension. All the women of the family have property of their own; some have property or building land in Sakaka, as their share of their father's estate, or they inherited it from their mother; and there are payments due to a mother when her daughter marries; or they sold their animals and bought land or gold. Two of my nieces are teachers, one in the girls' secondary school in Swair, and another used to be a teacher but now she lives in Qatar with her husband who is an engineer for Air France there. It's mostly our family that live in Nathaiyim. Apart from us, there's my sister and her husband, and a cousin; I've bought their land, they could see no use for it. I've put a centre pivot system on it, for wheat and barley but there's no real profit in it."

Some decide to sell land because they see their preferred livelihood to lie in other enterprises; others sell in hard times to ensure their survival, as with Beni Hamida in the 1940s drought when the going rate was a dunum of land for a sack of dates. Others sell land to raise money for compensation, like an 'Ata'ata who had killed a man and had to sell all his share of family land to raise the money due to the victim's family; he and his family now live on his wages from casual building work, the proceeds of share-working arable land, and a small goat and sheep herd. Some lose land from their commitment to fulfilling obligations of generosity, or from being unable to fulfil contracts of protection they have undertaken. Others lose land to creditors, often urban merchants who used moneylending to assure supplies of agricultural produce on favourable terms or to acquire the land itself. Others lose land from disputes over their ownership or preferential access, while yet others find the establishment of claims to land difficult to prove. An Amarin at Fainan has tried for several years to establish a claim on land on the north bank of the wadi, consulting elderly men of the area for examples of use by members of his group in the past, but has been unable to find any; he does have claims on family land to the south but his share is small and distant from his employment by an NGO. A Christian family from as-Salt owned a piece of inherited land in Fuhais; "my grand-father planted it with olives and vines, and because the land is steep he carried all the water for the plants on his back. All the trees got burnt in the Civil War, it almost broke his heart. We didn't know we had to re-register the land in 1972, and when we went to replant we discovered it had been given to one of the Princesses by her father the King." The inheritance of shares in land and other property, and the transfer of land at marriages, also results in sales and exchanges over time. This continual market in land reflects similar movements in flocks, buildings and all other productive enterprises.

Credit and money-lending were one of the two main forms of investment in the countryside, the other being share-partnerships. Rafeq (1992: 323) points out that the use of credit and money-lending were socially accepted, economically necessary and legally binding in early eighteenth century Damascus and its countryside; at this date, debt was 'legal debt' while the 'fair loan' of the early Ottoman Sultans rarely appears in the extant court records. In Rogan's (1992) examination of money-lending and capital flows from Nablus, Damascus and Jerusalem to the district of as-Salt in the late 19th century, both 'legal debt' and 'fair loans' were common, and "by definition, interest-free," while "a creditor was entitled to lawful gain on his/her capital as in any other lawful business deal" (1992: 242). Creditors had to remain within the law, and debtors had the means to extricate themselves from debts through family support networks and property. Creditors, whether urban merchants, artisans, villagers or relations, were under informal social pressures not to call in debts unless necessary. In spite of these safeguards, land belonging to tribesmen could move to ownership of urban merchants and moneylenders in relatively significant amounts. In al-Karak, the Beni 'Amr lost land to Karak town traders during the eighteenth century (Musil 1908: iii, 86); in the mid-twentieth century Gubser (1973: 28) reports Karak tribesmen losing lands to merchants from Hebron, Gaza and Damascus. Ben-David and Kressel (fc: 31), in their study of the market as the axis around which Beersheba developed in the Mandate period, describe grain merchants encouraging bedouin landowners to accept loans so as to acquire their land.

The extension of credit through money-lending and some forms of partnership enabled the development of the mercantile commodities of Jabal Nablus from textiles and cotton to olive oil and soap, with grain a constant important factor (Doumani 1995). Partnerships were crucial for trade, manufacturing and agriculture, and were made between merchants, merchants and landowners, merchants and peasants, landowners and peasants, merchants and bedouin, and peasants and peasants. The soap industry, based on olive oil from Jabal Nablus and Jabal Ajlun, *kilw* from the east bank of the Jordan local artisans for its manufacture, and on transport to Nablus and to markets in Egypt, Damascus and Hijaz for its distribution, illustrates the complicated networks of supply, credit and distribution involved. The soap industry was capital intensive as the fixed assets were expensive to purchase, construct and maintain, and soap had to be manufactured in large amounts. The industry was divided between the merchants who got the oil, commissioned cooked batches of soap and operated the regional trade networks, and the factory owners who had the buildings and employed the cooking teams. The merchants' investment was three times greater than that of the factory owners (Doumani 1995: 193), and two or three years passed before there were any returns. Partnerships were everywhere in the industry, and family *waqf*s were used to protect property from arbitrary confiscation, sudden downturns in family fortunes and fragmentation from marriage and inheritance. By the early 20th century, the average worth of a rich soap manufacturer was between 10 and 15 million piastres or £70,000 — £105,000 if 143 piastres equalled £1 (1995: 183–214).

Partnerships were important in all forms of economic activity (Firestone 1975). Sheep production using partnerships were noted for Qalamoun (Thoumin (1936: 150ff) and the Hauran (Burckhardt 1831: 17–8; Issawi 1988: 301–2). Partnerships continue to be important in livestock raising in the *badia*. Urban merchants investing in sheep use partnerships as do town-based large scale tribal owners. Details of partnership agreements vary in part according to where profits in sheep come from; at some times and for some management systems these are in milk, or in wool, as in the 1930s, or in lambs for meat. The effectiveness of a partnership depends on the degree of supervision given by the general partner. Most partnerships in livestock are for at least one rearing year, counted from shearing to shearing, but some owners make short-term partnerships where animals are consigned to a second party/ies for sale or grazing in areas inaccessible to the owner. Although these arrangements are spoken of as partnerships, they may not correspond to a strictly legal terminology; the consignee is responsible for the well-being of the animals under his care, he therefore contributes more than mere labour since it is his networks that provide for the grazing and water or the market outlet, and his share comes from a percentage of the profits realised at the end of the partnership. Métral (1989) refers to the multiple contracts between urban merchants, arable farmers, sheep-herding tribes, and Sba'a in production and distribution around Sukhne as long-standing and adapting to the new conditions laid down by the Syrian state. Partnerships between Syrian governmental agencies and foreign companies, usually Saudi, Qatari, Kuwaiti or Emiri, operate on the basis that the Syrian partner provides the land and fixed capital like water and electricity, while the partner provides everything else. The Syrian/Saudi Livestock Company southwest of

Palmyra produces sheep's milk products for the Saudi market, using a local co-operative for labour. Long-distance camel transport in 'Unayzah used *buda'a* partnerships (Altorki and Cole 1989: 71), where men provided camels to a transporter for a share in the money earned by the camel. Ben David and Kressel (fc) describe the three-sided partnerships and contracts between grain merchants, peasant cultivators and bedouin land owners in the Beersheba area under the Mandate; the movement of peasant farmers to becoming merchants, bedouin landowners to being merchants, and merchants buying land; and the integration of livestock and grain trading.

In urban enterprises, partnerships between the provider of a building and its user are an alternative to renting and tenancy, while some partnerships are re-negotiated as tenancies and vice versa. An enterprise may have joint owners as partners, who then employ staff. In al-Ruwaishid, one of the first shops was opened in 1933 by Abdullah al-Ramthawi in partnership with al-Aurens ash-Sha'alan of the Rwala; the partnership was re-formed by Abdullah's son/s on his death, and again following al-Aurens' death with a nephew. The current partnership of the son and nephew of the founders also own a money-changing business in the town. Another Sha'alan partner owns the main hotel, run by an employed manager. Another shop and restaurant is owned and run by Abu R., a Christian whose grandfather, from Anjara, worked for the Iraq Petroleum Company in the then H4. When he left the company in 1936, he started a shop and a small hotel, which later became the restaurant. Like the other trading families in ar-Ruwaishid, mostly from Ramtha or Ma'an, Abu R's family never lived there, residing in their towns of origin. R, with his brother, paternal nephew and maternal nephew are starting a company selling medical disposables; the maternal nephew will do the work, while the money comes from the other three, and profits will be shared. The family also own shops nearby, rented out on a monthly basis. An earlier enterprise was a share partnership in sheep between Abu R and his brother as one partner and a shepherd; "but it was a loss, it was always our sheep and lambs that died."

Similar multiplicities of partnerships, shares, tenancies and leases exist in all spheres of economic activity. Partnerships, since they embody significant contributions to the enterprise by both/all partners, reduce costs and risks. Each partner, by having to put up less capital, can spread risk by investing in further businesses. A tailor's establishment in ar-Ruwaishid is a partnership between the owner of the building and the tailor who owns his sewing machine, scissors and pressing equipment; the owner takes a larger share of the profits since he buys the bolts of cloth. A tailor's business in al-Qasr divides profits equally between building owner and tailor (who supplies the equipment), and customers supply their material. The owner of a furniture-making business in al-Qasr owns the building and machinery, buys the wood, and hires a carpenter and his assistants. A similar business in al-Qadisiyya is owned by a partnership, one of whom owns the building and the other the machinery, and the partnership employs a carpenter, while customers provide the wood. A car-hire business employs an office-manager, who is also a partner in a shop selling paints and wall finishes; he runs this shop in the evenings, while his partner provides the rent and stock. Many small businesses, small supermarkets, barbers, laundries, bakeries, restaurants, repair shops, metal and wood-working workshops, mechanics' workshops and so on in small rural towns, are started by men when they retire as retirement from government employment brings a pension and usually some sort of cash lump sum. It brings in some income, and gives the owner a place to go and a role in the community. Some are partnerships, others employ staff, often foreigners, Syrians, Egyptians, or Palestinians, and in Saudi Arabia, from the Indian subcontinent as well.

At ar-Risha, the buildings of the *sûk*, together with water by tanker and electricity from a generator, were provided by the shaikh to known traders in return for rents; as a known trader, protection and assistance with papers was assured, while credit was extended to members of the Sha'alan. One trader, dealing in tentcloth and general goods, commented that although he was owed several thousand pounds the debt was more than covered by the entitlement to trade and the protection he received. This position is similar to that described by Musil (1928a: 270) for the merchants camping with the Rwala in the early part of the century.

Women are quite often general partners in share partnerships, although the face-to-face dealings with the active partner is usually through a male agent. Women's shares in urban property, housing, shops, offices, gardens and orchards are relatively common, and known from historical records for Palestine (Ze'evi 1995: 166–8), around Damascus (Rafeq 1981: 665; 1992: 307), and in al-Kura (Mundy 1992: 70), while Doumani (1994: 154–5) comments that while in Jabal Nablus women inherited property they often did not press their claims. Women's ownership is unlikely to have been limited to these places. Money lending by women was known, and continues as in 'Unayzah (Altorki and Cole 1989: 158–161). Women putting money into large-scale family investments in arable agriculture are mentioned by Métral (1993: 208). Women's property in livestock and jewellery is wellknown (among others, Lancaster 1981:113 who notes the selling of jewellery by women to finance smuggling operations, and their ownership of urban land). Some Rwala women, like other tribeswomen, own urban building land; others have small businesses, dress-making or selling school materials, sweets, cigarettes and cosmetics from a chest in the tent or house. Many own animals and/or gold in their own right, and it is almost unknown for a woman to be without property and income. Several Sha'alan women have business enterprises. One owns a garden in the Ghuta of Damascus, inherited from her father, and plans to develop the garden as a cow-dairy, initially selling milk to the local co-operative but later setting up a dairy business to make ice-cream and other milk products. Another has boutiques, beauty parlours and a gymnasium in Riyadh. The first woman's father registered cars, pickups, buildings and land in the names of female relatives, who were unaware of these gifts until his death. A Sardiyya woman invested several thousand dinars in providing the well and electricity for her son's irrigated garden, and is a shareholder in the enterprise. A Majali woman makes *samn* and *jamîdh* for the market from the milk of the family flock; part of the profits are hers outright. Seasonal dairy businesses by women in the *badia* and the countryside are common, either using family flock milk or by arranging a supply of milk from another flock through a partnership (and see Abu-Rabia 1994: 75–8 for women's dairying businesses in the Negev). Some women in al-Karak and al-Ruwaishid have small dressmaking and embroidery businesses, and Altorki and Cole (1989: 196–7) describe a larger scale dressmaking partnership run by a woman and her uncle. In the past, women of Christian tribes in al-Karak and women in the Jabal al-Arab made pots for sale or exchange. Weaving rugs for the market is an activity of some Beni Hamida women, but most women who weave rugs intend them for household use or as gifts to sons and daughters; the rugs, dividing curtains, and cushion covers are assets that can be sold when cash or an alternative asset, like jewellery or animals, is needed. Some older women are traders, taking orders for women's underclothing, children's clothes, cosmetics and household textiles, or visit tents and village houses where they have networks with their goods. Altorki and Cole (1989: 142–161) describe the women's market in 'Unayzah. Some tribeswomen in the *badia* have reputations as finders of lost or stolen camels and other animals, and as healers; these women are not paid for their services but given presents of material, clarified butter, or a lamb, or feasted. Women owning and running garages in Sukhne, Qaryatain, and Qutaifa are mentioned by de Boucheman (1939: 91–92), and seen as inheritors to the tradition of women running *khans*. Many women are now employed in the rural small towns of Syria, Jordan and Saudi Arabia. In Saudi Arabia, most are teachers, while in Jordan and Syria, opportunities are more varied from teaching, nursing, security staff, and in the post office. It is rare for women to practice as doctors, dentists, architects, lawyers or accountants outside the cities.

Credits and debits are further aspects of economic enmeshments. Doumani (1995) and Rogan (1992) examine court records for credits and debts between merchants and peasants, and between merchants. Many credits and debits are less formal and unrecorded in legal documents. Altorki and Cole (1989: 147–54) show the importance of credit in the women's market at 'Unayzah, where the extension of credit to those who are trusted underpins both selling in the market by the women traders and their purchases of goods from male suppliers. The less-structured credits and debits of everyday activity enable much short term resource management, while share partnerships, themselves open to sale and purchase, are usually concerned with longer term investments. Z wanted to buy five tons of *alaf* from Abu A; Abu A had five tons but would sell only three as he needed the other two. Abu A set a non-negotiable price of 160 JD per ton. Arrangements for payment by Z took an hour of amicable and teasing discussion. Z put 120 JD down in cash, and transferred to Abu A a series of debts owed to him by four or five other people in money, sheep or goods valued at the balance of 360 JD. Both men were in ar-Ruwaishid, while at least one debt to be collected was in Turaif and another in Za'atari, a village east of Mafraq.

Abu A trades in sheep and trucks as well as *alaf*, making profits from his sources of supply in Saudi Arabia and from his proximity to the Iraqi border, selling to Jordanians without such access. Jordanian tribesmen ask him to find them a Saudi registered (significantly cheaper than a Jordanian registered vehicle) lorry, tanker or pick-up; his extensive networks extend deep into Saudi Arabia and he is a Saudi citizen. He is successful because he supplies a need, enhanced by his reputation and by his extremely wide range of acquaintance, acquired over time and starting when he was a young man employed by the late Emir Fawaz to vouch for Rwala using the al-Tinf border crossing. He is, of course, not the only person in this position, and similar patterns of trading in livestock and transport are followed by many.

At the moment such trading is usually profitable, and profits are invested in sheep, gardens, building land in towns and villages, or agricultural land in Saudi Arabia, once the traders' core resource is assured. Within Jordan too, payment for goods by transferring debts or shares is relatively common, since credit and debt between members of a wider domestic unit and the *jamâ'a* are widespread, and shares in cars, pick-ups and small scale enterprises are common – as were the shared ownerships of camels and horses in the past. Partnerships, credit and debt are and were as widespread in the *badia* as in the towns and countryside, and build networks intermeshing the three.

Investments in enterprises need not be large. It is possible to buy a crop for harvesting, the growing fruit of a date palm, a share in a car or camel, or to sharework the milk of a small flock. This scale suits many people; they have some extra time or a bit of spare money. Such enterprises may supplement the main livelihood, or be part of living from a multiplicity of small scale enterprises. Part of the attraction of small scale opportunities lies in some employment patterns; some work for two, three or four weeks and then have a few weeks off. This pattern of two or three week's work followed by leave enables many families with men in the army to cultivate land as arable, irrigated crops, orchards, or olive groves with the help of the retired and women. If serving in government abroad, a pattern of six month's away and six month's leave is usual. Abu B has a garden at Ain Lazhha, and works in army security in Jordanian embassies for half of each year; he grows fruit, supplying the family with fresh fruit and sufficient to dry, and he sells the rest, making about 500 JD a year. He share-worked 25 dunums of poor arable land in a dry year at Qadisiyya for Abu I 'Ata'ata, from which he expected to harvest 500 kgs. Half was his, which he would sell to buy flour, and he took half the straw for his sheep and goats. When in Jordan, he was paid 150 JD a month, which was adequate, and when serving abroad, he earned more.

People assume that not all enterprises will succeed, nor that all debts will ever be fully repaid. Three members of a Christian Palestinian family had a desk-top publishing business; they owned the equipment, bought with family money, and rented an office. The business survived for five years before closing for lack of work. They considered it had been a success as it had supported them for five years and they still owned the machines. A Syrian pedlar, with a pack of women's underwear, children's clothing and towels, travelled in the Karak area on foot. He stopped wherever he saw groups of people in gardens or women by their houses. He sold little, as his goods cost double similar products in Karak itself, but refused to lower his prices. When asked how he made a living, he replied that he had no expenses. The textile factory owner brought him and four others to Jordan, and collected them at the end of the summer; he walked or was given lifts; he was fed wherever he stopped, and slept with his evening's hosts; "the profit for me is that my family in Aleppo don't have to feed me. Anything I sell is profit for the factory-owner." SM buys Jordanian registered cars, takes them to Saudi Arabia where he re-registers them, and brings them back to Jordan for sale in the *badia*. The profits come from there being no import duty on cars in Saudi Arabia while Jordanian duties are high, although Jordanian rates of duty and Saudi registration fees each fluctuate at short notice. SM buys cars because he likes the look of the vehicle, sporty models or flashy pickups, rarely has a customer in mind, cannot afford to keep the re-registered vehicle to wait for a good buyer, and does not keep in touch with border officials to know current import or registration fees. He buys high and often sells low. This is a second business for him, "something to do", as his main income comes from a share in a tile factory in Jiddah. MK, like all his family, served in the army; unlike his brothers, he remained in the ranks. His elder brother succeeded in getting M a posting in the Emirates so he could benefit from higher rates of pay. On retirement, SK and MK bought buses, but M's casual attitude to costs and timetables gave rise to friction. The buses were sold, and M herded family sheep, and traded sheep, but showed poor judgement. When the sheep were sold, S and M settled on separate enterprises. A cross-border trading partnership was set up between al-AT, who supplied the cash to buy the goods, and four nephews who did the actual buying, driving and distributing. The partnership ended in disarray when the four young men found that their uncle refused to pay them their agreed shares in the profits, but only a wage.

Many people say having an enterprise or shares in a group of businesses that can be accomplished by their own labours or with a small group is what they find satisfying. Expansion into a situation where the enterprise would depend on workers not of known family and background is not attractive. Many families have sleeping assets, often bits of land or buildings unused. SR herds the family sheep in the *harra*. He and his sons own two pieces of unused land; "I'm looking for someone who knows about fruit trees to work them with us in a partnership. But there's no hurry, we don't have the money yet for a well." GJ has a successful business supplying and servicing computers, and providing training; "I don't want to expand any further, as I couldn't know my customers or staff properly. What would happen to the company if I employed staff who let me down with my customers?" Here, limiting expansion is tied to the need for a good reputation.

An alternative to limitation is to 'hand on' to sponsored others, either as sources of supplies of similar goods or services, or as developers of produce by processing or adding value. 'Handing on' can create an alternative 'self', either by employing an agent (*wakîl*) who acts for the supplier or producer in a defined field of action, or by recommending a member of the close family, wider domestic unit or *jamâ'a* as a suitable substitute. Developing a product or service may necessitate employing labour and agents who supervise and answer to the owner/s. The decision to develop on a large scale may come from a desire for personal wealth and the influence that wealth now brings together with a slackening of wider family ties, or the intention may be to create wealth to be used in asserting wider family and tribal interests. An outstanding example of the latter was the smuggling by Rwala (and others), seen as traditional trading and transport services transformed into smuggling by hostile actions of nation states (Lancaster 1981: 91–5, 112–4), and which provided livelihood for many tribespeople at a time of hardship and a political focus in a period of change.

While many tribal families content themselves with modest livelihoods, some set out to become both wealthy and good tribespeople. M from the Z took part in the smuggling, and invested in sheep herds; he settled in Sakaka and set up a well-drilling business, and now has several farms, and a large number of other businesses locally, in other parts of Saudi Arabia, and in Syria. He is said to be the richest man in Sakaka, after the Emir, and has a high reputation among tribespeople as he and his family are perceived to uphold the values of hospitality and generosity as well as providing a focus of Rwala participation in the changed economic and political arenas. The family economic management is based on a large number of separate autonomous businesses rather than on a unified hierarchic structure. For many participants, economic action using partnerships and face-to-face relations is preferred since this respects the autonomy of others, and ensures that initiators of a business partake in it as members of family, wider domestic units and communities.

Families and wider domestic units encompass individuals working in varied scales, while individuals have enterprises which they operate in different ways. Generally, people distinguish between enterprises using wider family/*jamâ'a* labour and capital, and those depending on inputs from outside the wider domestic unit. The first type provide livelihood, with a market or exchange component and is multi-resource; that is, there must be products for processing or sale, or additional related enterprises like seasonal or share-labour, or contracting. Many of these small-scale enterprises are currently supplemented by wages or pensions. The second type are on a larger scale, necessary because the enterprise demands additional inputs of capital, labour and skills. Irrigated gardens, multiple centre pivot irrigation systems for cereals and legumes, the larger commercial sheep flocks, cow dairies, garage complexes, and cross-border trading are examples in the countryside. The degree of market involvement is different, but each system has a market component. The crucial difference is the amount of capital needed, physical plant and skills. While most individuals are capable of acquiring access to land and water resources, maintaining these, having the skills to produce crops or livestock, and to distribute produce, it is rarely possible for one person to do all simultaneously.

An additional important factor for people participating in economic life is reputation as a good person, largely achieved by being competent and honourable. Having enterprises that depend on secure supplies of produce or services implies a wide reach of acquaintance and information, and the ability to ensure delivery or fulfilment through jural processes if need be. In forward buying contracts, the supply is secured by early payment, collected, stored and then sold at an unknown price to the market. FF suggested to WS that they should form a partnership to forward buy wool from herders; they would buy wool in the late winter when herders needed ready cash to buy *alaf* and collect the shorn wool in early summer, storing it in FF's buildings until they sold to Syrian or Turkish merchants; profit would be the difference between the price they paid and what they later received. The price of wool varies from year to year, so the enterprise was speculative but should have been profitable. FF would contribute one third of the costs, the knowledge of sources of supply and his personal reputation to assure delivery, together with the stores; WS's contribution would be two-thirds of the money. FF saw customary law together with his personal reputation as sufficient to secure the contracted goods, while Doumani (1995: 168–9) writing of the nineteenth century Jabal Nablus regards the expansion of the Ottoman Code and Islamic courts for the restitution of debts to encourage the use of forward buying by urban merchants of rural products. Acquiring information and reputation takes time and political and jural skills at their widest. FF, like some others, has these skills in a high degree; many have them to a lesser extent, while yet others may have them in a reduced form or rarely use them. Further options are to inform such political and jural skills from customary morality or a more modern individual ethos.

Some families say they have no interest in politics or government. The N family of as-Salt are one such; "we are not a political family, except for Dr AN, who was the ambassador in Chile, and his son is with the Jordanian delegation at the UN. Very few of us are in the army, ministries or civil service. We're interested in trade, we always have been, and in land. We used to trade between Jordan and Palestine, but we always lived this side, first in Rabba, then Hisban where we still have land, and we arrived in as-Salt by 1800. The family was strong enough to acquire and develop land in and around as-Salt – Wadi Shu'aib, Sirru, Fuhais, Muhais, Umm Jauza, and Umm Zuwaituna. We sold our own products in Jordan, and traded grain, grapes, wine and sheep to Nablus, Jerusalem, Haifa and Jaffa, bringing back clothes, sugar and rice to trade here. In the late 1920s, 'AN was employed as a driver, and then he bought his own truck. During the 1930s his sons joined their father and they set up the N long distance international transport company. We have continued to expand until the present. Another group of three brothers own the Honda agency in Jordan. Almost all our children go on to higher education in Jordan or overseas, and enter the professions. We have lawyers, doctors, architects, and engineers." A Christian Palestinian family concentrates on the electrical and electronics sectors, while its women work as secretaries, accountants and architects. This family, like others, observes the rounds of visits at religious festivals, and on appointment or promotion to members of government that they know as individuals from former occasions. This is not to maintain or establish a client relationship, but because "we were neighbours. When my father wanted me to go to university but couldn't afford the fees, he asked X (a prominent politician) as a knowledgeable neighbour if he knew any way of managing it. X suggested a scholarship. When he knew I played basketball for my school team, he recommended me for a sports scholarship and I got one. We continue to visit his *majlis* on occasions, like when he was appointed Minister, because we are pleased for him. It doesn't mean we support his politics or vote for him."

There are differing ideas of wealth in the Bilâd ash-Shâm. One set is concerned with wealth as material goods, enterprises run for profit, and cash in the bank; the other sees wealth to be in social relations and reputation as a good man and a good family, and the acquiring of livelihood in an honourable way. For holders of the latter, wealth and participation in economic activity have important moral components; those of the former regard the use of moral referents as old-fashioned, uneducated, tribal, rural, unrealistic, and as a cover for failure – while often simultaneously expressing admiration for these views. A wider domestic unit may have members operating both sets of ideas, those of reputation as a good family in the enterprises associated with the family, and those of profit in new businesses in cities or abroad. The understanding of particular economic actions and results by initiators and audience swings between the two ideas as each interprets and re-interprets the others' statements and behaviour, and negotiates between shifting demands. There is a constant interplay between the generation of economic activity for the present and future well-being of the individual's immediate family, and the necessity to fulfil requests and demands from the wider group which, while lessening immediate resources for himself and his children, provide reputation and insurance. The response to demands for money, credit, work and resources is not automatic but measured on the known needs and reputation of the asker, the probable response from other sources open to him, and the likelihood of success.

Money or labour used for investment in the enterprises of others is "spare money or labour or resources, after you have made sure you can support your immediate family. You always make sure, God willing, that you have your family's livelihood secure. And this is possible because we have land and animals and shares, and employment or pensions. It is difficult for everything to go wrong at once. If things do go badly wrong, we reduce our consumption, call in debts, sell investments, switch assets, and if all this isn't enough, we call on our wider family members. Because families are so widespread and have parts in different countries, there is always at least one section of the wider family that will be all right. This is one of the main reasons for the wider family, and for members of each part of it, to be scattered in different places. Then, as we are a good family, God willing, we support each other in need. We would ask them and they would ask us, we would offer and they would offer....... Listen, our economics are not so much about capital and income, but about liquidity and hedging. We aren't so concerned with owning things outright but about being able to use resources on which we have claims. Because we are all members of families who all have some sort of assets and access, there are no people without anything for ever. Someone might lose his job or his land or his business, but in a while he or his children will have something else. And that is why we put so much importance on being a good man, having a good reputation. Wealth is as much or more in social relationships as in goods and money." This attitude makes sense of Finn (1878: ii, 182–4) talking of rural Palestine when he says "pauperism, as we understand it, is unknown" even with the extraordinary impositions of government taxation for the Crimean War. Surplus went first through the hands of local chiefs who took a percentage, partly used for hospitality and local investment. People owned their lands and houses, bought silver or gold jewellery as a store of value for their women, and buried hoards of coins. Analysing current Syrian industrial and commercial sectors, Perthus (1992: 224–5) notes that even with a cumbersome and inefficient government bureaucracy where regulations change constantly and are regarded as unreasonable, and with no legal security, private businesses at any level suffer very few bankruptcies. He found this surprising, considering the apparent lack of turnover or profits, but makes more sense if seen in local terms.

Similar concerns are evident in attitudes to wealth and its distribution. Tribal society is often described as egalitarian, assumed to mean members are or should be economically equal.

Tribespeople in fact differ widely in their possessions of herds, flocks, houses, tents, clothing, weapons, jewellery and so on. Such inequalities have been taken to indicate that the concern with an egalitarian ideology is a facade for actual relations of power, dominance and inferiority. Rural society in Syria, Jordan and Palestine is often described as embodying precisely these power relations (Weuleresse 1946; Antoun 1991: 4–7, both about Syria; Gubser 1973 for Jordan). But tribespeople and peasants say that ideas of equality refer to jural equality before God, that all things come from God and He is generous. This could be regarded as a public statement that obscures economic and social reality. However, the ideas of participants are a framework within which many construct and measure many of their actions for much of the time. In such a perspective, a man who has little but is generous within his capacity has as much reputation as one who has plenty and is generous with that, and both have higher reputations than a very rich man who is mean.

In the countryside and small towns there are poor individuals and families, supporting themselves and their families and fulfilling social duties with difficulty. Such people are often those with limited family networks and/or, since each reacts on the other, poor reputations for meanness, incompetence and laziness; misfortune and illness play a part. However, no-one starves or is without shelter and clothing, although they may depend on the generosity of others. Appearances may be deceptive. A three generation family, living in two small, tattered goathair tents, wearing shabby clothes, without coffee and eating grain and dairy products, own seventy camels and three hundred goats worth between £50,000 and £70,000. They enjoy a good reputation with endless visitors and an extensive social network. Here, outward appearances obscure relative wealth. In small Saudi towns, the public faces of housing and clothing appear similar but may hide relative poverty; "you know your neighbour doesn't have a regular salary coming in and depends on help from his brothers and family. I give *zeka* (2.5% of profits at the end of the year) to this man, and other neighbours do the same." A Rwala living in a small Saudi village has two building sites; on one he has his tent permanently erected over a metal frame, on the other he has a two roomed house of concrete blocks. He is retired from his job as a school janitor, and has a pension of 800 Saudi Riyals a month – £1,920 a year. He asked the government for more and received another 1,600 SR a year, which brings this income up to around £2,250. While his immediate family is extensive, the children of his first marriage are adult, most are married and employed in the Post Office, the local hospital and the National Guard; those of his second marriage are in their teens. The family could manage on the original £1,920, but "I thought I'd ask for more and see what happened." A considerable proportion is spent on constant entertaining[27] and social expenditures since the family is a nodal point in a series of dense and extended networks. There are continual calls for help with marriages, housing and flock purchases from the many nephews, cousins, younger sons and grandsons. There are also elderly female connections who spend long periods in the household. Women continue to regard cloth and textiles as a necessary part of generosity to women guests. With a dress length costing from £30 upwards, continual long stay guests drain off money. The reverse is that guests bring gifts or return gifts when they are visited in turn, so there is in effect a constant circulation of cash bound up as goods. This family looks poor from its housing, lack of transport, and few resources of agricultural land or flocks; and others regard it as poor because its livelihood depends on semi-skilled employment and pensions, it has few developed assets of herds, land, or education and professional networks. Its members, however, have a good reputation as hospitable and as traditionalists.

Poverty and wealth, even in relative terms, are difficult to establish since how people talk about the two alter depending on a variety of contexts. If a government survey is undertaken to establish quotas for subsidies or relief works, many say they are poor since it is considered the function of government to provide for its citizens, and they want their share. These same people say on other occasions that they are not poor – poor here meaning they have a sufficient standard of livelihood to maintain themselves and fulfil their social duties. Material wealth and poverty is separated from social and moral poverty, so that people whom outside observers describe as materially poor see themselves as socially and morally rich, and their observers to live in conditions of moral and social poverty. The apparently rich with villas, cars, land and assets often

---

[27] See Abu Rabia 1994: 111, 123 where all 20 surviving male kids of a man's flock were killed for guests or for vows, while 3 male lambs were killed for guests and 3 given as gifts out of 22 surviving.

regard themselves to be living at the full stretch of their resources. This is due partly to the particular part of the family cycle they may be at, since a family with teen-age and young adult children is at its greatest demand on resources for the education, establishment and marriage of its junior members. In addition, these families have greater expectations for the futures of their junior members, and are subject to greater demands on family assets from within the wider family and from people attracted by their reputation for generosity or help in pursuing claims or injustices. Such families illustrate well the dilemma between limiting resources to immediate family members, and so not having a wide reputation, or expending resources more widely and acquiring a reputation, but possibly denying material opportunities to close members. The question exists in all families as to how best to forward its interests, and to tread the balance between material wealth and reputation as a good family.

Food, clothing and housing are both observable aspects of consumption, and crucial parts of internal distribution within and between families, especially at marriage and death. Everyday consumption, including casual visitors, is distinguished from the formal, public occasions, with feasting of a guest, weddings, funerals and memorials. People like to produce as much as possible from within family resources, so they know the quality of the food. Many say they are unwilling to buy meat, dairy produce, vegetables or processed foods because "I don't know how the animals have been fed, how the crops were grown, or what has been used in the making of bought ready prepared foods. You hear such things about artificial fertilisers, chemicals and additives. In the past we ate dates and milk, and sometimes bread or *'aish*, and meat at feasts. And we were healthy, much healthier than children now, who always have colds and coughs. People feel ill all the time now, they take pills and have operations. It must be all the bought food we eat, because we live in towns. Of course, our food comes from our garden and the sheep, so I know it is wholesome. Eat up, eat up....", urging one to eat grilled chicken and *hummus* from a restaurant, bean and tomato stew and salad from shop-bought vegetables, and bread from a bakery. For large feasts, one can hire a catering service "who provide all the equipment, everything down to the soap and towels for washing afterwards. You provide the sheep and rice and coffee and things. The young women don't know how to cook for feasts anymore, and cooking for large numbers (from fifty to five hundred men) is difficult now we live in houses, there's no room, and we don't have the numbers of big cooking pots." Other families, especially those who give feasts for visiting dignitaries from other tribes or from local or regional government, have storerooms full of huge saucepans, serving trays, dishes, coffee pots and so on.

Private family food is based on what is available seasonally, and in the villages and the *badia* what the family has in its stores and garden with some bought goods. People eat two or three times a day; breakfast may or may not be a meal, or it may elide with lunch; the other meal is at sunset or later. In the *badia*, breakfast may be remains from supper, or fresh *sâj* bread (or *masliyya*, a soft poured batter, in a few tents), with a selection from fresh sheeps' butter or *samn*, yoghourt or soft cheese balls, grape or date *dibs* or apricot jam, tomatoes or cucumbers, olives, dates with warm butter, fried eggs, *hummus* or *foul medamas*, or *halâwa*. Lunch may include some of the fore-going, with the addition of a vegetable stew, fried potatoes, or bread and *samn* or bread and tomatoes or fruit. Supper may be rice with a yoghourt and onion sauce or rice with hot *samn*, perhaps with spring onions. If someone has gone to the town to collect *alaf*, or to collect children from school, there may be shop bread, chicken or pickles. Village food uses more vegetables and fruit, olive oil as well as *samn*, olives and more lentils, less fresh butter, yoghourt and buttermilk, *burghul* instead of rice, and *tabûn* bread as well as, or instead of, *sâj* bread. There are regional, local and family variations. Dates are far more common in Saudi Arabia. In Fainan and the Wadi Araba, as in the Negev, *fatît* (bread soaked in liquid, cooked with a few lentils or onions and served with *samn*) is standard fare, whereas for the Rwala and the Ahl al-Jabal, *'aish* (rice, barley, *burghul* or millet boiled and served with yoghourt or *samn*) is the basic, with *fut-wa-kul*, bread soaked in yoghourt or *samn* and cooked with onions or lentils, and *khmai'a*, bread boiled in water, buttermilk or yoghourt and covered in *samn*, other standards. Some families use onions and/or garlic a lot, others do not; some use salt and pepper or cumin in abundance, others sparingly. Syrian influences are strong in some families, with *kibbeh*, stuffed *cousa*, and stuffed vine leaves appearing in season. Some families are actively interested in food and regard eating as a pleasurable activity, others regard food more as fuel.

Meat consumption is largely at formal or family feasts, but not always. A guest may be asked if he would prefer kid or lamb, and as *kubsa* or *mensif*, but he and his fellow diners might receive a tray of rice and *samn*. No one comments, or not openly, and the assumption is that an animal was not available or that the women refused to prepare it. Alternatively, some herding families eat meat in the early summers, or at other times of the year, because they have no or little grain for bread, or no rice; or because the family head feels like having meat. An Azazma family at Fainan had a kid every day for a week in December, "because we have little flour left, just enough for bread for breakfasts. My son went into Safi this morning, and he could have bought some, but he didn't. He had the money, he just didn't buy any flour. We have plenty of kids, and it's good to have meat when we have family guests (a relation from Israel)." Irby and Mangles, travelling in Karak in 1817, mention (1823: 366) how bored they were with eating meat all the time and how they longed for bread. A Rwala herding family has meat most days, because the family head prefers meat to *'aish* or bread, and loathes vegetables; another herding family lived on milk and dates for months.

Formal feast food is ideally meat; lamb, kid or camel, cooked as *mensif* in Jordan and Syria, and *kubsa* in Saudi Arabia. For *mensif*, the meat is boiled in fresh yoghourt, reconstituted dried yoghourt, or water with flavourings of optional onions and/or spices, and served on *sâj* bread, rice or *burghul*. If, as is usual, several trays are being served, variations of the cooking medium may be offered, such as stock and chopped tomatoes or stock with onions and spices. Fried pine nuts or almonds and raisins are commonly added to the finished dishes. *Kubsa* is a drier dish, and the meat is braised. Side dishes of *kibbeh*, stuffed *cousa*, stuffed vine leaves, fried aubergines and salads may accompany the main dish. *Ouzi* is a feast dish with townspeople, where lamb is cooked slowly with spices of cloves and cinnamon, and served with rice and fried almonds, and many side dishes of various seasonal stuffed vegetables, meat balls, and salads.

Family meals with visiting members have dishes such as *maglûba* (fried chicken and cauliflower line a large saucepan, and rice is put in the pan and the whole lot cooked in water and seasonings; it is served by turning out onto a serving dish so that the rice is hidden by the golden pieces of crisp chicken and soft cauliflower), or a variety of meat and vegetable stews on bread or with rice or *burghul*, or stuffed *cousa*, chicken *kubsa* with side dishes of chips, salads, and pickles. Sheep herding families have *uthun shâyib*, 'old men's ears', little dough envelopes filled with chopped meat and pomegranate seeds and poached in yoghourt, stuffed sheep's stomach, and trotters in yoghourt sauce. Soups, mostly lentil or with noodles, are served in a few houses and tents, while 'stir-fries' of green beans and carrots are new. Macaroni is fairly new, although noodles or *rishta* were always served in some families, usually mixed into rice and accompanying haricot bean and meat winter stews. Spoons are now part of the serving of a meal; people eat with a spoon or their hand as they feel appropriate.

A change in food in the *badia* is the increased use of vegetables, now available and cheap in the towns along the highway where the men collect children from school and *alaf*. Clinics, television and magazines tell women that vegetables and fruit benefit them and their children. Fewer wild foods are collected as people move less, although in spring people go out to collect *khubbayza*, and *akûb* are sold by the side of the road. In Jordan there is no commercial production of camel's milk, which has a high vitamin C content. Less bread and rice are eaten. Milk and milk products remain desired foods. In the seventies, dried milk supplemented fresh milk, because then there were relatively few sheep in the eastern *badia* since many families lived from smuggling. Dried milk consumption has declined with the increase of sheep numbers and the availability of fresh processed cows' milk products in shops in ar-Ruwaishid with the coming of electricity and refrigerators. Similar changes are apparent in the small towns of northern Saudi Arabia, where fresh vegetables grown in plastic tunnels appear in the markets of Sakaka and Turaif, and cow dairies provide yoghourt, soft cheese and 'breakfast cream'.

The most noticeable change in Saudi Arabia is the increase in meat consumption. At feasts it is no longer *de rigeur* to eat one's way through the rice to the meat in the centre of the tray; people start by eating meat. People now talk as they eat. Ready-cooked chickens from restaurants in towns sometimes supplement family suppers, as they do in Jordan. Some see the availability of food from bakeries, restaurants and shops as limiting a former range of traditional dishes. A middle-aged man in al-Karak said "when I was a boy, the women used to make different sorts of bread; some had onions in, or herbs, or dried tomatoes. But now they don't make them. They say they are too busy, with more children and all the washing and house-cleaning. It was a lot easier, living in tents." The most traditional place for food is the Fainan area, where people are a long way from shops and are not well-off in cash terms, although all have goats and sheep. The foods mentioned by Burckhardt (1831: 57–65) and Musil (1928a: 90–94), among others, remain extant. d'Arvieux, staying with the Arabs of the Emir Turabey at Mount Carmel in 1665, describes the food in detail (1735: iii, 269–282). Dairy products, bread, rice, beef, goat, mutton and chicken, honey, olive oil, seasonal vegetables and fresh and dried fruits were the main items. People drank water and milk, sometimes an infusion of fresh or dried fruits, and occasionally a tisane of water, barley and liquorice. Sherbet, "or as we say sorbet", was served only among the sheikhs on special occasions. Coffee was frequently drunk, from little earthenware, porcelain or wooden cups. Food for special occasions included kebabs, meat stews, and a whole lamb or kid cooked with breadcrumbs, raisins, salt, pepper, saffron, mint and other herbs. The daily dishes were rice or *burghul* cooked with yoghourt and/or butter, or in water with chickpeas, onions and raisins. Broad beans, lentils and peas were stewed in oil and served on rice or *burghul*. All these dishes were served on tinned copper trays, and eaten by hand. Figs, dates, grapes and water-melons were the usual fruits.

The distinction made between private and public in food extends to clothing and furnishings. Men's formal clothing is outwardly uniform in the countryside; all men own light summer *thaubs* and heavier weight winter *thaubs*, a black *aba* or fawn *bisht* for summer and a sheepskin cloak for winter, and a head covering (check or white) and *agal*. There are considerable differences in cloth quality, trimming and finishes, and the number of formal clothes a man may have. Regional variations show up in the cut of sleeves and body, details of cuffs, front openings and collars; in the finish of head coverings and *agal*, and the patterning on cloaks. Some of these change over time. Regional variations in everyday work clothes are greater; baggy trousers are common in the Hauran and Qalamoun, cross-over long shirts worn with a belt in mountain Jordan by middle-aged and older men in the fields, while herders wear loose *thaubs*. The young men tend to wear locally made or imported second-hand western clothing for work. Women's formal clothing emphasises newness and quality of material and finish, but is more varied through the region. Modest dress is a prerequisite, especially in public in the countryside and on formal occasions. Hair is always covered, and the body is hidden from view by a long black cloak or *abaiyya*, while young women in the rural towns of Jordan wear a long coat of neutral colours. Under this concealing garment, students and young women wear jeans or skirts and shirts. Long dresses or skirts and long-sleeved tops are always worn on formal occasions. Once away from the possibility of being seen by a man from outside the immediate family, *abâyyas* are cast off, and dresses are gorgeous in colour and material. Within the necessities of modesty, details of clothing change around necklines, shoulder padding, waistlines, fastenings, sleeve details, and finishings. Such changes are most noticeable among the young and the more urban and more travelled women. Older women wear what was the fashion when they were young married women, and hairstyles echo this convention. Rwala women who were married in the seventies wear their hair plaited, like their mothers, while their daughters wear their hair with bangs at the front, and coiled into a bun at the back. Older women move into darker colours, while black or indigo dresses were the standard traditional wear for women in Saudi Arabia and Jordan. However, such a fundamental garment as an *abâyya* does change. Twenty years ago, younger Rwala women wore gauzy *abâyyas* over their heads, kept in place by tension when the ends of the cloak were thrown over the lower arms, and held against the body, in keeping with the formal gliding walk; the face was veiled either by an end of the *isâba* wound across the face below the eyes or, in Saudi Arabia, by an *isâba* end wrapped across the face and head. Now, all *abâyyas* are opaque black material and worn like a coat, with the arms going through sleeves, while the veil is either a thin scarf wrapped round the head and across the face, or a black crochet face covering pinned to the head covering. There are also strong regional differences. In al-Balqa, al-Karak, and Shaubak, many women wear the *madraka*, a black unwaisted dress whose sleeves, once very long and tied behind when working, have become transformed into token sleeves held together by a cloth strap on the back. Ahl al-Jabal women wear long dark dresses, belted around the waist, and with a deep neck opening exposing the bosom; the under-dress, high to the neck, is always a pale or bright colour. Clothes of women in southern Syrian villages seem to differ between religious confessions and from village to village; but "in fact, it is impossible to tell accurately who a woman is by her clothes. A Druze woman might be wearing the very full skirt and blouse associated with the Druze or she might wear a loose velvet *thaub* that's associated with Haurani women.

They see clothes they like in shops, and so they buy them. And they give each other clothes." There are generation differences, where the young women are more likely to wear long skirts and tops than the loose dresses of their mothers. Outside influences come in from travel, television and magazines, especially the Burda publication with its paper patterns, on sale widely in the Middle East. Work clothes for women are best clothes grown old, or cheap bought or made clothes.

    Clothing is a main item of family expenditure, difficult to cost since many clothes are handed on to younger relatives or the poor, or cut up and resewn for small children. Clothes or material are common gifts from visiting relations. Prices cover a wide range for the formal clothes that every adult possesses. A decent summer *thaub*, made to measure, costs from 15 JD; a winter one from 20 JD; an *aba* or *bisht* from 20 JD, with most probably spending around 50–100 JD, although it is possible to spend several thousand for one in fine black cashmere with real gold embroidery. A herder's sheepskin cloak starts at 25 JD, while an exceptional *furwa* can cost thousands, depending on the type and matching of skins, backing material, and type and quality of trimming; few men have such expensive ones. A good sheepskin cloak is between 75–100 JD. Work clothes are much cheaper, because the quality of material is poorer. The range of prices for women's clothes is also large. Some herding families spend little on clothing annually, especially those whose men do not enter into any formal political relations with governmental agencies. Clothing for school is not expensive, but has to be purchased and washed frequently. The number and quality of clothes, particularly for children, girls, young men and elderly women, may be a fair indicator of the cash available to the household, and to the regard in which they are held. On the other hand, some regard expenditure on clothes and shoes as pointless; "so your shoes cost 70 JD; are they going to last any longer or protect your feet better than mine which cost seven?" commented a Rwala notable to his son.

    Tent and house furnishings in the public apartments range from the just adequate to the sumptuous. The differences are in quality, number and range of goods, and if electricity is present. Some tents of herding families used as home bases have generators powering television sets, with washing machines and freezers in the women's side. Expenditure on furnishings reflects income, extent of entertaining by men and women, and interest in furnishings as such. The traditional items were textiles and cushions, and the coffee apparatus. Textiles remain important for floor coverings, seating and covers, and curtains. Coffee making equipment may now include electric grinders and makers, and hotplates. Most houses and fixed tents have televisions, satellite dishes, telephones, fans or air conditioning, and cabinets displaying old coffee mortars, photographs, books, swords, graduation certificates, glasses, and trays. Most of the new houses, like many tents, are actively enjoyed by their owners, who take pleasure in their decoration. Some are outstanding in particular features; well stacked piles of homemade quilts and pillows, and cushions with home-woven or crocheted covers; sets of scrubbed shining cooking pots in order of size; sets of graduated coffee pots in hearths; painted adornments on porches or around doorways and windows; rugs woven by the women that appear for important feasts; storerooms with neatly arranged sacks of flour, *burghul*, lentils and chickpeas, bundles of herbs, and jars of soft cheeses, olives and pickles; framed drawings or paintings by children; laid floors of beautifully matched marble slabs, or the decorative window grilles of a villa; or the rugs, sofa covering and painted walls in a villa in Jauf co-ordinated and designed to give an air of cool freshness. This villa, completed in 1995, cost £300,000, of which a fifth came as a grant from the government; a one-roomed house in Fainan costs £1,000.

Many of these details of internal consumption get noticed on social occasions, times of reciprocated distribution through hospitality within the wider family and *jamâ'a*. Much hospitality is general: someone comes to ask for help in collecting money for compensation or to pay a hospital bill, and stays to lunch; neighbours or relatives drop in to chat or consult, and are given lunch or supper, or family members based in other areas visit for a few days or weeks and are feasted. Some has a more overt purpose, as when an official of the local Ministry of Education is feasted so that the community can put forward its desire for a secondary school, or when a new government official is posted to an area and he is feasted in turn by the senior men of the various groups. Some is obligatory, as at weddings and funerals which mark the movement of family members from one stage to another, and the feast ensures that these alterations are known.

Major distributions of wealth take place at marriage and after death. Marriage transfers wealth from the groom's family to the bride's. The amounts and types of property to be transferred, the timings and to whom, vary between groups and within them. Some groups transfer agricultural land to the bride's family, others do not or did not until recently. Since virtually all groups, with the exception of some Christian merchant families who marry within a circle of families like themselves, prefer to marry within the five or three generation *ibn 'amm* group, property largely circulates within and along groupings. Within a wider unit, various families are enmeshed together inside a series of exchange marriages, where brideprice is either very low and rarely paid or cancelled out; the occasional marriage to a girl of a family with whom there have been no previously known marriages means a very high brideprice officially to be paid in full before the wedding goes ahead. The groom must be able to support a wife and family, house her in a suitable manner, and give her in her own right a substantial present in gold. There is also a series of feasts over a series of three, five or seven days given by the groom's family. The bride's family provide the bedding and cooking equipment of the new house or tent, and in some families a daughter is given a large present by her father in lieu of a share in the inheritance. Where there is a high brideprice, and a house to be built, together with the feasting, contributions are expected from all close members of the groom's family, with lesser amounts from more distant members. The success in getting promises of contributions indicates the support there is in the family for the proposed marriage (Lancaster 1981: 52–4); if there is very little, the wedding may be called off. At a recent Rwala wedding between a girl of the ZN and a man of another section of the N, the feasting cost between £4–5,000 and the girl's present was around £8,000. Another Rwala wedding, between second cousins, was recalled by the girl; "I was covered in gold, I wore gold from head to foot. But I never saw it again, he must have hired it from the goldsmith's for the wedding. Well, I knew none of us had that sort of money."

Funerals are marked by feasts to commemorate the dead man (or woman but among family members only) but less extensive than for weddings. Inheritance makes a real transfer of assets down the generations, since property is divided among the children, with daughters having a half share to that of sons. Although the transfer takes place, it need not result in any outward division of assets or their re-registration.

Gifts to children of the family of income producing assets are common from parents and relations; boys receive animals, cars, weapons, sheepskin cloaks or money, girls get animals, clothes, jewellery and money. These gifts are made at their birth, at religious feasts, casually, or by visiting relations. The recipients are in control of their gifts. One ten-year old boy had received 25 JD after a summer of visiting relations, and got his father to buy him a case of cigarettes which was put into a smuggling run. The animals, lambs, kids, horses or camels, are put with the family herd, and any offspring or produce are kept in the child's name. Products, wool, hair or milk products, may be absorbed into the family's use, or an account may be kept; this is more so when the children are older and actively look after and work with the animals. A lazy child can find his or her animals being commandeered by the active members of the family.

Gifts are also made within the wider family unit to men who have, through misfortune, lost their livelihood in order that they may re-establish themselves, and outside the family, to those who need help. Within the *jamâ'a* gifts may be made to an individual or family on an occasion of general celebration.

Outside the family, further wealth distribution takes place through expenditures to providers of services. In the countryside, this used to be either for goods not produced within the family — cloth, sugar, coffee, spices, salt, cosmetics, jewellery, weapons and some craftwork – or for protection while using areas controlled by others; or for using the skills of arbitrators or judges. Protection contracts of *khuwa*, *rafîq* or *kafîla* all incorporated restitution and recompense of goods and person, so payments could go both ways. Other causes for expenditure were necessary social payments from compensation. Payments for protection and compensation, together with the jural processes that established these, and the ultimate right of self-help, contributed to the maintenance of dispute settlement within the region. This itself enabled the participants in this system of tribal administration to produce livelihood from the variety of environments and to be part of inter-regional exchanges. An additional distribution of wealth in the *badia*, countryside and small towns is and was the constant generosity to the needy, unknown travellers, seeking money for compensation, medical bills, education, or because they have had bad luck and are without other means.

The value of regional production at any date is difficult to establish, since most exchanges were unrecorded. Hutteroth and Abdulfattah (1977: 76–110) consider that there was a rural surplus of production after taxation in the early Islamic period, as does Bakhit (1982: 150–1) who mentions the importance of the collection of the desert plants, *kilw* and *shnân*, especially in Qalamoun for local industries and export to Europe. During the eighteenth century, the districts of Nablus, Ajlun, Gaza and Jerusalem funded the return of the Pilgrimage from agricultural taxation (Barbir 1980: 122–5). Owen (1981: 38) points out that agricultural production was high enough pre-1800 to supply towns, provide materials for local industries and sustain a low level of exports, while Ze'evi (1996: 102–8), Cohen (1973) and Doumani (1995), among others, indicate important regional and local manufactures and trade based on agricultural products of cotton and oil. Marsot (1984: 233–4) comments on the grain production and export of Syria in the early years of the nineteenth century. At the same time, the value of goods arriving at Damascus by caravan across the desert was valued at 18,528,000 francs (Issawi 1988: 159). *Badia*, countryside and towns were economically interlinked, as Ze'evi (1996: 980–108) notes for sixteenth century Jerusalem, with the *badia* supplying *kilw*, milk products, sheep, horses and camels, as well as transporting and military services. Burckhardt (1822; 1831) depicts the standard of living for herding families in many regions, and the economic enmeshment between *badia*, countryside and town. Again *kilw* was important, Doumani (1995: 192) finds that *kilw* was two-thirds of the total production costs of soap, while Burckhardt (1822: 354) noted that three thousand camel loads of ashes went to Nablus from as-Salt each year. Over the century, the demand for *kilw* tripled. In 1851, sheep raising in the Damascus region gave a profit of 25% on capital invested, "very close to the return on money in the Damascus market (Issawi 1988: 301). Merrill (1881: 474–5) considered the trade of Damascus, Jerusalem and coastal towns with "the Bedawin tribes .... must be considerable", with sheep, goats, some grain, horses, camels and ashes for soap being important, although he travelled almost twenty years after the introduction of caustic soda to soap manufacture. The export of grain to Europe between 1874–86 was reckoned by Schumacher (1886: 23–4) to be on average between 100,000 and 120,000 tons annually, although by this time profits in the grain export trade were falling with the opening of the Suez Canal and the advent of the Great Depression (Schilcher 1991a: 53).

That the regional economies produced a surplus has been established. Distribution of this surplus to administrative purposes switched over time between agents of central states and local, tribal notables. Both categories could and did see themselves as providing security and as guarantors of dispute settlement (either through an Islamic state code or customary law), so that production and distribution of surpluses could be achieved. Although the two systems appear as polarised alternatives, which is how they can be portrayed by some members of both, notable individuals from tribal systems were agents of central states when such states decided to incorporate the tribes or to decentralise their rule in parts of their Empire. The history of Jordan during the Ottoman period and to the present illustrates the changing methods used to acquire surplus from the Bilâd ash-Shâm by the state, and how local individuals and families managed the state to acquire income for themselves.

Taxation by the conquering Ottomans started the day after Sultan Selim arrived on the outskirts of Damascus in September 1516, and in 1521, the Sultan sent an official to survey the land and divide it into state domain, *timars*, *waqf* and privately owned lands (Bakhit 1982: 143–4). Local notables were incorporated into the system of *timar* and *za'ama* land grants, as were tribal leaders (Bakhit 1982: 189–91, 200), and tribal leaders were given administrative and tax-farming posts (204ff). The official administrative divisions at subdistrict level were not primarily meant as an effective grid in a centralised political hierarchy but as flexible fiscal units, maximising revenue at the least political cost; "the government, in other words, read the existing local political map and then drew boundaries around the actual relations of power" (Doumani 1995: 36). In seventeenth century Jerusalem, Ze'evi (1996: 145–54) ties the increasing decentralisation of tax collection to the abandonment of the military role of the *sipahis*, and the development of direct links between tax payers and receivers, and increasing struggles to acquire tax exempt status; the result was that "profits and property amassed though decades now started to dwindle rapidly". In Jabal Nablus, from 1657 the leading families who had military *sipahis* holding timar land grants shifted to being tax-farmers holding tax-farms and later to merchant entrepreneurs collecting taxation and customs duties as members of the Advisory Council. Eventually the area, like the whole region, was incorporated into a capitalist world economy dominated by Europe. Doumani concludes (235–6) "The socio-economic transformation of Jabal Nablus, like that of many of other interior regions in the Ottoman Empire during the eighteenth and nineteenth centuries was, therefore, neither a linear march into the modern period nor predicated on a sharp break with the past......many of the features associated with capitalist transformation had indigenous roots that were clearly evident before they were supposedly initiated by outside forces, and ingrained modes of social organization and cultural life...... proved highly resilient and adaptable."

The urban Advisory Councils, set up by the Egyptian authorities between 1831–40 and continued by the Ottoman *tanzimat* governments, were intended to extend the role of central government in the appropriation of surpluses from local production and its distribution. Doumani (1995: 241) describes a new configuration of political reference points taking place through "dozens of separate negotiated deals concerning specific issues, the outcomes of which spilled over into an ever-widening political and cultural space" and "in each bargaining session, their (the Council's) responses to requests and admonitions from the central authorities were designed to facilitate their own objectives and, at the same time, to secure the state's recognition of their own legitimacy." Two basic contradictions underlay the intention of central government towards control of local surpluses. One, that the council members implementing the new policies of tax collection and conscription were the people who would lose by them. Secondly, while council members needed the legitimacy, administrative authority, and control over the militia conferred by the state to secure rural surplus, the state could successfully eat away at the notables' share of surplus and regional independence.

Similar political struggles are evident in other areas of the Bilâd ash-Shâm in the late Ottoman extension of rule, as in Rogan's description of the situation in as-Salt and al-Karak (1991). Wetzstein (1860: 138) recalls a conversation with a Rwala shaikh who blamed the expansion of grain cultivation because of rising grain prices for competition between the Weld 'Ali and Rwala for pasture. Schilcher (1981; 1991: 51–4) describes the changing situation in the grain producing region of the Hauran where "an informal cartel of Damascene merchants and rural-government sanctioned political and fiscal brokers emerged in the course of the bloody Hawran conflicts of the 1860s." This cartel alternated sales of grain between different markets, "wherever profits were highest or government pressure strongest", so that, depending on how the cartel was operating, the Hawran could be described either as a relatively remote region with small peasant units of cultivation or as a region pulled into commodity production for world markets. The cartel needed political control and economic incentives for continued success, and with the price falls in the late 1870s and the extension of direct taxation by the state, it increased its pressure on the peasants. By 1887, world grain prices were so low that the government could find no notables interested in farming agricultural taxation in the Hauran, so it had difficulty in collecting enough grain for its troops (Schilcher 1991: 61), and in the following years had to send out gendarmes to raid villages and seize livestock to extract revenue. Demands for taxes caused peasant populism, bedouin uprisings, and resignations by members of the Damascene Administrative Council, culminating in a unified Haurani uprising in 1897, with alliances between the peasantry, bedouin and Druze. In 1900, the Sultan ordered a general amnesty, dropped demands for tax arrears, and allowed some measures of autonomy. Musil (1927: 426–33) records Rwala views on Ottoman demands for taxes and baggage animals, and restrictions on sale of grain and cloth by the settled population to them, Ottoman support for tribes opposed to the Rwala, and how they saw Ottoman actions as influenced by English policies in the region. There were continual restrictions on tribal freedom of movement and action by government at the same time that tribal economics were affected by a lessening demand for camels, and tribally provided services being superseded by state provision. Both political and administrative restrictions and changing economic patterns continue under the Mandates and ibn Sa'ud.

The transformations come with the great shifts in production of surplus and therefore the sources of state income from wealth produced by individuals and families from agriculture, herding, and processing and manufacturing based on these, and transit and local trade carried by animals, to royalties from oil and mineral extraction in which the state has a direct stake, imports funded by oil wealth or aid, and aid to governments from geopolitical considerations. Although the beginnings of these are visible under the Mandates, agricultural produce and animals remained the foundation of taxation, and urban merchants, whose wealth was based on land and trade, were the financial backers and supporters of Emir Abdullah of Jordan (Amawi 1994) and ibn Sa'ud (Field 1984: 105–17). By the mid-sixties, Jordan had switched taxation away from land to transactions, especially on imports, and by the mid-eighties could be described as a rentier state (Chatelus 1987), depending on income from aid, mineral extraction and tourism. The oil economy transformed the economy of Saudi Arabia by the mid to late seventies. In Syria, agriculture remains the single most important sector (Perthus 1995: 26), although less than 30% of the population work in agriculture, and since 1986, the country has been a net exporter of oil. The state capitalism introduced by Asad in the early seventies was transformed since the eighties, by decisions seen by Perthus (1995: 7) " as a complex collective process of action determined by institutional structures and by conflicts and collusions of interest."

Incorporation in a world economy had happened earlier in the 'long thirteenth century' from 1250–1350 analysed by Abu Lughod (1989), until this ended with the closure of the trade routes from Central Asia, the withdrawal of the Chinese from the Indian Ocean, and the arrival of the Portuguese. Within the region, animals for transport and draught remained essential. Agricultural and pastoral commodities were essential for industries, at home and for export. Soap and textile industries had to adapt to techniques and products from the west from the early part of the nineteenth century, while the grain exports of the mid-nineteenth century were hit by production in the Americas and Australia. The internal combustion engine and oil and the hydrocarbon industries spelt death to the camel's role in transport and draught. Production strategies are resilient, and people adapt new technologies to their needs. But the control of the new technologies of production is from outside the countryside and the people of the countryside can only get access to its productive potential through participation in the institutional processes of the states and their agencies.

CHAPTER 8

INTEGRATION INTO MODERNITY

The view that states, of themselves, produce the security of property and person necessary for production of agricultural, pastoral and industrial surpluses has been questioned recently by scholars working from documents and from archaeological survey and excavation. Local people in the Bilâd ash-Shâm consider that states move in to regions only when surpluses are already in production. The state is seen as a predator from outside, cunningly hunting down opportunities for removing surpluses out of the control of local producers and consumers for its own benefit and at the same time seeking to constantly enlarge its field of operations. Local opinion holds that surpluses in production on the scale to attract states, over and above local requirements, are generated from external inputs and demand. The development of the oil industry and economy is frequently cited as an obvious example, where the West provided the motivation and the technologies, and its continued demand for oil affects the economics and politics of the whole region. People also cite the grain market at some periods as driven by outside demand, as during both World Wars and regional wars at various dates in the nineteenth century and earlier. The transformation of herding, with the camel now as a commercial meat or dairy animal rather than having a major function of transport and draught, and the consequent rise of sheep herding for urban markets expanded by oil money, is attributed to the introduction of modern forms of transport and pumping equipment.

People make a distinction between government or *hukûma*, and state or *daulat*. States govern, but so do other political associations, such as tribes, village or peasant associations, or urban based tribal or merchant polities. In local logic, any reputable individual has the ability to offer the specialised functions of government of dispute management and protection which permit the general function of "allowing people to live their lives"; the ability to pursue disputes for restitution and recompense, ultimately by self-help, is also seen as a basic aspect of *hukm*. *Hukm* comes from a root meaning arbitration, used by disputants who cannot settle their differences by mediation, and who therefore agree to accept the verdict of an arbitrator accepted by both. Being unable to reach a settlement through mediation is somewhat unusual. In the past, this would have indicated a complex situation or personality clashes. Since the abolition of customary law and the involvement of state authorities in dispute settlement, the state through its agents is arbitrator, through ratifying an agreed mediation by the disputants and their mediators within terms laid down by the state. The reference points of government as arbiter and as enabler of the pursuit of livelihood and social practice necessarily involve ideas of consultation between participants and action coming out of consensus, such consensus being continually redefined and restated. The protagonists of government are known persons, responsible for their actions on behalf of government and answerable to the governed and governors. Government in the sense of the administration of affairs by a group from the community implies the authority to carry out its decisions. Authority comes from the adherence of governors and their agents to the moral premises of the community, and their ability to deliver suitable actions and conditions of livelihood. Their power comes from forces that they command, and increased by restrictions they are able to impose on local holdings of weapons or other instruments of force.

Daulat or state has opposed connotations of being outside the local community, physically imposed power legitimized by ideology into authority, with decisions taken at the centre and transmitted by fiat. Its agents, as agents, are not known persons and perceived as answerable only to superior officers of the state. There is little sense of the relations between state and rural populations assumed to exist between government and people. Its authority is seen to come from its own ideology, although in the Bilâd ash-Shâm these ideologies, such as Wahabi Sunni Islam, democratic socialism, or Sunni Islam and democratic capitalism, have a wider resonance than the state itself. These ideologies are imposed on populations as 'better' than those of the past, and seen by the heads of states as relevant to their populations. Some members of the general population may already support these wider ideologies, while specialised groups close to the head of state will be adherents.

This apparent opposition is rather a way of speaking, a way of explaining ideas, than an exposition of actuality, although participants hold both concepts to be valid. While the two ideas are distinct, each holds within it aspects of the other, since any kind of government has relations outside itself, and any state also governs.

In the history of the Bilâd ash-Shâm, states have invariably delegated functions to local groups, and local groups and individuals use the state as a resource. Nor does the appropriation of rural surplus under delegated systems of tax collection or *khuwa*, the tribal alternative, remain entirely with the central tax collecting agency, either state or tribal. Al Rasheed (1989) describes the nineteenth century Shammar polity of Hail in central northern Arabia as founded on the payment of *khuwa* to the amir by weaker tribes, oasis dwellers, merchants and pilgrims. As Hail was on a main trade route between Mesopotamia and the Hijaz, *khuwa* revenues allowed the amir to employ a permanent military force whose role was to make the peace respected, protect property and to punish wrong-doers, and to legitimise his regime by generosity or subventions to tribesmen. A proportion of tax revenues stayed in the area, to be used for local welfare as generosity and hospitality, and local good works. In other tribal arenas, customary surplus was used for building and supplying guest houses, (as in Salt and Karak), shaikhs cleaned out *ghadîrs* and other water catchments, and guaranteed markets and distribution. State agents from outside the region, for example Mamluk governors and Ottoman Pashas, built schools, bridges, mosques and hospitals. Not all 'good works' came from state agents, others were funded by private endowments. But in the present, most state income comes from the resources the state creates and controls – oil exports and royalties, customs, import and export duties, agencies for the preceding, mineral exports, and aid. There is less local contribution with less representation in local areas for the production of state income. This leads not to a situation where local economic activity is self-sustaining, free of state appropriation, but to one where local groups, having lost land to state bodies, compete for employment in such bodies. Any sense of partnership goes, to be replaced by fiats from the centre. State agents assist in the distribution down or outwards but procure this agency through appointment as unattached agents, not as representatives of local groups and of known families.

Participation by tribal and rural groups or notables in the activities of decentralised or re-centralised states is often seen by historians and political scientists to be as clients of the patron state/ polity or as being tributarised by the state (e.g. al-Azmeh 1986: 82, 86; al Rasheed 1989: 232; Kostiner 1991: 225: Velud 1995). Central government may intend such a dependency. Locally, payments of subventions by the Ottoman Empire, the Mandate governments, the early and modern ibn Sa'ud rulers are seen by tribal leaders and tribespeople as a return for their help in keeping the peace. Both sides contributed to the ideal purpose, and both receive benefit, which on the rural side included monetary payments and adjustments to central rulings inappropriate in the regions. Honorific titles are regarded with a degree of polite scepticism. Delegated responsibilities in state systems, such as tax collection or protecting roads, were seen to assure access to markets for grain, cloth and other necessities. The facts of state activities in the real world often meant that these negotiated arrangements between state and decentralised authority were breached, resulting in disputes considered legitimate by tribal and rural groups but recorded as predatory incursions by state documents and consular officials. As Singer points out in her analysis of peasant and Ottoman officials relations in sixteenth century Palestine (1994: 130), "peasant interests were determined by an entirely separate schedule of concerns from those of the state. The two might intersect, but they were not congruent." Similarly, the interests of tribespeoples and merchants had further separate agendas. Doumani (1995) considers those of merchants in Jabal Nablus, but the concerns and interests of tribespeople are rarely considered because of a lack of source material and of understanding of social processes. The history of the Hauran in various decades of the nineteenth century has been analysed by Schilcher (1981; 1991a; 1991b) from the peasant perspective.

Tribal views remain to be written. Musil (1927: 429ff) quotes an analysis by Nuri ibn Sha'alan and his son Nawwaf of their current difficulties with the Ottoman state, where although taxes had been paid, and conscripted animals delivered, the Rwala were not allowed access to markets; in addition, the Ottomans were supplying ibn Rashid with money and supplies to fight the Rwala. Velud (1995: 65) sees great shaikhs, like Nuri Sha'alan and Mijhem Mhaid, were made loyal to the Mandate by subventions, honorary titles and electoral mandates. Lewis (1989: 154ff), however, considers the Mhaid shaikhs realised that the Mandatory powers needed them as intermediaries in a French system of indirect rule, and the rewards and honours they received were in return for their services. The Sha'alan themselves view the subventions as part payment for services rendered in keeping the peace in the Bedouin Control region of Syria together with use of markets and summer grazing in areas to the west by some tribal sections; their seats in the National

Assembly recognised the need for participation in the tribal economy by tribal and other regional groups in an advisory capacity. The decision by Sha'alan to live in Syria rather than Saudi Arabia or Jordan at that date was based on assessments of which state was more compatible with what they regarded as important for themselves and the Rwala in general, and where livelihoods could be obtained (Lancaster 1981: 126). By that date, many Sha'alan assets were in Syria, while several tribal sections had wells, gardens and markets in Saudi Arabia. The Jordanian *hamad* was used by the Rwala, but tribespeople owned no assets there; it lay between resources in Saudi Arabia and Syria, and the British Mandate government took to itself the role of protection, banning *khuwa*. Ibn Sa'ud also abrogated all aspects of rule and peace-keeping under the logic of Wahabism. The French Mandate government in Syria allowed a degree of tribal administration in the *badia* to continue; while *khuwa* payments were reduced, they were paid (Thoumin 1936: 153), and tribal shaikhs collected payments for guaranteeing markets. The continuation of tribal economic income through traditional means was a direct factor in the Sha'alan and Rwala decision to live in its Syrian areas as well as its Saudi bases in al-Juba and the Wadi Sirhan. In addition, Syria was larger and richer than Jordan, and at that time, richer than northern Saudi Arabia.

The modernisation of Saudi Arabia during the sixties, seventies and eighties transformed its society economically and politically. The Saudi government actively used sedentarisation to undermine tribal processes (Fabietti 1982). Fernea (1984) and Kostiner (1991: 245) see such policies leading to tribespeople being absorbed into social classes, although "tribal values compensated both for the creation of a formal and unfamiliar bureaucracy and for the absence of political parties in the kingdom" ...... through networks which ultimately "created a large clientele dependent on the royal family." Tribespeople, unless they adhere totally to Wahabi tenents, have difficulty in accepting the royal family's control of all fields of action in the kingdom. The Syrian state has been explicitly anti-tribal since 1963 with the coming to power of the Ba'ath party (Seurat 1980: 111–2), who points out that integration into a new national society cannot be achieved merely by decrees and bureaucratic apparatuses. This anti-tribal stance, together with the seizure of assets, was the cause for the Rwala, like other tribes, leaving Syria for Jordan and Saudi Arabia.

Given these disparite comments, how tenable is it to say that tribal leaders and tribespeople see state benefits and employment as elements of a working relationship between tribespeople and state agents? Just idealism? One aspect, granted; but underlying this is a genuine view that just as relations with other polities, whether state, tribal, or urban/merchant, in the past were conducted by negotiations between parties who became jural equals for the occasion, so should the relationship between state and citizen, especially where tribesmen or citizens provide necessary services as many saw themselves as doing. This view was widely held in the early seventies in the countryside of Jordan and Saudi Arabia, where many served in the armed services, National Guard and Badia Police and specifically equated this employment with the former defence of tribal interests. Tribal interests were best served by participation in the security forces of governments where tribespeoples' livelihoods could be pursued and assets maintained, and whose rulers regarded tribal shaikhs as quasi-ambassadors who could offer real services or present obstacles (Lancaster 1981: 89–90). Such interests were pursued at the personal level, whether at the centre or more local foci of authority. Since then, oil wealth has increased, aid from a variety of sources has grown, the geopolitics have shifted, and ideas of state and government have altered.

States, like the alternative of tribal or village political and governmental communities, continually redefine themselves. Moral premises and ideology are one focus, while the provision of the physical and/or social infrastructure of government is another and legitimised from the first. The modern nation states of Syria, Jordan and Saudi Arabia are often seen as rentier-states, whether or not they are oil producers, since the allocation of oil-related aid in non-oil producing countries is as significant as the direct allocation of oil revenues (Chatelus 1987: 206; 1990: 101). Although Jordan has had periods when it has not received aid from oil producing states, it then had aid from the west because of its position in regional security. Similarly, Syria has received substantial aid, mostly from the Arab oil-producing countries, and public investments are financed from this source (Perthus 1995: 34–5). Saudi Arabian revenues increased greatly after the rise in oil prices in 1973, and a series of five-year development plans have been implemented for physical infrastructure, and economic and social development. The primary aim of development in Saudi Arabia is to "maintain the religious and moral values of Islam" (Koszinowski 1981: 209).

The distribution of surplus, whether from taxation or from direct or indirect oil revenues, is administered by agents of a state who at one level or another are also regional or local figures, and subject to pressure from local sources. In the past, such local leaders retained a proportion of surplus locally rather than transmitting the total to the centre. At present, local leaders wish to direct as much as possible from the centre to their areas. In her discussion of state centralisation and rural integration in the Hauran during the 1880s and 90s, Schilcher (1991a: 74) comments that agricultural depression contributed to a struggle for survival among powerful provincial and urban families. Those who continued as members of the political elite achieved this by increasing their identification with and their dependence on the central state. "For those who were excluded or who rejected overdependence on principle, the concepts of state decentralisation and Arab nationalism represented new political alternatives," a point echoed by Lewis (1989: 150ff) in his discussion of the Muhaid shaikhs of the Feda'an under the Mandate. Later ideological and religious fundamentalisms, such as the Ba'ath and the Muslim Brotherhood, became attractive to others. Among many, a dislike of any central government is common, and the preferred government is one that leaves the most control of local affairs at local level. The participation possible in local affairs is given by many in the *badia* as the reason for living in Jordan and taking Jordanian citizenship. This view does not prevent complaints that there is too much government. Such an attitude might be expected in rural areas, but it is also common among educated, urban, travelled Jordanians; a high official in the Bank of Jordan said "There is too much government all through Jordan, at every level. It simply isn't necessary."

Does the redefinition of states imply a redefinition of the social groups of society? Have tribespeople, merchants, farmers, artisans and herders become members of social classes? Do they identify themselves in such a way? Some scholars answer in the affirmative, and the fact that many would not so describe themselves is seen to be irrelevant. Kostiner (1991: 244), writing about Saudi Arabia, sees class as a *sine qua non*; he writes "Tribal chiefs who acted as mediators between the central government and individual tribal members became large landowners and joined the upper class. On the other, many rank-and-file tribal members formed the bulk of the Saudi lower class." Fernea (1987) regards the processes of social differentiation occurring among bedouin in northern Nejd as the bases for a developing class society, although Fabietti (1990: 245; 1993: 141) sees these changes as possibly prefiguring the emergence of a class society but not suitable to describe the situation of the majority of the bedouin. Differences in wealth between members of wider domestic groups have existed over time, as has individual property, but this does not necessarily cause class stratification. If one nuclear unit consistently refuses to be generous with its surplus wealth, distancing occurs between the constituent parts of the wider group, as though a lack of expected generosity, and therefore participation, delegitimises the offending party from inclusion. They are distanced, they are not subsumed into a class structure. Antoun (1991: 1–12), analysing the work of several scholars writing on Syria, presents class as one model which "assumes that modern Syria can be best understood by the assumption that deep-seated social dislocations have pitted one class against another" (1991: 3). However, the diverse economic activities of wider family units as discussed by Métral (1984; 1993), Khalaf (1991), and Longuenesse (1980), the way in which family units bridge social categories (Hinnebusch 1991), and the flexibility and mobility of social practice make "the application of the usual categories of class difficult" (Antoun 1991: 7). Local emphasis on multiresource economics, the relevance of the wider family group, and the use of embedded networks provide an alternative base for analysis.

What are the views of change and development held by people in the countryside? Some comments are common in the eastern *Badia*, al-Karak, the Wadi Araba, in small towns, and in al-Juba, and whether the speakers are herders, farmers, army officers, small shopkeepers or artisans. Increased control and regulation from the centre is seen to accompany development to the detriment of the exercise of personal autonomy; "now we need papers for everything and anything, before we went ahead and did things." Roads, police posts with radio, phone, and computer links, and local government offices are the concrete evidence of this extended control. There is more material wealth and goods; cars, villas, televisions, furnishings, bathrooms, washing machines, refrigerators, satellite dishes, telephones and faxes are common. Some say these have affected marriage arrangements; "a bride is no longer prepared to share a house with her husband's family, she wants a villa of her own with just her and her husband. And she wants a salon, bedrooms, bathroom, kitchen, furniture, gas stove, fridge, washing machine and I don't know what else. It makes it so expensive." Some desired goods are largely dependent on state provision of electricity and piped water. People are aware of the ratchet effect of the demand for consumer goods, pushing up the need for income. Population increase is an additional cause of change that local people comment on. There are far more people, partly from immigration, but also through the provision of health care by the state and private agencies. Families are larger; children live, women rarely die in childbirth, epidemic disease is greatly reduced, the old live longer and the young leave home later. Urbanisation is understood to come largely from the influx and natural increase of Palestinian and West Bank refugees without land or herds in Jordan and from the movement and settling of local populations employed in government services, and their need for education. In northern Saudi Arabia, the growth of towns and villages resulted from the decline of camelherding as a major resource, the partial substitution of employment in TAPline and local government, and the need for education for children. These towns then generate demands for meat, dairy products, fruit and vegetables which affect production in the countryside.

A coherent view of change was given by a local government official in central Jordan; " In the past, being a trader was shameful, because he made his living by taking advantage of people. The honourable way of living was from your own efforts. Being a manufacturer was shameful too, because he depended on other people for his livelihood. Most of the big merchants and manufacturers who dealt here came from outside the region. This would be until 1948, and there were always Palestinians coming across for farmwork, herding work or as building labourers. In 1948, there weren't a large number of refugees. There were three small camps, and the rest were absorbed without difficulty because they brought skills and they were better educated. They were the pressure for development. Their skills and labour helped development in the fifties, and they settled in the towns. In 1967, it was a much bigger flood of refugees. They brought skills but Jordan couldn't accept them all. Initially, the same effects of development were felt, but the numbers were too great, and they had to be supported by aid from outside. Jordan could have absorbed them but not at the standard of living they demanded. The ones who couldn't be absorbed got taken up in the Fedayeen, and this problem came to a head in 1970, when they were a state within a state. Fortunately Jordan won, and all those registered as Fedayeen were thrown out. Having got rid of this dead weight which absorbed time and money, the economy took off, partly from aid, partly from remittances from the Gulf and Saudi Arabia. Everything boomed and that was when everyone took to trading – in land, goods, transport, everything. Before 1967, a man could afford to employ only one person; afterwards, he could employ five, because Palestinians were desperate for work. So there were employees and agents. It was no longer shameful because the rules of subsistence had changed, as well as the rules of political action by individuals. In the tribal system, incomers are accepted and given space and they provide their own subsistence. Now, the state accepts them and gives space, and its citizens provide subsistence for incomers." This acute analysis focuses on the substitution of employment for what was portrayed as a subsistence economy with a market component based on agricultural and pastoral production. Now, wealth creating production is ultimately dependent on oil which is in the control of states, and accessible through education, skills, and citizenship.

The inhabitants of villages and small towns give the main reason for their growth as the need for education. Education is seen as the entry to participation in current economic and political activity, of which a large part revolves around employment by the state, in whichever country the person is living. The state in its governmental aspect is on the one hand a supplier of surplus and on the other a resource to be managed by its citizens. The surplus a state has available for infrastructure and services, and employment in these, now comes from outside. Much of this is earned from sales of oil, gas and minerals, but mostly from oil. Aid from the oil economies of the Gulf and Saudi Arabia to those of Jordan and Syria is seen either as reciprocal payment for continuing regional stability or in return for services in supporting Palestinian refugees, or, alternatively, as a religiously sanctioned generosity and a due, not a charity. The growth in government, funded directly or indirectly by oil wealth, is such that its resources as a supplier of livelihood in wages, pensions and benefits is seen as necessary by a majority of participating citizens. While most groups feel they are entitled to access to government resources for contributions to livelihood, these groups also feel that to back up their claim they must be seen to participate in government activities. The two aspects of claims to access and participation justify present action, ratified by past actions and will be used in the future to legitimise choices of action.

The way to the resources of government employment and service is initially by being a citizen, registered with the appropriate authorities. Then it depends on educational attainment; success in national examinations is essential for university entrance, recruitment for army officers, grades in the civil service and so on. Family networks are important but do not replace educational achievements. Participation in government, the armed and security services as an officer, and in the ministries as a relatively high ranking civil servant, has a political aspect for the families with members so employed. At one level, the families and their wider groups visibly support the regime; secondly, participating individuals are able to direct government benefits of employment or services to their area and thus enhance their reputations as effective men; and thirdly, some can make known local views and requests to central agencies. Government service, especially in the army, gives access to pensions, payments for children, and aid for further education. Pensions and gratuities finance much rural investment in housing, gardens, flocks, and workshops, as well as adding to livelihood. Access to these desired resources being through education, most people say they settle to give their children education.

All villages have primary schools. Further movement to a village or small town with a secondary school that provides education to university entrance level is quite common. The most rapidly growing villages are those that have such secondary schools for boys and girls. Education is also valued for its own sake, in that being educated includes the person in a wider community of knowledge, "and this is important now we travel outside and watch television and meet other people. And we have to know our religion and our history." Wahlin (1982) discusses two village schools in central Jordan, and the progress of two cohorts of pupils.

Not everyone sees education as relevant, including some who are educated. Herding families find educating their children difficult, and those who think it worthwhile have a variety of strategies. Some men marry a second wife; one is with the flock in the *badia* for milk processing and household duties, while the second has a house in the town where the school-age children stay during the week. More have a base camp within driving distance of a school, and the flock and shepherd range far away but are supplied from the base. Others largely ignore education; "We don't see any point in education. All the boys are going to herd, and they don't need school teaching for that, they need to be in the *badia* with the sheep and goats. They have to learn the plants and where the grazing is, and how to herd, animal diseases and cures. Even if they don't all herd all their time, they'll know how to drive and mend trucks so they could do that work. And they will have got to know people and be known. They can write their names and work numbers, that's enough." Cole (1985: 295–6) "observed no change in educational status among the al-Azab Al Murrah I revisited in 1977 after a seven year absence: none of these long distance camel nomads have gone to schools, nor have they sent any of their children to schools, nor do they have any plans for doing so". He notes an ambivalent attitude to education, which is 'a good thing', but 'no use to herders'. Among Rwala in the north, every section of the main tribal sections has families who herd camels and do not educate all their children. A Rwala said "education is essential to live in the system organised by the state, and it's difficult to live outside the state because you have no political base. Some regard those who do with admiration, others are neutral." A Sa'idiyin in the Wadi Araba said that "we don't have many educated people, we herd or go in the army. But education is important for its own sake, not just for getting jobs. With education, you can get knowledge of the past, so that you can properly claim your rights." Many tribeswomen are ambivalent in their attitude to their children's education; "it's good that the young men study at universities, because they can get good jobs, and they widen their minds. They know about things that are important now and that we don't understand. But at the same time we lose them, they go too far away from us, especially if they study abroad for three or four years. Sometimes they marry abroad, they stay there and work. If they return here, they get a job in a city or a region far from us, and they visit maybe once a year. So we don't know all our grandchildren as we should. Their father and I are glad that A is married to a good man from our *jamâ'a*, and that their marriage is happy because she was educated at university as well, but we wish they didn't live the other side of the country. But that's where his work is. We talk by telephone at least once a week; it isn't the same but it helps. Lots of women are in this position. But the future is the children's. That we are sometimes lonely, and our knowledge and skills are less valued is a part of modern life."

The increase in student numbers in further education means that some graduates do not find the professional or white-collar employment once assumed to be normal. Jordanians joke that the young men running small restaurants selling *falâfel* and *hummus* are all graduates, and it is not uncommon to meet a graduate share-working tomato or other crops. Older men often remark "if young men have studied at a university, they do nothing but sit in offices. Some of them do nothing but read the newspapers and drink tea. They don't want to get dirty and sweaty working with sheep or the land. That's all right if there are proper jobs in ministries and companies. But now so many have degrees, and there aren't enough proper jobs. Those who can't get those jobs should be prepared to get dirty like the rest of us." "Their idea of gardening is to bulldoze all the terraces and make new ones in concrete. They're not gardeners who understand soil and water and rocks, they're engineers who don't." Some young graduates choose to work in their fathers' businesses. One is a highly skilled mechanic, whose own enterprise is rebuilding four-wheel drive vehicles; another runs his father's farms, "because I like making everything work and making each enterprise profitable." Other graduates set up their own businesses. Two young men, one an engineer and the other with a business qualification, run a thriving building and electrical materials shop in a village. Another with vocational training has set up a village garage and repair shop, as well as working family land.

Some young sheep-herders in the *badia* at peak work times are on vacation from school, university, army, or professional work; part-time herding resolves for some families problems of labour for herding and education. There are stories of boys who were herders and become educated professionals. "The brain surgeon who operated on my sister is of bedu origin. He used to herd in the *badia* as a boy, and his father, who was himself illiterate, bought him books and he would read while herding. Aged thirteen, he had a chance to attend school, but the teachers wanted him to start at the beginning and he refused. His father had a relative who knew Prince Muhammad, and the Prince said the school should set an exam for him. So, given permission to buck the system, the school set the boy an exam for his age, and he got 97%. They set him another one, and he again did very well. So he entered high school, and matriculated a year early. He always wanted to be a doctor and he studied in England and America, and now he's a top brain surgeon. And he isn't the only boy from the *badia* who's done this."

Some who are well educated question its relevance. A tribeswoman in Saudi Arabia said; "I went to school, I did well, I went to University in Riyadh and I have my degree. The only job I can do is teaching. I teach girls to get educated so they can go to university and learn to teach other girls...... For what purpose? There isn't one. I'm less happy than my mother and she can't read or write. My older sister can't and she's far happier than I am. So I have a degree; I can't weave or spin; I don't know about local plant medecines; I can cook and make bread and sew, but I can't herd. I'm dependent on a town and the state, and I don't like it." Other women and men think girls' education is good because it leads to paid employment and more financial independence for women, although they criticise the limited choice of careers for women in Saudi Arabia. Many men and women in Jordan and Saudi Arabia enjoyed their education and feel they put it to good use, as teachers, computer scientists, in banking, as dentists and doctors, lawyers, accountants, managing farms and herds, in government service at home and abroad, as pilots and aircraft engineers, to mention the careers of just some from tribal families.

Altorki and Cole (1989: 240) report that in 'Unayzah, "local people opted to abandon manual work in favor of state employment or managerial/ownership positions in private sector enterprises......... The new work ethic ....places value on acquiring wealth fast from 'clean' work that involves little financial risk and minimal physical effort." State employment had short-term attractions for some Rwala (Lancaster 1981: 105–6), largely because of the information and connections that could be gained. Many found more risky occupations more rewarding, financially, intellectually, and emotionally. Risk is valued by many young men, not only because the rewards can be high, but because risk management is seen as skilled, with responsiblity for success or failure on the heads of the participants. A frequent complaint about living in Saudi Arabia is that "it is so boring. Once you've got your herd together, and the truck and the pick-up, and you're married and you have a villa in town so the children can go to school, there's nothing else to do. Any business you think of developing, you have to get a licence. And they can just say No. There needn't be a reason, they just say No."

Some say there is too much money around. The rich rarely say this, but those who give the appearance of at least relative poverty frequently voice this opinion. In Amman, the south of Jordan is held to be a poor region, and many in the south would agree.Yet some in the south insist that part of what is wrong with their world is that there is too much money, which works against the living of 'the good and moral life'. One of its strongest upholders explained; "it was because there were surpluses that states were tempted to move into the regions that governed themselves. Then states had to get more money from outside to sustain themselves and provide livelihood to their citizens. But because the state did not have enough to trade to earn money, it had to ask other states for aid. So it was not free to make up its own mind on a course of action, the aid corrupted. And that corruption works on into local communities." A secondary teacher from an old *hâdhr* Sakaka family considered that "in the past everyone lived on milk and dates, which are healthy in themselves. People were fitter because they walked everywhere. We were healthier then even though it was dirtier and there were no hospitals or clinics. And we were better people. What we had we got by our own efforts and we paid for everything we had by labour in the gardens or by money from our produce. There's no value in things like health services or education if you don't pay for them. We prefer to go to hospitals in Jordan, it's much cheaper than going to a private hospital in Riyadh, and better than going to a free state hospital". A Jordanian saw the state as eating away at private civic enterprise; "there is nothing left for people to do for themselves. There's no energy left in Jordanians, everything is done for them. Electricity, water, roads, schools, health centres, everything comes from the government. There's no room for private civic enterprise by a group. We as the wider family wanted a *madhâfa*, a meeting-house. But every part of the family wanted it on their land, and as each bit could afford to build one, every one has their own." Many describe life in Saudi Arabia not only as boring, but arbitrary and through its arbitrariness, unjust. A young tribesman resigned from the prison service, although pay and pension rights were good, because he saw the system as unjust; "What finished me was the case of a man whose neighbour kept repeatedly moving the boundary stones between their plots of land over the years. The man couldn't get the government to do anything, because his neighbour had better connections. At last, they had a fight, and he knifed him and killed him. He got eighteen years for trying to defend his own interests when the government would do nothing. It's the government's function to defend the interests of its citizens. But they won't do it, and they don't let you do it yourself. It's got to be one or the other. It's like that all the time, and I couldn't stand it." Another small gathering in the *badia* said "living in Saudi, there is no freedom to speak your mind. There are informers everywhere. You have a new pickup or your wife has a new bracelet, there's someone asking you 'how did you get the money?' The subsidies distort how people evaluate their enterprises, you don't make real choices. We don't approve of the state system of government on principle. There's no need for it, because it limits freedom of speech and freedom of action. Saudi is worse than Jordan, and Syria's worse still". (This was in 1993. In 1997, opinions were Syria offered most freedom and opportunities, Jordan some freedom but had little money, while Saudi Arabia had money but little freedom.)

Nation states are seen to restrict actions necessary to the livelihood of those based within their borders. Many groups in the Jordanian countryside have found their livelihood strategies in herding or trading affected by borders. Winter migrations to the Wadi Sirhan by tribes whose summer bases were in Jordan were early disrupted by the new borders between the kingdom of Saudi Arabia and the Mandated territory of Transjordan, Bocco and Tell (1994: 111, 120), see this disruption as a reason for many Beni Sakhr sections registering agricultural land in Jordan during the thirties. Trade with and migrant labour to Palestine were important in the economies of the Wadi Araba, al-Karak and al-Balqa, as well as for other East Bank areas (Antoun 1972: 28–33). After 1948 these opportunities largely ceased, especially in the south where the borders of the new state of Israel extended to the Wadi Araba and Dead Sea. Certain groups lost trading outlets, grazing lands, wells and agricultural land. Herding, trading and labour movements into Syria were also affected by border controls, especially after the establishment of the Ba'ath government in 1963. Again, certain groups lost wells, markets, grazing land and agricultural land. As states became more nationalistic and had differing and often opposed politics, control over their populations became more important. In the seventies, it was common for individuals who used two or three states to have passports from all. Now this is not possible, and people who formerly moved between states must decide of which state they will be a national. People make their choice on where they see their assets as being and which state, to whose nationality they can substantiate claims, will offer them and their children the best future. Members of tribal or family groups decide which state offers them the best option, but all groups end up with members in each state now occupying the areas traditionally used. Nationality is seen as a state confirmation of the right to residence and work, access to state services of education and health, and a passport. Many believe that these rights are inherent in being a member of a named and known group that lived in an area and used others, and that such rights can be earned by those who move to other areas for whatever reasons, as long as they work through customary practice and are accepted. The reluctance of Kuwait to give nationality rights to people who had lived and worked there for years, and were often members of tribes who had always used Kuwait for grazing or markets, was the main reason for unwillingness to support Kuwait in the Gulf War.

A Sardiyya said "I'm a Jordanian national, I have a Jordanian passport, but my heart is Syrian. My brother runs a transport company in Syria, and he has a wife in Damascus and a wife in a Sardiyya village in Jordan, and he has a Saudi passport. All his adult children now live in Saudi Arabia. What does nationality mean? What does being a Palestinian mean? Our ancestor (pre1600 AD) had a brother, and they quarrelled. His brother joined the Saqr tribe at Baisan, in the northern Ghor of Palestine. His descendants came to a refugee camp outside Irbid. They've left now and got land. But when they arrived here they were Palestinians." Many Sirhan are Saudi citizens, while others remain in eastern Jordan. Most Sharafat of the Wassamet al-Bahl Ahl al-Jabal are in Jordan, with a few in Syria, but the Hassan section of the Ahl al-Jabal are all in Syria. Most Rwala are Saudi citizens, with some in Jordan and a few in Syria. Family units have individual members and their families in each state, although tribal sections tend to focus around their traditional assets of wells and access to markets, now often transformed into irrigated gardens, and access to state employment and entrepreneurial opportunities. Many Azazma left land in Israel to come as refugees to areas of Jordan they had earlier used though never owned, while others remained. In the winter of 1995, an Serahin Azazma from Israel was visiting close relations who have been herding around Fainan and the upper part of the Wadi Araba since the early fifties. There were conversations about rainfed agricultural land (*shamsiyya* land) owned by the Serahin pre-1948 on what has become the outskirts of Bir as-Saba'. Some land was expropriated, and more has been legally removed from them by the authorities insisting on registration by individual owners; if the land was not then used within a period of time, the owners lost their rights which were given to Israeli settlers. Other land was rezoned as building land; as they never had the money to build, they lost that land, although in some cases monetary compensation was paid. Some Sa'idiyyin stayed on their land in Israel, while others lived only in Jordan, from where they continue to claim former land in Israel. Contacts are maintained between the composite parts of groups in different states. Access to former areas varies from country to country, and also for what purpose it is required. Visiting is not usually difficult. Investment and supervision of business enterprises is normally possible, with greater or lesser facility. Tribal or group held agricultural land is often lost. Herding is sometimes impossible, as between Jordan and Israel, and always difficult with restrictions on cross-border mobility except at official crossing points and with papers of sale and sometimes medical certificates.

Borders and nationality are greatly resented by rural populations of all present states of the Bilâd ash-Shâm. The French and British Mandate governments are blamed for the imposition of borders. Kostiner (1993: 188–9) describes how in the early days "both ibn Sa'ud and the Sa'udi tribes had difficulty adapting to the principle of permanent, defined borders, for these would clash with free tribal movement and Saudi control. In the early 1920s, under the pressure of British authorities who were seeking to demarcate the borders of local states, the Saudi perception of boundaries began to change....... In the mid-1930s, Saudi territorial perceptions focused on regionally recognised, permanent, and demarcated boundary lines". A widely held local view is that "Arabs should not have different nationalities. We all know we have different traditions and so on, but all Arabs should be able to move freely. Governments shouldn't have this power, movement should come from market forces. People would move to where wealth is, just as always happened in the Arab world. In the Islamic tradition there are no boundaries, and people moved all the time. Now those who move are employed and without protection and rights, under the control of their employers; they should be *musharriq*, share partnerships, so both are equal and free people. We Bedu have managed this up to a point........ The borders are a given at the present, but they came from historical events and so they can be changed." Local arrangements can be negotiated, as for example along the south-eastern Jordanian/Saudi border, and on the Jordanian/Syrian border.

The current givens of nationality, borders, currencies, duties and registration are all perceived to extend state control. Some living in the countrysides can profit from them, rather as peasants 'caught' between state and local power bases in the past were sometimes able to use the conflicts of power and its legitimization for their own ends. Rather than consider faceless categories of peasants, herders and state officals, it is more useful to note how local people assess and utilise the 'givens' in their environments.

State land reforms often do not deliver all the results they were established to achieve. The Wadi Sirhan project of 1960s in Saudi Arabia (Hamza 1982), the Ghab project in Syria (Métral 1984), and al-Jafr (Bocco 1989a; 1990b) and the Northern Ghor in Jordan all developed in ways not apparently foreseen by their developers, rather like the agricultural projects of Midhat Pasha in the 1870s at Dmayr and other places (Musil 1927: 381–2). All were based on state ownership and management of land and water, with the state distributing land and water rights to families. The Ghab project aimed to break up existing local village communities so, while kinship relationships were respected, the reform committees tried to mix people from different villages and faiths. "The cement to bind this created mosaic was supposed to be a new ideology, that of the socialist revolution, and the process generated by a unifying framework" (Métral 1984: 75) to be expressed by the new organization of landholdings and agricultural cooperatives. This did not happen because peasants found gaps in the apparently rigid state-run agricultural system which allowed them to develop private initiatives supported by traditional social practice. The state operated a 'socialist' system of development, owning the means of production (though not exclusively, private land ownership below the level set by the state continued), but encouraging private investment and allowing the private sector, including peasants on state land, to realise substantial profits. These apparent contradictions are an inherent part of methods of land management used over the centuries regardless of state ideologies. The Northern Ghor situation is somewhat similar, a state directed programme which redistributed confiscated land in 30–40 dunum plots to former owners, former users, other users of the area, people from outside the area, and Palestinian refugees. These last two groups rarely moved to the Valley (Layne 1994: 47). The Jordan Valley Authority controls water provision and distribution. The JVA concentrated on vegetable crops using the new technology of drip irrigation, with piping from the new plastics factories. During the seventies and early eighties, the project was hailed as a dramatic example of progress; Layne (1994: 45) says "state subsidies have guaranteed a level of financial success, and the prevalent pattern of family farms grants tribespeople a degree of personal/familial autonomy that is highly valued." She notes the continuation of the multi-resource economy among every family and by individuals within each family, and an expansion into transport and services, as does Métral in the Ghab, and many observers of rural and small town economics. From the late eighties, the export market to Saudi Arabia and the Gulf collapsed, as they developed their own produce and restricted Jordanian imports. Some Valley farmers could not afford to continue and sold out to urban buyers, who work the land using share-working agreements with the former owners and other locals and hired Egyptian labourers. Many local people prefer to share-work for urban or larger landlords since they do not want the responsibility of arranging for and paying for water allowances, subsidised fertiliser, and marketing to the big wholesale markets or canning plants.

The modern example *par excellence* of using borders as a resource by local people and state power groups is smuggling. Smuggling has several facets. It can be used to make political statements – "these activities would be trading if the borders and different currencies and regulations weren't here; as they are, it's smuggling." A participant may see smuggling as allowing an autonomy of action denied by states who do not fulfil their function of allowing goods to pass freely. The young and others without assets smuggle to get livelihood and profits for investment from their own activity, rather as they formerly raided. Particular power groups in a state can use their position (Perthus 1995: 149–50) to benefit from the supply of goods not produced by the state for whatever reason but desired by the population. Again, this is reminiscent of a form of raiding, moving goods from areas of surplus to those of deficit. The smuggling of cigarettes, electrical goods and quality cloth into Syria in the seventies was run as raiding had been (Lancaster 1981: 91ff). This trade moved into the hands of the Syrian army during the eighties. People transport drugs as individual operators for profit and risk-taking. Weapons are traded across borders as only Jordan has a free market in guns, and "they are useful things to have"; guns are traded in the same way that cars and pick-ups are. Bringing in sheep from Iraq is semi-official and profitable because of the discrepancies in the exchange rates between Iraq and Jordan. Some see the enablers of this alternative system as profiteers, "because he is charging too much to circumvent his own government, he's double-crossing virtually his own family, and he doesn't need the money." Others, while disapproving, pay up as "it saves me having to do the paperwork and from answering awkward questions, and it's cheaper than the official rates."

Face to face relations are important for achieving personal and group aims. Many need to call on third parties at some points in their lives. Third parties are used either as middlemen/mediators (*wasît*), or as sponsors/guarantors (*kafîla*). A middleman mediates between two persons unknown to each other but both known to him; he introduces them to each other as reputable persons. His ability to perform this service is the *wusta* of the person seeking to be made known to the unknown. This process has been taken to designate the securing of favours, since a usual reason for someone to wish to be made known to another is to request something, but it more truly denotes the process of mediation and the intermediary himself. The *wasît* makes the two parties known to each other, any action is up to them. Farrag (1977), discussing *wusta* among Jordanian villagers, found the term used particularly as a recommendation by a respected member of the community concerning a younger and therefore less known man to a potential employer, usually the army and government services, that the applicant is politically reliable and suitable. *Wusta* is not patronage, since the unknown and the middleman are in a symmetrical relationship; the *wasît* may be a senior man with a respected reputation, or a friend and contemporary with different networks, but for the process of the transaction they are both free and autonomous individuals who choose to make it. Tribesmen of known reputation present in one state are frequently called upon by fellow tribesmen and others of different nationality with problems in herding, trading or transporting between the two. Often, a telephone call to the appropriate bureaucrat or police station enables the tribesman to put his case to the relevant official and the problem to be sorted out between them. Tribesmen of reputation are often asked for help in getting a Saudi passport for someone formerly using Jordanian or Syrian papers; here, the *wasît* will often tell the applicant which office to go to first, what he needs, and to say who sent him; elderly applicants may well be accompanied by the *wasît* and introduced to the relevant officials, as the old are usually illiterate and confused by bureaucratic procedures.

A sponsor or guarantor makes himself responsible for the acts of the person he is sponsoring in his area. All foreign nationals employed in Saudi Arabia must have a *kafîla* responsible for his/her actions, and paid for his services by the foreigner's employer. The *kafîla* may earn his money dearly, with much expenditure of time and money in tracking down absent or absconded workmen; government penalties for missing sponsorees and the costs of repatriating unsatisfactory workers are considerable. The *kafîla* must provide recompense and restitution for damage or injury done by one whom he has sponsored, and he must look after them if they are injured. Foreign companies doing business in the Kingdom have a *kafîla* from the Royal Family and their associates. Once a foreign national is in, and finds that his job does not come up to expectations, conditions are poor, or work ends, he can change *kafîla* if he can find another to take him on. A *kafîla* acts like a labour exchange with social provisions, fitting job seekers to work available, and dealing with the employees' paperwork, renewing residence permits and sorting out problems.

A *kafîla* may also be used as a *wasît*, and the two categories blur into one another when someone makes the initial move in getting a problem sorted out. A respected Rwala, AZ, in one of the northern towns in Saudi Arabia was visited by a fellow tribesman whose sister, a Saudi national, had married an Umuri, a Syrian national, who had recently died. There were now problems over the children's nationality, technically Syrian, although their parents had lived in Saudi Arabia for years, and the children were born there. What could be done? AZ said he would look into the matter, and tell the man what should be done next. If the man and his sister should see the relevant authorities themselves, AZ will pass this on with a name to ask for, and will have acted as a *wasît*; if the problem can be resolved by AZ guaranteeing the information from the Rwala and his sister, and so enabling the issuing of papers, he would be a *kafîla*. The Rwala with the problem had alternative sources of information and help from the umda/local government official, responsible for relations between the local population and state authorities; the reasons for his choice of interlocuter are unknown. People using a *kafîla* may appeal to a third party if their man seems slow or unsatisfactory. Three Sulaib went to see a Rwala shaikh in Amman when trying to move their sheep from Jordan to Saudi Arabia after a nine years' stay. They had a Rwala *kafîla*. At that date, people could take out of Jordan only 10% more sheep than they had brought in. This group had arrived with 600 and were leaving with 2,500. Sa'udi regulations allow a returning flock to have increased by 50%. But are these increases each herding year, or *in toto*? With outstandingly good management, good fortune, and keeping all ewe lambs in the flock, the total of 2,500 could have been reached over the nine years. The Sulaib had been back and forth between the customs and the badia police three times with various papers, and felt their kafila was not furthering their interests. The shaikh told them to see a named official in the customs, and to tell their *kafila* "I will be getting in touch with him." On another occasion, the shaikh received a telephone call from the border post, asking him to guarantee 29 Rwala tribesmen, which he did, and they were allowed to enter.

In this society, people expect to be answerable for their actions and to pursue claims and rights against injury and loss through accepted social practice and ultimately by self-help. The twin questions of responsibility and answerability exercise many. At one level, women in Saudi Arabia do not like hospital nurses being veiled and not wearing name tags on their uniforms; "you don't know who they are. What happens if something goes wrong? How could I make a complaint if I don't know who anyone is?" The lack of answerability by 'the authorities' also irritates; "some women in the town wanted to have a club for sports, like aerobics and so on, and perhaps a writing and poetry circle. We found a site, and we were supported by our fathers and brothers. But we couldn't get a licence. There was no reason given, we don't know why we didn't get permission, the men's clubs do." Projected business enterprises are also refused licences; "we wanted a licence for a juice pressery for grapes and pomegranates, but we were refused. We couldn't get one for a drying plant to make raisins, either. Or for a factory for making tomato paste and canning tomatoes, or for an olive oil pressery or a bottling plant. And no one in the area has been allowed a licence for a potato crisp factory, although potatoes do well here, because someone influential at Court has the licence for importing crisps. There's never any reasons given. And then we see, after a couple of years, other people close to power building an olive oil pressery." In some instances, the reasons for refusals are thought to have political overtones in that local enterprises are restricted to protect the profits of those close to the centre of power, or that potential businesses are directed to those within the ruling elite. Some consider they are discriminated against so that they remain dependent on central funds. It was said that the government had refused to buy someone's wheat crop as a mark of political disfavour; other sources said there was a free grain market where prices were lower, used by those in arrears with the repayment of government loans, since the government can seize a crop in lieu of loan repayments.

In Jordan, complaints – in addition to those about not being able to cross borders for grazing and supplies, and the difficulty of herding with educating children – focus on democracy and the state system (*nizâm al-daulat*). The two are interlinked and interdependent, and seen to block communication and redress. A *mukhtâr* (headman of a village or a tribal group in a village) in the *harra* said "it's no longer possible to go and speak to the people you know and sort things out, everything has to go through the state. Disputes have to go to the courts rather than to mediation. When the piped water and the electricity took so long in coming I went to the local offices in Mafraq, and to Amman, because I knew people there who could help. Now I can't do that, we would just have to wait. Being *mukhtâr* is nothing now." A Nusayr in al-Husn took a similar view; "I don't like written contracts or courts. There's no need for them, they only lead to bureaucracy. You should be able to do everything yourself, and if matters get beyond you, you go to the *shuyûkh*, men of good reputation and expertise. Going through the courts to settle a dispute has no advantages for anyone except the lawyers you have to pay. I don't see why we can't use the old system, where we did things for ourselves and talked to people we knew and who knew us, and that was how it worked. A *mukhtâr* does nothing now except be a rubber stamp for identity papers, and at weddings and land transfers. He doesn't settle domestic disputes anymore, that's done by the police; he has to be there when the police search a house, that's all." These could be taken as the complaints of middle-aged men about a system they do not understand, but the dislike is deeper than misunderstanding, and more widespread. An educated young tribesman and university lecturer considered "democracy must be at the very least consulting with people, listening to them and having discussions. It isn't telling people what they are going to get. Proper government has to come from inside the person, not be imposed from outside." The only people met who rather approved of increasing bureaucracy were a family of Communist Christians, who saw it as "more democratic." However, they used a middleman or mediator frequently, "not to contravene legal requirements but to speed things up and because using a *wasît* is more convenient. We use a *wasît* for building permits and getting electricity and water. The *mukhtâr* is just a rubberstamp. If there are problems in the village, or between one of us and someone from outside, we never involve the police or the government unless we absolutely have to. We keep it all quiet, and get any respected older man, any *kabîr*, to sort things out."

A respected retired Jordanian administrator analysed state government as corrupt and unnecessary of itself. "Ruling works where it is through face-to-face relationships, and where everyone is responsible for what they say and what they do, and where they are answerable to complainants. It might be slow and unlike western nation state bureaucracies; but everyone knows where they are, and how to get access to mediators, not only for settling disputes but also for getting the means to support themselves. People really administer themselves and sort out their problems between themselves. And because there are so many networks between all the different social groups and between regions, it is an open form of government. Because of the way in which resources are owned, there are lots of ways of getting to resources by people who don't own them. Most people are good people, honourable people, who defend their resources and assets, but are also generous and know that everyone needs livelihood. People know how they should behave for the benefit of their community, and on the whole such a system works well. If there are disputes and of course there will be, these can be usually be reconciled. Once ruling incorporates a rich elite on whom it is dependent for money and who have to be rewarded with opportunities for more wealth making, rule is lost. This always happens. It is how states get established. Look at Sa'udi Arabia in the early days, look at Oman, look at Kuwait, Jordan. The Ottomans were the same, so were the Mamluks, all of them. .... There are two strands: one is wealthy merchants who supply money and then are rewarded with monopolies or agencies; the other is the people from nowhere whom the state makes its administrators, and they have to be rewarded. Sometimes, like under the Ottomans, local notables got incorporated into administration, but these had responsibilities to their groups, they couldn't be just bought, there had to be accomodations by the state. This was the case here, but now the state brings in more institutions, more bureaucratisation, more technocrats, elections, and they call it democracy. It isn't our kind of democracy, where we work through people we know about and who are answerable to us, good people, people with honourable standards of conduct, people of family. The reason the rich are rich is because they are not of family, they have no standing, no reputation, except through the rulers. It's for the very reason they are not of family that rulers pick them, they can be bought.

The rewards that are given now may be in heading NGOs which provide prestige, endless opportunities for aid money from abroad, contacts, the chance to build a private empire, jobs for relations and connections .... And the ruling families here and in the other Arab countries, they use NGOs in the same way. There are families who do not participate in these, and these are *asîl* families, the well-thought of, the honourable. Democracy is a face-saver. The elections make it look as if people have a voice, and as if the old families and the honourable have a function as representatives of their local areas, but the members of parliament have no meaningful authority. The house of representatives just looks good on television.

It's more difficult for people to get redress for wrongs now. Customary laws were abolished in the seventies, but that wasn't too bad because in fact the law authorities ratified decisions that people had made through mediators or arbitrators.[28] And new situations had arisen in the country that needed modern laws. And of course, in many, many cases people didn't go near courts but sorted things out themselves. But now everything and everybody and every enterprise needs papers, the courts get involved all the time. Papers don't tell everything about a case, and local *mukhtârs* and shaikhs knew this, but courts can't. Local affairs aren't in the court's official knowledge, and those who know how to use papers and laws rather than rights and obligations benefit."

The facelessness of government ministries and NGOs is perceived as a major reason for discontent by rural populations in their relations with state institutions. One arena is the extension of control over state land (and see Bocco 1989a). Control of land by the state is an integral part of Islamic law. *Miri* state lands can be rented for cultivation and may be released for purchase, but the state can also withdraw land from rent and re-allocate it to itself as *kharaj* land for forests or reserves. Many rural areas have been caused hardship. Antoun (1972: 26) describes the effects of creating

---

[28] Ghazzal (1993) discusses *qâdi*s in Damascus who largely ratified decisions made by local leaders when settling disputes through customary restitution and recompense from *qanûn* and *shâri'a* law.

a government woodland reserve of one thousand acres of land in 1939 from land formerly used for grazing and crops. Both landholdings and stock raising by households declined while the population was increasing. Twenty years later, half the village families were landless, and most of the remainder needed part-time employment for a living. Generally, expropriation of forest land used by villagers for grazing and firewood by the Lands Department was common, with some villages losing nearly half of their lands (Fischbach 1994: 96). Popular resistance forced a halt in 1938, when over 4,000 hectares were registered as state forests. Glubb (1959: 173) considered that "there was less inequality in wealth and social position in the old insecure chaotic time that there was under the new theoretical 'democracy'. The establishment of law and order resulted in the rich becoming richer and the poor growing poorer ..... The establishment of public security deprived the farmer of the power to threaten the usurer with violence."

Further increase in forestry areas after independence had two approaches (Kingston 1994). British aid programmes favoured a gradual approach, and emphasised training, establishing a legal and administrative framework, and some limited afforestation, but in 1954 were outbid by the American Point 4 plan with a greatly expanded programme. Point 4 was not successful in its forestry projects and left the Jordanian Department of Forestry "an administrative and developmental mess" (Kingston 1994: 208). Over the years there have been more initiatives to increase the amount of forest lands, while the clearing of forest for building does occur through Royal Edicts. Villages are required to have a certain amount of land as forest reserve; the Ministry of Agriculture takes in state land for grazing reserves, experimental forage reserves, and for forests; the Royal Society for the Conservation of Nature takes land used for grazing and firewood for nature reserves; the Ministry of Tourism restricts access to and use of large areas of land, Petra being the most notable example; the Department of Antiquities fences off sites; and the National Resources Authority takes land for development. The aims at the centre may be admirable, but local inhabitants see an extension of state control at the expense of their interests, with the beneficiaries from outside the area. Local people are not usually against the protection of archaeological sites, wildlife, or places of natural beauty. They are against control by those from outside the area who do not have to live with the results of restricted access and a diminution of resources.

The 'Ata'ata villages of Dana and Qadisiyya and the Sa'udiyin village of Busaira have been subject to restrictions on land use and loss of land to several state bodies, including various departments of the Ministry of Agriculture, the RSCN and NRA. Local employees in an office of the Ministry of Agriculture said "We have lost a lot of grazing land and arable land. In the past we didn't use all the *miri* land on which we have claims, but now all the village land is cultivated and developed, and we need the *miri* land because the number of people in the village has grown, and will continue to grow. We're short of land. And then the Ministry of Agriculture takes land out of the miri category into *kharaj*, for forest and forage reserves. They don't seem to recognise that people have a right to subsistence. We can see no need for the RSCN Reserve. There aren't many wild animals, and the ones that are there use far bigger areas than the reserve. The trees aren't in any danger because our goats don't damage the trees if they're herded properly, and they only really eat some fruit and fallen leaves. We've lost seasonal grazing to the Reserve, and they try very hard to extend it over patches of arable land we've been using for always. These arable lands are between the rocks where *butm* trees grow, and you can tell we've used these because there are pounding holes in the rocks for getting oil from the fruits. The land that the cement factory is built on used to be farmed by five families, and it's some of the best land for apples in Jordan."

Young men in the *baladiyya* in Qadisiyya were resentful of the loss of winter grazing land, and of restrictions of access to forest land included in the RSCN campsite. "We don't mind tourists staying at the RSCN campsite, and we don't mind the President of the RSCN using it for his guests. We do mind not having anywhere for our picnics, or for putting a tent in the spring. We can't use the RSCN site, and we can't use the Goethe Forest. Everything is for tourists and nothing for us and it's our land.[29] We can't use the Wadi Dana for winter grazing anymore so some families have to pay Huwaitat herders to take their sheep east to beyond al-Jafr for the winter. It's too hard for the sheep up here on the mountain, they have to move down." A guard at the Goethe Forest complained "We've lost grazing land to the Ministry of Agriculture's Goethe

---

[29] There have been two recent disturbances against tourists in area; in one, a shot was fired at a hired car, and in the other, the RSCN campsite was wrecked.

Forest and their forage reserve, and to the RSCN reserve, and to the cement factory. The Ministry's forage reserve has sheep on it now, but the sheep belong to the Ministry." People living in Dana remarked "We don't mind tourists visiting the village for short periods. It's a beautiful area and we're proud of our gardens and old houses. The Friends of Dana (who supply the money) put difficulties in renovation of the old houses. We're only allowed very small windows and not many of them. And the Friends want to buy a house for a museum, but the family won't sell. Why should they? We're not a museum, we're a village of living people. And the Friends don't like a family taking in a few tourists for bed and breakfast."

At the western end of the Reserve, in the mountain foothills, herding is permitted, largely because of the arguments put forward by herders and transmitted through third parties. The middlemen used archaeological and written sources to support the local view that the area was not a wilderness but had been used and was used by tribal families with rights in customary and Islamic law. The herders' arguments established that they managed their herding practice through the seasons and the years to maintain vegetation, that they protected trees and wildlife because they valued them, were careful over their firewood collection and use of timber, that firing tamarisk and willow allows for regeneration, that coppicing acacia and other trees is sustainable, and that some of the assumptions about plant resources were not based on knowledge of the area. A second line of argument was that they had rights of use for livelihood, that removal of these would be tyrannical and they would appeal to the King through a local mediator. They won over the herding but lost the right to have tiny gardens using runoff water for barley and local tobacco.

Similar attempts by state bodies and NGOs to restrict use of the area around Petra by the Bdul and others with customary rights have been contested in a variety of ways, reported in Jordanian national newspapers. It is possible to negotiate use of *kharaj* lands by individuals and groups. A man from Sirfa negotiated use of government land, formerly planted by an agency with trees that had died from lack of care, on the grounds that he would be responsible for the garden, replanting, and would pay rent. Local groups do negotiate with local ministries for access to grazing in forest reserves in drought years, although permission is granted centrally. Local groups can negotiate grazing regeneration schemes with the Department of Forests. They must produce detailed plans of the types and number of trees and perennials, planting methods and after-care, grazing management, and costs of setting up and maintenance. An official said "trees and grazing are not incompatible. We can arrange for a group to have access to *kharaj* land, like we have near Zarqa Ma'in."

Parallels exist in the *badia* areas of Jordan. The RSCN Shaumari reserve was originally taken out of grazing use for agricultural development which was abandoned. Then it was given to the RSCN for the re-introduction of oryx, gazelle and ostrich. Wadi Shaumari was one of the best grazing areas in the region and local tribespeople, while accepting its use for the re-introduction of oryx and gazelle, thought the scheme not well-founded in animal behaviour. When the reserve was extended, local herders were angry at further loss of grazing and eventually took matters into their own hands, repeatedly cutting the fence in several places. After some years the fence has been withdrawn, although the Royal Edict extending the reserve has not been. The important water pool at Burqu' is reported to become a biosphere reserve for the RSCN, who say that the area will not be fenced and that herders will be allowed to get water. But local people are both suspicious and dismissive of their intentions.

In principle, herding tribesmen are not against conservation of grazing. They point out that traditional tribal practices, *khuwa* and *sohba* and *qosra* contracts, worked to manage grazing by restricting access and because people had to pay in areas not their own, herds were reduced (Lancaster 1981: 123–4; Lancaster and Lancaster 1997). Raiding also worked to reduce grazing pressures, since raided animals were the unguarded and so above the labour capacities of the owner; raiding of camels moved surplus animals to deficit individuals and groups. In one sense, tribespeople consider over-grazing is the responsibility of states for banning traditional practices, restricting cross-border movement and trade in animals, and taking grazing land for a variety of state developments. "This area of eastern Jordan is too small for reserves to work, either for wild or domestic animals. The wild animals need to move with seasonal grazing and water, just as we do with our sheep, goats and camels – which is why the borders are so irritating. In Saudi Arabia, there is a huge reserve, more or less from Tabuk to Jauf, and no-one is allowed in. Westerners patrol in little planes to check no-one is there. The state doesn't consider herders from tribes like the Shararat and us and others have to live. It should be the people who live in and use the area that have the responsibility to conserve grazing and plants and wildlife, not people from towns employed by the state. But it won't last for ever, nothing does. In Syria, the *hîmas* and *mahmîyyas* (types of reserve) provide for a sort of rotation of grazing." Some who have joined Syrian government co-operatives or societies which give access to rotated grazing in the *hîmas* and *mahmîyyas* see them as yet another form of control. A group of Ahl al-Jabal herders discussing reserves made a distinction between nature reserves and areas reserved for particular groups of people. Nature reserves were considered unnecessary since the militarised Jordanian-Syrian border acted as a reserve for gazelle and other animals. The conservation of areas for future grazing for particular groups was a good idea, since plants would spread from the conserved areas outwards; but the social mechanics of making a conservation area work "is very difficult since there are four groups of Ahl al-Jabal who use the whole *harra* regularly, and people from the Beni Sakhr, Beni Khalid, Sardiyya, Sirhan, Shararat and Rwala. It wouldn't be any good organising it through shaikhs, they'd only register the area in their own name. And the Ministry would be worse, useless."[30] At the same time, many deny that there is over-grazing in the long term and over the whole *badia*. According to them, bare areas result from temporary intensive use and/or lack of rainfall, and the vegetation will be restored once appropriate amounts and timings of rain have fallen.

Restrictions on hunting are regarded with some cynicism in the rural areas; people say "Who were the hunters? People from outside the area, from the cities. And they're the ones who stop us hunting. We hunted and we killed what we could eat, not like those from the cities who took photographs of themselves with mountains of birds or gazelles all piled up. Hunting with hawks and a saluki is fine, or shooting a gazelle or sand grouse in season for food. And shooting a wolf that's killing your lambs is right."

Badia development projects are regarded with scepticism and disinterest. "We can't use the *badia* properly because state governments won't let us, with their borders and crossing points, and

---

[30] A suggestion to use a traditional *faqîr* to mediate claims of access and use into contracts, witnessed in front of a *qâdi* and the head of the *badia* police was considered possible, together with registering such a co-operative with the Minsitry of Co-operatives in Amman.

passports. And the development programmes the state sponsors, they're no use. The state doesn't know how we used and use the *badia*. I don't know if they even understand there is anything to know. All those programmes are a waste of time, and no-one on them ever says anything interesting, even if intelligible. To get your living in the *badia*, you have to be skilled. We know how to get a living here; what do they know about the *badia*? There's no way a townsman could survive here, and yet we survive in towns pefectly easily." Another man complained about social survey teams; "Why do these people come out here asking me questions about how much salt we put in our food and how many blankets we have? They take up a lot of time, and for what? We don't have any blankets, we have proper wool quilts made by the women. They didn't ask about quilts made at home, only about shop-bought blankets, so I said 'none'. Have they put me down as without bedding? or poor? I spend my money on spares for the lorry and feed for the sheep." "These projects have no purpose; they are not for the *badia*, they are for jobs for people from the cities, who don't understand how our systems work. They talk at us, they never listen to us. We develop our own enterprises from profits from herding in good years or employment in Saudi Arabia, and ours work." "Some of the people are quite nice, but the projects are stupid. We women don't have time to weave rugs for tourists, we're busy. It takes a long time to make a good rug, and mine are for the family. If I wanted to weave rugs for sale I would; there are men who come round and buy them or I could take my work to merchants in the town." A shaikh said about a grand project for the *badia*, "We proposed digging out entire *ga'*, maybe a kilometre or two long and a kilometre wide, for rainwater storage. It would be really cheap, the Ministry of Works could supply bulldozers and the diesel would not cost much. It would take about four years for the silt to line the *ga'* to make them waterproof, but it wouldn't cost anything. But the project turned down our idea, they said it wouldn't work. They said the silt would never make them waterproof, and that the water would evaporate quickly. Rubbish! We know how the silt in the *ga'* works, we see how long a rainwater pool can last. A big one can hold its water for weeks, two, three or four months. A water expert from Britain agreed with us. We think they turned it down because there were no jobs for them in it, they wouldn't be employed for years making surveys and planning and so on, trying to find out what we already know. They want to build a big dam almost on the border with Syria, damming the whole of the Mingat and Ruwaishid waters. We think this wouldn't work, the dam would break under the strain. They don't have any idea of the force of the waters that can flow down. When we make dams to hold water for growing barley, we build little dams and place them so the water can flow past as well as being held back. They don't know how the water flows, they don't ask us who do know because we live here and use the area all the time, they just make another survey and go back to their offices and their computers in Amman. The whole project was to give work to engineers and agricultural experts, nothing for us. They had another idea to fence in the whole of the Dumaithat, Dibadib to Anqa and the border, for growing perennial shrubs. This is some of our best grazing! It has plenty of perennials and dormant annuals. When there is rain, everything will regenerate. It is the nature of the *badia*. We (the senior respected men) stopped that project. If governments really wanted to help us and the *badia* environment, they would reduce the border restrictions on grazing and trading."

Local knowledge is often regarded as an asset that should not be transferred to outsiders willy-nilly. Many travellers commented on accusations by local people that the travellers wished to take away for their own purposes information acquired from them, while at the same time appreciating the sharing of knowledge between themselves and their hosts and guides. The ambivalence appears to reflect, the the one hand, an awareness of knowledge as a capital asset that can be turned against its owners if acquired by those who wish to move into the area or in some way control the local population and, on the other, the pleasures derived from transferring knowledge to those in some way under their protection and who are appreciative of the information or can be construed as needing to know by the holders of items of the information. State agents and those without a relationship to the holder of information fall into the first category, where requests for information are met with courteous formulae, non sequiturs, denials, or silence. These enquirers can get round the impasse by negotiating themselves into an alternative role compatible with the second category, but many are not prepared to do so, or indeed realise this is available. 'Knowledge is power' is relevant in the rural Bilâd ash-Shâm, but the knowledge confers 'power to' rather than 'power over', although 'power to' can be converted into 'power over' by those coming from arenas of hierarchical power relations. There is therefore an awareness of the opposition between local knowledges and official knowledge where, in the eyes of the holder of official knowledge, local systems are to be incorporated and subsumed or to be denied. Partly this opposition is a function of a context in which official and local are necessarily opposed; partly it reflects the reality of urban highly educated personnel having no awareness of rural landscapes, economics or social processes except in the terms of their education. The possession of local knowledge is intimately bound up with identity as a person of the area, so the transfer of small parts of this to one from outside implies social relations between them, where the recipient acknowledges the possession and transfer of the donor. Knowledge may be widely known and general, widely known but largely unimportant, or known to a few but deeply relevant in certain fields. The routes, methods and recipients of transfer are similarly layered and negotiable.

While viewing development programmes with scepticism, local groups see them as a potential, though often unsatisfactory, resource. Archaeological excavations provide short-term work for labourers, short-term rents of local houses and employ a permanent guard; these benefits should be distributed among those from the local population who 'own' the site and need income.[31] Achieving these to the satisfaction of archaeologists and locals is mediated through the local representative of the Department of Antiquities, the director of the excavation, and local notables. Conservation programmes hire local guards who gain an income and are usually in a difficult position between their neighbours and their employers; other workers may be local, but most are from outside the area and have used personal networks to get their positions since the qualifications needed are held by many; managers and administrators are invariably urban-based and educated while technically well-qualified and enthusiastic. Big development projects, like the cement works at Rashaida in the south of Jordan, are major regional employers. The cement works was set up to provide cement and to alleviate poverty in the south. Each major local tribal grouping has a proportionate share of jobs, and over-manning is specifically seen as a contribution to social welfare. The relatively few specialised

---

[31] This line of reasoning is also used by those who make part of their living from finding and selling antiquities who argue that they have the right to profit from their excavations since they live in the area, have always lived there and are entitled to make a living from their own efforts.

and skilled jobs are well paid, although employees complain of boredom. Other development projects provide local employment, usually for unskilled or semi-skilled jobs, while technical positions are filled by nationals, but rarely from the area. A fully local resident will not have the formal educational qualifications and language skills to fulfil the employment criteria, while those who do cannot know the area, population and their enterprises in any great detail.

What specialists on development programmes hear from local people is discourse aimed at managing any potential government action arising out of the programme; "everyone lies to government officials and tells hard luck stories about how poor they are and how difficult it is to manage. It's natural. They might change the regulations and we'd get something out of it. I know the development project has foreigners on it and they say it's research, but they're from the government. They've got Ministry cars and there are people from the Ministry with them. How else would they get here? Of course they're government."

There are two disparate attitudes to the functions of government. On the one hand, government should provide for its citizens as an enabler of livelihood, i.e. 'power to'; this leads to the corollary that government is itself a resource to be managed. On the other, government maintains itself by appropriating local surplus, by fiscal measures or restrictions on freedom of resource management, i.e. 'power over'; the corollary is the development of elite families who become richer while others become poorer. 'The rich get richer and the poor get poorer' is as common a remark in the Arab Middle East as it is in Britain. People complain about the way state government acts, but explain disliked actions from current political realities and the greed and lack of responsible behaviour by members of the elite. "There are so many people now that it isn't possible to support everyone from land, even with trade and manufactures and services. We have developed mineral resources and manufacturing, and the service sector, which includes the transit trade, banking, private healthcare and educational establishments, and tourism. But their development takes a great deal of money, so we needed aid for that. It is almost that the government hunts aid money to get wealth into the country. It's necessary for the people to survive. Of course, some private money comes in for development, especially for the private hospitals and universities, and some manufacturing projects, and for the businesses that people working abroad set up in their villages for when they return, but most of it is government organised. To succeed in attracting aid, there must be locations for projects. As the state owns the land, the government can provide locations although it might mean local people have to give up land or stop using it how they want. Local people don't like this because they see their rights being restricted while really the government is trying to benefit them, or other sections of the population." It sometimes seems as if there is a policy of pursuing aid for any reasonably plausible scheme. Conversely, an NGO gives permission, for a substantial fee, to a foreign NGO or company to assess some particular project the foreign party has in mind for development, while knowing the viablity of such a project is virtually none.

If government and its branches are seen as a resource for local people, and if surplus wealth no longer comes from the products of land, why do people not concentrate on government employment and on the new sources of wealth in development in tourism, banking and so on? Why does land remain important? There is a change but perhaps not a transformation in the relations between wealth creation, landholding and identity. Education has developed from the situation where every village had a primary school to the present where all centres of over 5,000 inhabitants have secondary schools that prepare for university entrance, and state universities in the north, centre and south of Jordan, as well as several private universities. The relation between population growth and urbanization is clear. Not only are more people available for technical and managerial government posts, but private businesses in the service sector have greatly increased. Wealth, for many, is to be made in the big towns rather than in the countryside whether by providing professional services or by investing in and developing building land. Less land per head, from the increase in population, in itself implies a reduced income from land if it is not developed beyond customary agricultural production by becoming used for industry, commerce or dwellings. People work land for all these productive uses by arranging share-working agreements, hiring labour, part-timing, or when they have retired. The retired often have 'a little business' or 'buy land for a little garden', producing income and family supplies. Arable land provides income and profits from sheep-herding that fits with grain land; irrigated land is more profitable growing stone fruit crops than vegetables as most think the export-led boom in fruit and vegetables of the early eighties is unlikely to be repeated. Many continue to regard sheep as one of the best investments; those who make significant profits are those who herd fulltime with their family or with a shepherd and supervise fulltime, or who herd and trade fulltime across borders. Land ownership and land use as an individual or as a member of a local land-holding group is important to virtually everyone in the countryside because it confirms identity. Individual ownership of land confers identity as someone with income producing assets and some degree of independent security. Longstanding group landownership confirms continuing group identity and assets, and thus local political standing and reputation independent of the state. The state may have confirmed ownership through registration, or through a gift of title, but the claims to ownership precede and are independent of the state. For many, ownership reinforces claims to government employment.

In northern Saudi Arabia, Rwala have changed their attitude to land since the late seventies and early eighties. At that date, government was urging land settlement on nomadic tribespeople. Tribespeople considered the Wadi Sirhan agricultural project of the late sixties unsuccessful as it gave a lower income and less security than herding and employment, and imposed living under a government sponsored project. Rwala had farms in al-Juba by the late seventies, but no-one expected to make money from them; they provided a base where the elderly lived with the families of the young who herded or were employed (Lancaster 1981: 109), and from where schools could be attended. A few saw farming and employment as the future. Rwala considered that few markets and competition from Jordanian, Syrian and Lebanese produce inhibited any real farming in the area (Lancasters 1986). There were traditional wells, and seismic maps showed there was deep water in some areas. These waters were made more available for agriculture by the provision of national grid electricity and the import of pumping, storage and irrigation equipment. Government development plans installed electricity and water storage to villages, provided grants for irrigation equipment and farm machinery, and established buying and pricing policies. It is said by Rwala that "ibn Sa'ud put all this money into agriculture rather than pay America for wheat when the oil price rises produced so much wealth. It was a way of giving some of the money to people in the rural areas." Since many pockets of land in al-Juba, al-Busaita and the Wadi Sirhan were now capable of growing crops, the land became potentially valuable. In the logic of the state, such land should be developed, and if those who traditionally claimed preferential access did not do so, then the land could be offered to others. In the event, Rwala took up their land "because if we hadn't, others would." The opportunities for wealth from agriculture were ultimately dependent on the state, since the state set prices for electricity, water, diesel, seed, fertiliser, and crop prices. Inputs were priced below market prices, and harvest prices well above, so profits were made in the early years of development. By the mid-nineties, government support had fallen and some were turning towards seasonal feed crops for sheep as more profitable, or growing vegetables in plastic tunnels. At this date, Rwala in many areas spoke of al-Juba, Busaita and the Wadi Sirhan as "where the Rwala are". Earlier Turaif and 'Ar-'Ar had been as important for their resources of TAPline and National Guard camps, but they do not have water reserves for agriculture.

We have spoken as if local and regional government officials come from outside local communities whereas it is members of local communities that staff local and regional government at their lower levels. It is the duty of government as *hukûma* to provide or ensure livelihood for its citizens. Local governments are relatively effective in supplying people's perceived needs. Working in local and national government is sought-after; in addition to job security and pension rights, participants are useful to fellow community members. Service in the army and police had similar benefits, and visibly useful functions. Army pensions, at least, are heritable so that service is seen as an investment for the future and the family.[32] Participation is necessary, locally to be seen as of the area, and from outside to indicate that the group has local standing. Government as *hukûma* is legitimised locally by participation.

Local government in Jordan is organised at three levels, explained by the *mukhtâr* in a village on the Karak plateau. "There are large towns, the capitals of districts, like Karak. Then there are small towns, like Qasr. Then there are the ones like us, groups of villages who join together to make up a population of about 5,000. All levels get money from the Ministry, and have income from

---

[32] There is a reference to Amir Abdullah of Transjordan recruiting army officers from the former Ottoman officer corps, where his acceptance of transferable pension rights played a large part in accepting the new employer.

other sources. Amman gets a lot of its income from local taxes which it spends on street cleaning and lighting, road repairs, bills and wages. Qasr gets a little from taxes, licences for business premises and so on, but most comes from rents of buildings owned by the *baladiyya*. Small villages like us don't have any income because the people are too poor. It's my job to protect the inhabitants from the necessity of *baladiyya* taxes by wheeling and dealing and coaxing so that all our costs – street cleaning and lighting, wages, roads, telephone and so on – are borne directly by the Ministry. They will do this. I negotiate with the Ministry and I have to be realistic in what we want because Jordan is short of money. All *baladiyyas* at all levels are different, there are no two the same."

Official regulations can be modified in certain local situations. Rather as a judge's decision in court often ratifies agreed mediations between accuser, accused and police, the governor of a district has the authority to ratify agreed modifications for local conditions. A widely commended example from the late seventies which still holds today is the arrangement for *badia* vehicle registration worked out by Shaikh Nuri ash-Sha'alan with the relevant authorities while opposed at every turn by the local Ministry of the Interior official (Lancaster 1981: 88–9). Vehicles had to be registered, but could legally use Saudi plates if not entering the urban areas west of Mafraq; the point at issue was the cost of registering a vehicle in Jordan as opposed to Saudi Arabia.

Serious crimes, killings, adultery and violent theft, have become the prerogative of the state's legal institutions, the police and the civil courts, who recognise tribal and family traditions. A killing in the southern Ghor in 1991 illustrates the combination of official and local positions. M became engaged to A's sister with the agreement of both families. M was visiting the girl, with other members of her family present, when A arrived and began beating his sister. M remonstrated, and A replied 'She's my sister. Get out of the house.' M continued to remonstrate and was shot dead. A and his entire *khamsa* to the fifth generation fled to somewhere north of Zarqa Ma'in where they waited for three days. Very early on the fourth day, those of the *khamsa* not immediately liable to blood vengeance (i.e. second cousins and beyond) came to the police station at al-Qasr in fifteen pickups to wait for the *mutaserrif* of al-Qasr, the head magistrate and chief civil authority in the town. They stated they were happy to pay compensation and put themselves under his protection, or more precisely, under that of the government he represents. In this case, tribal practice is elided with the presence of the state. In 1992, a young Beni Sakhr killed an ex-slave of the family. The young man was put in prison for his own safety, and all men of the wider family liable to have vengeance taken against them came to al-Qasr to take protection from the Majali. The victim's family were forbidden to come to al-Qasr by the government. "The state looks after the killer and the victim, but the tribal system looks after the rest of the family while compensation is worked out, which might take six months or a year or more. The women stay to look after the land and the sheep, they are in no danger of vengeance". Wider Beni Sakhr opinion held that this killing "should have been dealt with within the family. The father of the killer got few contributions from the wider Beni Sakhr for his son's compensation to the victim's family because we felt the killing was dishonourable and the responsibility was on their heads." The protection seekers stayed in al-Qasr for over a year while compensation was agreed, collected and paid. Sorting out disputes within the group inclusive of the disputants is considered honourable. A Ministry of the Interior official commented "if there is a killing within the Azazma, the police never hear of it. They deal with it themselves. They're very honourable people." A police officer remarked "in the *badia* the bedu are responsible for security, they vouch for strangers. In the towns it is the *mukhâbarât*'s (internal intelligence service) duty to check up on unknown people", although in many tribal villages the inhabitants feel they vouch for accepted outsiders and bureaucracy has no business interfering.

The extended networks so important to the functioning of society stretch into those founded on shared service in army and security units, university, or ministries. Khalaf (1991: 71ff) discusses the changing networks within local Ba'ath party groups in the Euphrates Valley after the 1966 land reform and its deradicalisation under Assad since 1970.

More formal groupings are the private voluntary organizations and clubs found in the larger towns and cities. Those in al-Karak in the sixties are discussed by Gubser (1973: 130–5), where they provided a focus of social life for "low-level government employees, teachers and middle-level employees, and professionals". Altorki and Cole (1989: 113–5, 130–9) comment on those in 'Unayzah which "manifest community spirit," funded by contributions from residents and people from 'Unayzah living in other parts of the

Kingdom, and contributing to town cultural and social life, and amenities. The Ibn Salih Centre of 'Unayzah appears to provide similar facilties to the Sudairi Foundation in Sakaka. Community works on this scale demand considerable financing and are found only in cities where there are extremely rich citizens. One reason for their development, mentioned by Altorki and Cole, is that while the state is acknowledged to have provided much, it is good for people to provide for themselves and the poorer members of the community, and to improve communal amenities. In rural Jordan, some NGOs have the function of improving rural life through women's societies, marketing rural crafts and products, and encouraging smallscale enterprises.

It is often held that the difference between state and non-state government is that the state creates institutions whereas non-state government operates by cronyism, patronage or power groups. Seale (1991: 103) sees Assad as "a man of order who values the concept of the state. Before he assumed power, Syria could not claim state institutions worthy of the name." Do these institutions have any independent existence or "are they mere camouflage for the personal exercise of power by the president?" Seale concludes a return to government controlled by a small group of Sunni urban notables is improbable, since any successor would come from a rural or smalltown background, and the political and security foundations of Asad's system are too firmly entrenched at all levels of Syrian life. A Jordanian politician and political scientist considered that government works through face-to face relations and family connections; "we've been trying to put institutions in the middle, with a proper civil service, so that there is delegation between the King and the people, but it doesn't work. I don't understand why it doesn't work, except that there aren't enough people from the educated middle class to staff the institutions properly. We were aiming at a meritocracy. But the middle class is shrinking as the rich get richer and the poor poorer. It is the middle classes who should be interested in governmental institutions like the civil service, but many of them are not. They would rather have their businesses and be professionals."

Yet people in the countryside regard face-to-face relations as essential to government but disappearing, replaced by faceless elites in the capital, faceless bureaucrats in the ministries and local government offices, and faceless masses on the streets. Is this hyberbole or is it a situation to be expected when government has no clear policy or, rather, many conflicting policies and attitudes, and sending out mixed signals? Many see the government (*hukûma*) becoming a state (*daulat*), with no recognition of partnership in ruling, without room for mediation between free parties, where fiats come down from on top in spite of – or even because of – a public commitment to elections and the democratic process. "This is not how an Arab government should be. Government should be between all the partners, all people who are participants. Government officials should be responsible for their actions and answerable to the people their decisions affect. There should be redress. And the government must be legitimised by Islam, or by Islam and a secular ideology like the Ba'ath." Both sides complain that there is a lack of institutions or that institutions mean little and largely a facade behind which the leader wields power, or that customary face-to-face relations are no longer permitted: these complaints are not new. They appear in material from the forties and fifties, and from the late Ottoman period, and probably earlier as well. They are perhaps a part of a dynamic between state and people, exacerbated by the increase in state revenues from outside and originating from state efforts.

Who are the relevant state personages in the countryside? In Syria, Seale (1991: 101–2) says there are three; the governor of the area, the party secretary and the intelligence chief, with the exact balance between the three depending on "their personalities, their sectarian background, their access to the powers-that-be in Damascus, and on the nature of the job being done in that particular governate." For example, in Deir az-Zor, near the border with Iraq and Syria's oil fields, security would be a priority and the intelligence chief important. At Raqqa, a development area, the governor would be the centre. Governors are the senior representatives of the state in Syria, assisted by a local government structure introduced by Asad in 1972. Each governor has a hundred member council, elected by universal suffrage every five years. The council has real powers, but is subject to constraints. Fifty-one of the members must be 'workers' or 'peasants', who thus have a built-in majority over business and professional groups. Damascus ultimately controls all projects and budgets. Lastly, the governor and council are scrutinised by a parallel party structure, responsible for laying down broad policy outlines and watching over implementation of policy by the governor and his council. In Jordan, the country is divided into governates, sub-governates, and villages. Governors and subgovernors are officials of the Ministry of the Interior and come from outside the area; heads of local government offes in large villages are also officials of the Minstry of the Interior but local men. The Police and Internal Security appointments parallel this structure, as do officials from other ministries, such as agriculture and forests, education, and health. In Saudi Arabia, the local governor, often a member of the ibn Sa'ud or the Sudairi family, is appointed by the Ministry of the Interior, from where his budget comes. He has local advisors from the well-known tribal and urban families of the region, and there are local officials in each village appointed and paid by the Ministry of the Interior. Ministries have offices in the regions. While senior officials come from outside the regions to which they are appointed, more junior officials are local, as are the officials in small communities. Senior officals from outside will be known to some local people, either through tribal or family networks, university education or army service.

Within each local office, each group of the community has a share in the employment opportunities. Local official positions where staff are appointed for a period are rotated among suitably qualified members from local groups. In Karak, the Agricultural Development Bank had a newly appointed head. Several members of his tribal village received loans for development of new gardens, whereas new gardens were hardly to be seen elsewhere. On asking about this, the answer came that "the head of the bank is from them, so of course they do well. His appointment is for three years, so it doesn't matter. It will be some other group's turn then." The sharing out of extra opportunities among the *jamâ'a*, with those deemed the most needy being pushed to the forefront, was seen in action at Fainan. Short-term labour was needed for an archaeological excavation season, and a permanent guard. The guard chosen by the Dept. of Antiquities was from the Rashaiyida, the tribal group that 'owned' the land on which the site was. He was one of the few more or less permanent residents as he had school-age children who went to the local primary school, and had only an army pension and a small goat flock for subsistence. He put forward his interest in the job, his need for the money, and his suitability to visitors with links to the Department and to the archaeologists, and to his tribal shaikh. In pointing out his claims to the position to a senior official of the Department of Antiquities, the senior archaeologist revealed that it was the leading man of the Rashaiyida who protected his, the official's, own tribal group when they moved from the Wadi Musa area north to Karak some three hundred years earlier. The *jamâ'a* included members of the four tribes living in and using the area. The *jamâ'a* compiled a list of possible workmen of members who needed work rather than necessarily on their ability. A Huwaitat from further east had arrived, down on his luck, and included in the list as a member of the group he was closest to; "we'll say you're one of us." He was not chosen, as he was elderly and suffering from diabetes, but he remained in the area all summer and built up a goat-trading business. In Karak, when every village got a Post Office, some family heads spent considerable amounts of time seeing a local member of Parliament who was the Minister for Communications. The Post Office provides jobs for women, in short supply except as teachers, and it was women's employment that was wanted. The result of these mediations was that while a man manages each village Post Office, there are several part-time positions filled by girls from different tribal sections. The new village health centres also provide some women's jobs which go to village people if qualified or trained. Not every village has produced a doctor, let alone a doctor who is prepared to work in government service, but most come from the wider region. Teachers in small towns are usually local, while in more remote areas this is not possible, and the teachers come from the urban areas.

The spread of education and the increasing numbers of young people with full secondary and university education have produced a move towards the professions and technical services and away from service in the military and police. Governments are reducing recruitment to the armed services, except for the more highly educated. Expatriate workers in professional and technical positions are becoming fewer because there are qualified nationals. Local government services in communications, health, education, agriculture and banking are expanding.

From the situation where most people could do everything with some formal specialists for particular difficulties, the present has specialists for most functions, and only informal general abilities. State agencies arrogate the practice of defence, law and protection from local sources, although they recruit personnel there. Education, medicine and religion formerly had a largely local content, where rural inhabitants had knowledge and skill but also called on specialists learned in their field.

In the sense of livelihood and social skills, education was the domain of the family, while as a link to a wider arena of knowledge it meant literacy by learning to read the Quran, writing, religion, and sometimes history, geography and some mathematics. Altorki and Cole (1989: 92–7) found a long tradition of Quranic schools for boys and girls in 'Unayza. The first Wahabi state sent out religious teachers to the towns for the study of the Quran and the dissemination of religion to counteract heterodox beliefs and practice. Wallin (1854: 146–7) mentions that reading and writing with a knowledge of religion was common among the youth of al-Jauf. Teachers had been sent by Ibn Saud, maintained after the dissolution of the first Saudi state and continued in the town until the 1960s. The second Saudi state established a public education system for boys in Najd in 1936, extended to al-Jauf governate in the 1950s (as-Sudairi 1995: 175–8). The first school for girls opened in 1962–3, and schools for girls expanded rapidly during the 1970s. Further education colleges for young men and women became established in Sakaka during the 1980s. In Jordan, Wahlin (1982: 15) notes the first school in Salt was established in 1850 by the Greek Orthodox patriarchate. By 1880, there were seven schools, Christian and Muslim, in Salt. Orthodox, Latin and Protestant Christian schools were founded in Madaba and Karak during the 1870s and 1880s (Gubser 1973: 62) and in other towns of Palestine, Syria and Jordan. The Christian missionaries found their co-religionists sadly ignorant and to partake in some Muslim practices. In response to this burst of Christian missionary activity, seen as a further sympton of western expansionism, the Ottomans sent Muslim religious teachers to the newly re-incorporated areas of the Bilâd ash-Shâm (Rogan fc). An 1883 report, quoted by Wahlin, noted that the shaikh of the 'Awazim tribe in al-Balqa provided a school for boys of the tribe. Other tribal shaikhs and village leaders provided local schools (Gubser 1973: 62; Wahlin 1982); the eleven-year old Sultan ibn Nawwaf ash- Sha'alan "had learned how to write in al-Jauf" (Musil 1927: 452, noted in 1915). Government education was in rural areas from the early forties and was initially compulsory from the late forties for boys only; towns and larger villages had secular government schools earlier. Government education for girls started in the countryside in the mid-fifties, and expanded rapidly; in the towns, Christian mission schools had been providing girls' education from the turn of the century in Madaba and Karak (Musil 1908: iii, 97). The association of formal education provided by external religious bodies or government employees with incorporation into centralised state systems is clear.

In either case, a system of identity and rules for living, wider than those of family and tribe, are presented as advisable or essential, while local identities and social practice are held to be insufficient, at best misguided and at worst actively destructive of the greater system. The inhabitants of the countryside had religious identities as Muslim or Christian, were part of networks linking them with other regions through markets, and had roles in wider political systems. But with a decentralising state, local identities, practices, and processes provided functions of education, medecine and religion. A centralising state makes itself the official provider of all these functions, thus establishing its own credentials and maintenance; local provision and techniques continue but as an almost invisible alternative.

Before the reincorporation of the countryside, its religious practice was largely informal and local and centred around shrines, rather than formal and official and taking place at mosques (Jaussen 1927: 141). Dervish orders existed in Nablus (Jaussen 1927: 186) and in the villages of Palestine (Canaan 1927; de Jong 1984) and north Jordan (Antoun 1989: 77, n. 26) going back to the Mamluk period and continuing to the mid 1950s. De Jong (1984: 44–5) reports a certain revival in Palestine after 1967, while Antoun was told in 1960 that "the Shari'a had clamped down on the Sufi orders (the use of snakes, fire, drums and banners) but not the dhikhr, repetitive mention of God itself."

Religion and its practice in the countryside has changed with the development of the nation state. Antoun (1989) has analysed the process in a north Jordanian Sunni Muslim village from 1959 to the mid-eighties, the post of preacher being held by one man from the village. Antoun sees the preacher as a 'culture broker' who interprets and passes on a message for his audience at the same time as he deals "with the overarching political and religious hierarchies whose norms and aims often differ from those of both the culture broker and his audience." In Jordan, as in the region as a whole, religious consensus exists among the majority, while religious and secular state officials live in areas where village and tribal local traditions are upheld by many. Eickelman (1989: 262) argues that the wide range of traditions and practices encompassed in Islam cannot be reduced to 'essential' practices and beliefs. He quotes Asad (1986: 15) that "a practice is Islamic because it is authorized by the discursive traditions of Islam and is so taught to Muslims "; there is a clear relationship between belief and authority. The village preacher has to accommodate the teaching of qualified learned scholars with the ideas and practice of fellow villagers through his dealings with the state religious bureaucracy and in his Friday sermons where he expresses "a concern for the problems and policies of their fellow-Muslims in the Islamic community.......... and symbolically .....represents the Islamic community in each local context" (Antoun 1989: 71) while aiding his audience in the quest for salvation.

The preacher's sermons analysed by Antoun use as sources the Quran and the Traditions, familiar and known to his audience. He also draws on learned Quranic commentaries, conversations with his son, a former student of the Islamic University at Medina, discussions with local religious figures, and his collection of books and cassettes of sermons by Egyptian anti-establishment scholars. The content of his sermons is primarily ethical, while legal implications are dealt with in the lesson prior to the sermon. His sources parallel his education; village school, then studying with and working for a neighbouring preacher, and studies with an itinerant preacher who had studied at al-Azhar in Egypt. He then became the village preacher, being paid in sacks of wheat by the villagers. In 1961 he was appointed, on success in the relevant examination, as marriage official for a number of local villages, and in 1971 he became an official guide for local pilgrims to Mecca. In the seventies, the Department of Religious Endowments (*Awqâf*) began to pay salaries and to supervise village preachers. Following the oil price rise in 1973, a Quranic school was established in 1977, a new mosque in 1983, a second preacher was hired in 1984, and more people went on the annual Pilgrimage.

Islamic teaching in towns and village has changed from small private groups in the forties and fifties to religious studies classes in the state primary school in the sixties, with private classes again in the 1970s and 80s. Formal further study is at state and private religious colleges in Jordan, or religious universities in Saudi Arabia. Antoun (1989: 267) sees a two-sided underlying attitude to religious knowledge in Jordan. On one hand, man's reason enables him to understand the meaning of the Quran, and on the other, the truths are so deep that they are beyond man's understanding. Thus Muslims of different ages, education and occupation can be interpreters of belief, while all interpretations are open to reinterpretation.

In a Muslim state, government and religion imply each other. The idea of political parties is antipathetical to many villagers and tribesmen, since they disturb the idea of consensus and reconciliation within the local community that Islam and customary community values uphold (Antoun 1989: 202ff). Jordan has had a two-chamber parliamentary system since independence in 1946. Parliament was suspended in 1974, and reconvened in 1984 with new elections, analysed by Layne (1987b; 1994: 108–127) and Bocco (1989b) in respect of questions of identity as citizens and as tribespeople for Abbad and Beni Sakhr respectively. Political parties, with the exception of the Muslim Brotherhood, were banned in 1954, although there were covert party involvements in the elections before 1967. Layne argues that since most candidates supported issues that all voters were likely to support, they attempted to build alliances to gain the support of their own group and that of others, relying on being able to influence the votes of group members. These attempts at influencing voters' decisions backfired. In the 1967 elections, discussed by Gubser (1973) for Karak and Antoun (1979) for Kufr al-Ma, votes were seen as valuable services that tribesmen could offer to tribal leaders who in exchange would procure goods of schools, roads and local emplyment. In 1984, similar approaches were made to men, but never to women who now had the vote. Layne (1994: 120ff) sees voting as "identity making", where "tribespeople celebrated the 'arab value of autonomy as they exercised their aleatory capacity to move, to position themselves vis-à-vis the candidates and other members of their family and tribe, in potentially new and surprising ways." Many voted orally, although many who did so were literate; Layne sees this as "a verbal, although embodied, performance through which people construct themselves ...... in a concrete social arena." Others regard the democracy of a centrally administered state with a certain scepticism.

Traditional medical care of people and animals in the countryside may be separated into three. One, that known to most, the household cures using plant or mineral materials for minor illness and accidents to people and animals. Two, the treatment of incurable, chronic or hard to diagnose conditions, and mental or psychological illnesses. Third, the application of particular techniques either for conditions in the second category or for specific diseases; burning is an example.

Many elements of the first category are widespread. The basic treatments by households and specialists use infusions, snuffs and fumigation, ointments and oils made from plants. Plants growing in many areas of the peninsula are used for similar illnesses, seen by comparing Morris' (1988) work in Dhofar with Mandeville's (1990) in Eastern Arabia, and information from Rwala and Ahl al-Jabal. Three widespread plants are *ja'ada* (*teucrium polium*), *hanthal* (*citrullus colocynthis*), and *shîh* (*artemesia sieberi*). The dried leaves of *ja'ada* are widely used as an infusion for fevers and feverish conditions, including malaria; it is also smoked in a pipe to relieve rheumatism. The seeds and pulp from the yellow gourds of *hanthal* are used as hot poultices to treat inflammatory swellings, especially those from thorns embedded under the skin, and rheumatic joint pains. *Shîh* is drunk as a tea as a general tonic, and the smoke inhaled also as a relief to a variety of ill-defined and minor symptoms. Horses were stood in the a smoke as a treatment for glanders. These are a minute part of rural plant use for healing, known and used by a very large part of the populations in the peninsula. These three plants are some of the most widely seen on sale. Traditional pharmacies in the towns and cities sell stones and minerals like rock salt and sulphur; oils; gums like myrrh and frankincense; aloewood, dried roots, leaves and seeds; some of their stock is imported from the Indian sub-continent, and is roughly similar whether in Sur in Oman, Sahab in central Jordan, or Damascus. In Sahab, the proprietor was not a pharmicist, knowing nothing about the properties of his stock; he sold to people, many of them women, who knew what they wanted. In Sur, the owner, a known specialist, was continually engaged in consultations with buyers. Traditional drugs, like asafoetida, salts of zinc or copper sulphate, dragon's blood, and spikenard were also traded far into the *badia* and used by the bedouin (Musil 1928a: 667–70).

In local thought of sufferers and healers, many of the second category are caused by the evil eye or *djinns*. Jaussen (1927: 225ff) came to know well a curer of illnesses and an exorcist of *djinns* in Nablus. This healer considered that "all maladies can be produced by the evil eye and in reality nearly all are ...... At Nablus, envy reigns in all hearts and secret hatred poisons all consciences ..... the evil eye is the plague of our town." A similar position is taken by Ahl al-Jabal, many of whom consult tribal healers on this kind of illness and the treatment of which is aimed to cure the social dissension that has caused it. The Nablus shaikh was consulted by townsmen, villagers and bedouin. His diagnosis was based on what the patient told him of physical symptons and on the laying of hands on the patient's forehead. This told the shaikh whether the suffering was from natural causes, the evil eye, or possession by a *djinn*. Since patients came to him after failing to be cured by ordinary methods, he dealt with the evil eye and *djinns*. His treatments for the evil eye centred on the writing of a *hijâb*, some of the ninety-nine names of God in a certain order, carefully folded and put by the patient on the afflicted part of the body and kept there. In addition, the rubbing of the afflicted part and prayers by the shaikh over five or six days were a part of the treatment, which usually worked. The shaikh could not treat palsy or convulsions, but was successful with partial paralysis. He used sympathetic magic in curing sciatica, using a root of the *'irn* tree, of the same length as the sciatic nerve. For the exorcism of *djinns*, the shaikh had to ascertain the *djinn* was in the patient's body at the time. He then wrote the names of God on the forehead, and the nails of the hands and feet of the patient, thus entrapping the *djinn*. The patient was enveloped in the fumes given off by the burning of scented plants, while the *djinn* was interrogated for the cause of possession. This could take days. When the reason was divulged, often love, the shaikh wiped the name of God off the patient's little fingernail, which allowed the *djinn* to escape. The writing of *hijâb* by religious shaikhs (Antoun 1989: 82) or itinerant religious figures continues.

An alternative approach by sufferers or their carers is to visit the tomb or a memorial of a holy person or *welî*. Some of the fifteen in the Nablus area were associated with particular illnesses (Jaussen 1927: 141, 162, 168–70). Tombs of Ahl al-Jabal *faqîr* are visited by tribespeople for the curing of paralysis and mental handicap or illness; a sacrifice is made, prayers said, and the patient kisses the grave.

Burning (*khiyya*) is a common technique in the *badia*, usually with a red hot tent pin, for people and animals. Although everyone can do this, and many apply it to themselves, there are known specialists who understand far more of anatomy and disease, and so are more able to place the iron in exactly the right place for a particular symptom. These specialists are often highly skilled, dextrous and conscientous, searching into the history of symptoms. The technique uses a knowledge of nerve pathways to block the spread of disease or pain along them. For many years, Western and western trained doctors have decried the use of burning, and its use is said to be illegal. Some now are showing a greater interest; "the specialist who treated my sister asked her how long she had

had the pain. It started when she was eight or nine and now she's twenty two. They were terrible head-aches, she would wake screaming with pain at night. I could hear her and I was sleeping right the other end of the house. Our father took her to hospitals who took lots of X-rays, but they couldn't find anything. After a year of this, our father gave in to his grandfather's suggestion of burning. He was really knowledgeable about it and a good practioner. He burnt her on one of the nerves on the right side at the back of the neck. It cured her for thirteen years. Then the headaches began to return but not so badly. Her husband took her to the University Hospital where she had a brain scan which showed a tumour and a sac of water on the brain. The pressure was making her squint and her sight was bad. Now everything is fine, thanks to God. The surgeon acknowledged my great grandfather's skill in curing her for thirteen years, and he said that burning was a system of medecine. That really surprised me, but it's good."

Traditional practitioners are consulted by those whom the official doctors have failed to help. Some offer alternative medecines, others offer different diagnostic techniques and therapies. Most appear to offer their patients a far closer and more personal consultation, often extending over several sessions to establish a patient's history. Sufferers often consult both systems in a short space of time, whether their illness is relatively straightforward or has become chronic and difficult to cure. Comparisons between their reception by the practiner, the amount of time he or she spent on their case history, and the treatments offered are made on the return home. The patient is usually sceptical, while his or her associates exhibit more enthusiasm. There are also those who refuse to seek treatment beyond a reliance on those mentioned in the Quran, like camel's milk, honey, and burning, and put their faith in God; "if God wants me to die, I shall die; if He doesn't, I shall recover."

Morris (1989: xxii–v) in a discussion of Dhofari traditional medecine comments on the combination of treatments used by healers and patients, herbal and practical, and more esoteric. Like those of Nablus mentioned in Jaussen, of the Rwala (Musil 1928a: 666–9) and current today, they have close ties with medicinal practice described in the Quran and use elements of classical Islamic medicine. Traditional medicine is successful in the treatment of psychosomatic illness and simple trauma, and those where the patient recovers from infection in the course of time. People appreciate the skill of Western medicine in reducing deaths and pain from childbirth, childhood diseases, cancers, heart, liver and kidney conditions, and so on. They are also appreciative of spectacles, hearing aids and the like in improving the quality of life.

The state expands from its centre into local areas, becoming closer rather than distant. More people in the small towns and villages of the countrysides are employed in enabling the functioning of the state in its role of providing services and livelihood. The financial resources for these roles come largely from outside the countryside and certainly indirectly; most money comes from state organised external sources. The state imposes its own physical and social infrastructures on those formerly provided from local resources, and is enabled to do so because of its own incorporation in wider global economies and political structures. The state's incorporation into an outside world and its provision of local infrastructures affect its members, but to what extent does this supersede or replace existing practice? Such a phenomenon is not new but inherent in the history of the region. People live their lives in both systems simultaneously, in both the infrastructure of the state and in those of local communities, switching functions, aims and signals according to context and their readings of the signals of those with whom they have relations at the time. It is comparatively simple to find examples of conflict between local arenas and the centre, the mindset of relations based on closeness and distance as opposed to those of power hierarchies, of action from within and that which is imposed, or behaviour based on individual autonomy and responsibility before God to that portrayed as deriving from state institutions and bureacracy. The contenders do not line up in neat defined boxes, but shift and change sides. Some individuals protest against what they see as wrong with current government functioning by moving towards fundamentalism, others maintain customary and tribal traditions. Each reaction of protest contains contradictions with what the individual sees as preferable to the existing and the means to achieve this. The ideas people have of the structures around which they and others function, and the ideas of the other participants are never totally congruent, although there may be close similarities with some and disparities with others.

A general consensus appears to be that an individual acts in several arenas, behaving in each code demanded, and that each arena and code has significance. The local arena of family or tribal grouping has greater depth and the one from which identity and assets are drawn. That of state is newer and has a more superficial relevance. Like the local, it can be managed and it makes claims on the individual; but the local arena both claims and provides more. However, people are free agents and can choose to live within a local arena or to move, to a city or foreign country where new 'colonies' may be established or to live outside these in a new situation.

Many authorities see a breakdown of traditional tribal social structures from settlement, education, employment and citizenship. A breakdown need not be a collapse but a re-fashioning, a reconstruction. The resilience and adaptive qualities of customary social practice, based on an entree through tribal or family identity, are remarkable when state authorities have been actively hostile to formal tribal processes as in Syria (Seurat 1980: 112) and Saudi Arabia. Layne (1994: 113) says that in Jordan "each clan is now led by its own shaik. Meanwhile the King has taken the role of 'shaikh of shaikhs' ." Certainly it is common to hear that "we don't bother with shaikhs anymore", but it may be that the late Ottoman and Mandate governments emphasised a structural position for shaikhs with their need to incorporate tribes. The 'solution' of settlement to 'the problem' of nomadism, actively pursued by the same and later independent governments, may be a re-fashioning of the means to livelihood under greatly changed conditions by independent and autonomous individuals who, with their families and networks, are capable of reading economic and political signals, and readjust their actions accordingly. The appearance and growth of small towns and villages does not mean people stop being mobile in pursuit of livelihood and social practice, but that they have a base, a preferred location where news of their whereabouts may be acquired and where things of value, the means of access to resources, are stored. The means of access to resources now, for most, means state registration and its education. The rules of the game change, and players must take account of these. Parallel to the resilience of social practice is that of the rural landscapes, themselves the creations of their users in the different natural environments. It was, and remains often the case, that the newly in dependent states of the region saw local customary practice as both the source of environmental damage and of under-productivity. Such perceptions by the various states permitted the extension of control over the rural areas and the creation of development plans for the conservation of natural resources and to increase productivity. Users of the various landscapes, as has been discussed earlier, consider their landscapes capable of infinite renewal since God is generous and will provide rain at the right times in due course. Their landscapes are resilient, like their social practice; like the animals, they move to sources of livelihood; like the wild plants, they can respond to favourable conditions and withstand hard times.

APPENDIX

Plants seen during the last week in March 1995 in various locations in the eastern *bâdia* of Jordan.

Identifications. This is not meant for botanists, it is a record of what local people know about their landscape and an indication of different plant communities. Plants were identified from previous knowledge and local names collected from herders. Plants were photographed and checked using Mandaville's *Flora of Eastern Saudi Arabia*: the photographs were further checked by Prof. Da'ûd al-Eisawi, Professor of Botany at the University of Jordan, to whom we are deeply grateful for all his time and help. (Numbers refer to first identification).

Location 1. (Southeast of Jâwa)

1 & 2.	Erodium sp.	*bakhatrî*
3.	Senecio glaucus	*rijlet al ghurab, jirjîr* (eaten)
4.	Gypsophila sp.	*salîh*
5.	Malva sp.	*khubayza* (eaten)
6.	Roemaria hybrida	*didhân, daydehân*
7.	? Hypocoum pendulum	*umm ath-thurayb*
8.	Speedwell	?
9.	Medicago laciniata	*nifal*
10.	Anthemis melampodina	*arbayyân*
11.	Carduus pycnocephalus	*shadd al-jamâl*
12.	Brassica tournefortii	*khafsh*
13.	Astragalus annularis	*gafâ'a*
14.	? Red campion	?
15.	Hyoscyamus pusilus	*hîshî*
16.	Centaurea sp. (Pink)	*amrâr*
17.	Herniaria hirsuta	?
18.	Alyssum sp.	*drayhma*

Location 2. (Pools east of Jâwa)

19.	Aaronsohnia factorovskyi	*jurrais, gurrais*
20.	Filago desertorum or Plantago ciliata	*gutaina*
	Medicago laciniata	*nifal*
20.	Seidlitzia rosmarinus	*shnân*

21.	Erucaria sp. or Cakile sp.	Slîh
	Erodium sp.	ba<u>kh</u>atrî
22.	Aizöon hispanicum	mlîh, mulayh
	Anthemis melampodina	arbayyân
23.	Picris babylonica	hawthân
	Brassica tournefortii	<u>kh</u>af<u>sh</u>
	Roemaria hybrida	deydehân
24.	Matricaria aurea	babûnij (drunk as tea)
25.	Caylusea sp.	thanabât
	Gypsophila sp.	salîh (Rwala), na'îma (Ahl al Jabal).
26.	Gymnarrhena micrantha	kaff al-kalb (Rwala), gutayna (Ahl al-Jabal).
27.	Rumex vesicarius	hamaitha (eaten)

Location 3. (South of Deir al-Ginn)

Much as above plus
28.	Paronychia arabica	buwayda
29.	Scabiosa palestina	umm ar-ruwais

Location 4. (East of Azraq, at the beginning of TAP–line)

30.	Notoceras bicorne or Lepidum aucheri	a<u>sh</u>bet umm salîm salîh, ra<u>sh</u>âd
31.	Trigonella hamosa	nifal
	Anthemis melampodina	'arbayyân
32.	Stipa capensis	samâ'
	Gymnarrhena micrantha	gutaina
33.	Calendula sp.	henwa, a<u>sh</u>bet al ghurab
34.	Artemesia sieberi	<u>sh</u>îh (used to flavour butter and tea)
35.	Achillea fragrantissima	gaysûma (drunk as tea)
	Aaronsohnia factorovskyi	jurrais, gurrais
	Malva parviflora	<u>kh</u>ubayza
	Senecio glaucus	rijlet al-<u>gh</u>urab
	Carduus pycnocephalus	<u>sh</u>add al-jamâl
	Astragalus annularis	gafâ'a
36.	? Cenchrus ciliaris	?
37.	Schismus barbatus	<u>kh</u>afûr
38 & 39.	Two more grasses.	
	Scabiosa palaestina	umm ar-ruwais
40.	Launea sp.	huwwa'
41.	Diplotaxis acris	<u>sh</u>iggâra

Appendix 399

42. Poa sinaica — *naza'*
43. Large thistle with large pale leaves
44. Astragalus spinosus — *shitâde*

Location 5. (Between Milgat and Burqa')

    Achillea fragrantissima — *gaysûma*.
    Notoceras sp.
      or Lepidum aucheri — *ashbet umm salîm or salîh*
    Aaronsohnia factorovskyi — *jurrais, gurrais*
45. Diplotaxis harra — *khafsh*
    Medicago sp. — *nifal*
    Artemisia sieberi — *shîh*
46. ?Stipagrostis ciliata — *sulayân*
    Schismus barbatus — *khafûr*
    Malva parviflora — *khubayza*
47. Horwoodia dicksonia — *khuzayma*
48. ?Atractyli cancellata — ?
49. A yellow medick — *nifal*
    Alyssum sp. — *drayhma*

Location 6. (By track at Burqu')

50. Matthiola ?longipetala — *hemhem, shiggâra*
51. Salvia lanigera — *'ajdayyân*
    Plantago ciliata — *gutayna, gurayta*
    Picris babylonica — *hawthân*
    Schismus barbatus — *khafûr*
    Aizöon hispanica — *mlîh, mulayh*
    Centaurea sp. (Pink) — *'amrâr*
    Malva parviflora — *khubayza*
52. Ifloga spicata — *zunayma*
53. Scorzonera tortuoissima — *dhu'lûk*
    Another 4 grasses
    Trigonella hamosa — *nifal*
54. Helianthemum sp. — *arja*
55. Pulicaria undulata — *jathjâth*
    Anthemis melampodina — *'arbiyyân*
56. ? Mandrake — ?
57. Cleome arabica — *ufayna*

Location 7. (Between Burqu' and *Hamad* Basin dam)

58. Papaver rhoeas — *deydehân*

	Diplotaxis harra	_khafsh_
	Artemesia sieberi	_shîh_
	3 grasses	
	Aizöon hispanica	_mlîh, mulayh_
	Caylusea sp.	_thanaybât_
	Plantago ciliata or Filago desertorum	_gutayna_
	Trigonella hamosa	_nifal_
	Medicago laciniata	_nifal_
	Diplotaxis acris	_shiggâra_
59.	Anastatica hierochuntia	_kaftah, kaff Miriam, qunaifida_
	Erodium sp.	_bakhatrî_
	Anthemis melampodina	_'arbayyân_
	Papaver rhoeas	_deydehân_
	Stipa capensis	_samâ'_
	Atragalus spinosus	_shitâde_
60.	Haplophyllum tuberculatum	_furayta_
	? Silene sp. (pale)	?
	Aaronsohnia factorovskyi	_jurrais, gurrais_
61.	Barbarea arabica	_sufayra_

Location 8. (Further east, near track to ar-Risha)

Anthemis melampodina	_'arbayyân_
Plantago coronopus	_ribla_
Brassica tournefortii	_khafsh_
Notoceras sp. or Lepidum aucheri	_ashbet umm salîm or salîh_
Artemisia sieberi	_shîh_
Trigonella hamosa	_nifal_
Matricaria aurea	_babûnij_
Diplotaxis acris	_shiggâra_

Location 9. (Further east, increasing flint overlaying sand on limestone; rolling country)

	Aizöon hispanica	_mlîh, mulayh_
	Schismus barbatus	_khafûr_
	? Rostraria pumila	_sajil, shu'ayyira_
62.	Astragalus sieberi	_gafâ'_
	Trigonella hamosa	_nifal_
	Plantago ciliata or Filago desertorum	_gutayna_

	Plantago coronopus	*ribla*
	Malva parviflora	*khubayza*
	Artemisia sieberi	*shîh*
	Anthemis melampodina	*'arbayyân*
	Gypsophila sp.	*salîh*
	Picris babylonica	*hawthân*
63.	Launea nudicalis	*hawwa'*
	? Calendula sp.	*henwa*
	Gynandris sisyrinchium	*sa'id*
	Calendula sp.	*henwa*
64.	Astragalus haurensis	*gafâ'*

Location 10. (*Sha'îb* and *ga'* west of ar-Risha)

	Anthemis melampodina	*'arbiyyân*
	Artemisia sieberi	*shîh*
	Caylusea sp.	*thanaybât*
	Malva parviflora	*khubayza*
	Erucaria sp. or Diplotaxis acris	*shiggâra*
	Aaronsohnia factorovskyi	*jurrais*
65.	Launea mucronata	*adhîd*
66.	Pteranthus dichotomus	*ru'aysa, na'îma*
	Notoceras sp.	*ashbet umm salîm,*
	or Lepidum aucheri	*salîh*
	Schismus barbatus	*khafûr*
	Brassica tournefortii	*khafsh*
	Plantago coronopus	*ribla*
	Gypsophila sp.	*salîh*

Location 11. (Southwest of ar-Risha)

	Artemisia sieberi	*shîh*
	Achillea fragrantissima	*gaysûma*
	Anthemis melampodina	*'arbayyân*
	Hordeum sp. (barley)	*sha'îr*
	Brassica tournefortii	*khafsh*
	Picris babylonica	*hawthân*
	Launea mucronata	*adhîd*
	Caylusea sp.	*thanaybât*
	4 grasses	

Location 12. (Southeast of ar-Risha)

| 67. | Astragalus kahiricus | *udhun al-himâr* |

	Astragalus hauarensis	*gafâ'a*
	Trigonella hamosa	*nifal*
	Erucaria sp.	
	or Diplotaxis acris	*shiggâra*
68.	Arnebia linearifolia	*kâhil*
	Plantago coronopus	*ribla*
	Schismus barbatus	*khafûr*
	Stipa capensis	*samâ'*
	Brassica tournefortii	*khafsh*
	Gymnarrhena micrantha	*kaff al-kalb*
	Lepidum aucheri	*rashâd* (eaten).
	Herniaria hirsuta	?
69.	Hippocrepis bicontorta	*gurayna, qurayna, umm al-grayn*
	? White campion, ?Silene sp.	*'ahheym*
	Centaura sp. (pink)	*'amrâr*
	Malva parviflora	*khubayza*
	Papaver rhoeas	*deydehân*
70.	Pimpinella sp.	*kusaybira* (eaten)
	Calendula ? sp.	*henwa*
	Plantago ciliata	*gutayna*
	Almost prostrate feathery grass	
	Anthemis melampodina	*'arbayyân*
	Caylusea sp.	*thanaybât*
	Plant with very large soft leaves	
71.	Haloxylon salicornia	*rimth*

Location 13. (*Sha'îb* between ar-Risha al-<u>Gh</u>arbî and ar-Risha a<u>sh</u>-<u>Sh</u>arqî, ploughed in last few years)

	Artemisia sieberi, regenerated and self seeded	*shîh*
	Erodium sp.	*bakhatrî*
72.	Neotorularia sp.	*khushshayn*
	Plantago ciliata	
	or Filago desertorum	*gutayna*
	Brassica tournefortii	*khafsh*
	Plantago coronopus	*ribla*
	Aaronsohnia factorovskyi	*jurrais*
	Stipa capensis	*samâ'*
	The prostrate feathery grass	
	Anthemis melampodina	*'arbayyân*
	Foxtail type grass	

	Another grass	
	Picris babylonica	*hawthân*
73.	Leptaleum fililium	*huwayira*
	Erodium ?lacianatum	*ba<u>kh</u>atrî*
74.	Erodium ?glaucophyllum	*dab<u>gh</u>a*
	Caylusea sp.	*thanaybât*
	Calendula sp.	*henwa*
75.	Allium sp.	*<u>kh</u>arît* (eaten)
	Gypsophila sp.	*salîh*
	Trigonella hamosa	*nifal*
	Medicago sp.	*nifal*
	A different foxtail grass	
	Ifloga spictata	*zunayma*

Location 14.(*Sha'îb* further east; ploughed ?1987, ?1991)

	Neotorularia sp.	*khu<u>sh</u><u>sh</u>ayn*
	The prostrate feathery grass	
	Erodium laciniatum	*ba<u>kh</u>atrî*
	Aizöon hispanica	*mlîh, mulayh*
	Anthemis melampodina	*'arbayyân*
	Matthiola longipetala	*<u>sh</u>iggâra*
	Caylusea sp.	*thanaybât*
	Artemisia sieberi	*<u>sh</u>îh*
	Matricaria aurea	*babûnij*
	Brassica tournefortii	*kha<u>fs</u>h*
	Helianthemum sp.	*'arja*
76.	Launea angustifolia	*? mrâr*
	Achillea fragrantissima	*gaysûma*
	Hordeum (volunteer)	*<u>sh</u>a'îr*
	Trigonella sp.	*nifal*
	Calendula sp.	*henwa*
77.	Stipagrostis plumosa	*nasî, nizza*
	Calendula sp.	*henwa*
	Horwoodia sp.	*<u>kh</u>uzayma*
78.	Erodium ciconium	*tumeyr* (tubers eaten).
	Astragalus haurensis	*gafâ'a*
79.	Anthemis ?pseudocotula	*'arbayyân*
	Filago desertorum	*gutayna*
	or Plantago ciliata	
	Foxtail grass	
	Alyssum sp.	*drayhma*
	Roemaria hybrida	*dey<u>d</u>ehân*

80.	Symbricum sp.	?
	Notoceras bicorne	a<u>sh</u>bet umm salîm

Location 15. (Ar-Risha a<u>sh</u>-<u>Sh</u>arqî)

	Picris babylonica	hawthân
	Erodium ?glaucophyllum	daba<u>gh</u>
	Schismus barbatus	<u>kh</u>afûr
	Brassica tournefortii	<u>kh</u>af<u>sh</u>
	Erodium lacianatum	ba<u>kh</u>atrî
	Pulicaria ?guestii or incisa	<u>kh</u>uzayma
	Anthemis melampodina	'arbayyân
	Lepidum aucheri	ra<u>sh</u>âd
	Trigonella hamosa	na'îma
	Medicago lacianata	nifal
81.	Echinops sp.	<u>kh</u>a<u>sh</u>îr
	Aizöon hispanica	muth'âth
	Gypsophila/Erucaria/Caile	slîh
	Roemaria hybrida	deydehân
	Diplotaxis acris	<u>sh</u>iggâra
	Plantago coronopus	grayta
	Plantago ciliata	gtayna
82.	Koelpinia linearis	athwa
	Stipagrostis ?ciliata	sahma
	Stipa capensis	sama'a

Location 16. (Basatîn)

Neotorularia sp. or Cakile sp.	
	slîh
Picris babylonica	hawthân
Artemisia sieberi	<u>sh</u>îh
Erodium laciniatum	ba<u>kh</u>atrî
Plantago ciliata	gutayna
Plantago coronopus	grayta, ribla
Papaver rhoeas	deydehân
Aizöon hispanica	mulayh
Gypsophila sp.	slîh
Matricaria aurea	babûnij
? Rostraria pumila	sajil
Poa sinaica	naza'

Location 17. (By <u>kh</u>abra in Basatin)

Gypsophila sp.	slîh

Appendix

	Neotorularia sp.	*slîh*
	Artemisia sieberi	*s̱hîh*
	Diplotaxis acris	*s̱higgâra*
	Medicago laciniata	*nifal*
	Schismus barbatus	*ḵhafûr*
	Anthemis melampodina	*'arbayyân*
	Plantago ciliata	*gutayna*
	Astragalus annularis	*gafâ'a*
	A grass	
	Picris babylonica	*hawthân*
	Red campion	?
	Hyocyamus pusillis	*hîshî* (smoked)
	Plantago coronopus	*grayta, ribla*
	Herniaria hirsuta	?
83.	? Altheae ludwigii	?
	Hordeum sp.?	
	? Lotus sp.	*na'îma*

Location 18. (Track to Traibîl)

84.	Arnebia hispidissima	*kahal*
	Grasses	
	Astragalus spinosus	*s̱hitâde*
	Gynandris sp.	*sa 'id*
	Haplophyllum tuberculatum	*furayta*

Location 19. (*G̱hadir* near Jisr al Ruwaishid)

Plantago ciliata	*gutayna*
Medicago lacinaita	*nifal*

Location 20. (Small *sha'îb* south of Jisr al Ruwaishid)

	Erucaria sp. or Cakile sp.	*slîh*
	Gynandris sp.	*sa 'id*
	Astragalus spinosus	*s̱hitâde*
	Medicago laciniata	*nifal*
	Trigonella hamosa	*nifal*
	Calendula sp.	*henwa*
	Plantago boissieri	*ribla*
	Plantago ciliata	*gutayna*
85.	Carduus pycnocephalus	*s̱huwwayḵh*

Notoceras bicorne a<u>sh</u>bet umm salîm

Location 21. (*Sha'îb* further to south)

  Gynandris sp. sa 'id
  Trigonella hamosa nifal
  Calendula sp. henwa
  Plantago boissieri ribla
  Schismus barbatus <u>kh</u>afûr
  Launea mucronata a<u>dh</u>îd
  Aizöon hispanica mulayh
  Plantago ciliata gutayna
   or Filago desertorum
  Another grass.
  Artemisia sieberi <u>sh</u>îh
  Astragalus annularis gafâ'a
  Herniaria hirsuta ?
  Aaronsohnia factorovskyi jurrais, gurrais

86. Reichardia tingitana halawla
  Diplotaxis acris <u>sh</u>iggâra
  Papaver rhoeas deydehân
  Another grass.
  Gymnarrhena micrantha gutayna
  Centaurea pseudosinaica mrâr
  Astragalus haurensis gafâ'a
  Hippocrepis bicontorta gurayna, umm al-grayn
  Helianthemum lippii arja

Location 22. (Big *sha'îb* southeast of ar-Ruwaishid)

  Stipa capensis sama'
  Schismus barbatus <u>kh</u>afûr
  Plantago boissieri ribla
  Aizöon hispanica mulayh
  Gymnarrhena micrantha gutayna
  Trigonella hamosa nifal
  Artemisia sieberi <u>sh</u>îh
  Astragalus spinosus <u>sh</u>itâde
  Gynandris sp. sa'id
  Calendula sp. henwa

Appendix

Filago desertorum *gutayna*
or Plantago ciliata

Location 23. (Further south, on a *sha'îb*)

Lotus ?halophilus	*na'îma*
Gynandris sp.	*sa'id*
Aizöon hispanica	*mulayh*
Trigonella hamosa	*nifal*
Calendula sp.	*henwa*
Erodium laciniatum	*ba<u>kh</u>atrî*
Astragalus spinosus	*<u>sh</u>itâde*
Artemisia sieberi	*<u>sh</u>îh*
Stipa capensis	*sama'*
Schismus barbatus	*<u>kh</u>afûr*
(On top of slope, flinty cover)	
Medicago laciniata	*nifal*
Aizöon hispanica	*mulayh*
Artemesia sieberi	*<u>sh</u>îh*
Stipa capensis	*sama'*
Gypsophila sp.	*slîh*
? Mandrake	?

Location 24. (Next *sha'îb* south)

Schismus barbatus	*<u>kh</u>afûr*
Trigonella hamosa	*nifal*
Plantago ciliata	*gutayna*
or Filago desertorum	
Gynandris sp.	*sa'id*
Gypsophila sp.	*slîh*
Aizöon hispanica	*mulayh*
Stipa capensis	*sama'*

Location 25. (At old *mahfûr*, almost due south of gravel workings at Jisr ar-Ruwaishid)

87.	Lycium sp.	*awsaj*
	Malva parviflora	*<u>kh</u>ubayza*
	Schismus barbatus	*<u>kh</u>afûr*
	Hordeum sp.	*<u>sh</u>u'ayyira*
	Erodium laciniata	*ba<u>kh</u>atrî*

Plantago ciliata or Filago desertorum	*gutayna*
Artemisia sieberi	*shîh*
Picris babylonica	*hawthân*
Trigonella hamosa	*nifal*
Medicago laciniata	*nifal*
Aizöon hispanica	*mulayh*
Cakile sp. or Erucauria sp.	*salîh*

Location 26. (Second *mahfûr* and huge *sha'îb*)

	Artemesia sieberi	*shîh*
	Arnebia sp.	*kahil*
	Trigonella hamosa	*nifal*
	Erodium lacianata	*bakhatrî*
	Aizöon hispanica	*mulayh*
	Picris babylonica	*hawthân*
88.	Farsetia burtonae	?
	Plantago boiserri	*ribla*
	Four grasses	

Location 27. (South of ar-Ruwaishid, sandy *sha'îb*)

Astragalus spinosus	*shitâde*
Picris babylonica	*hawthân*
Hordeum sp.?	*shu'ayyira*
Schismus barbatus	*khafûr*
? Rostraria sp.	*sajil*
Artemisia sieberi	*shîh*
Anthemis melampodina	*'arbayyân*
Gypsophila sp.	*slîh*
Trigonella hamosa	*nifal*
A quaking grass	
Gyandris sp.	*sa'id*
Plantago cilaita or Filago desertorum	*gutayna*
Matthiola longipetala	*shiggâra*
Calendula sp.	*henwa*
Roemeria hybrida	*deydehân*
Erodium laciniata	*bakhatrî*
Another quaking grass	
Nine other different grasses	
Centaurea pseudosinaica	*mrâr*

Appendix

	Pimpinella sp.	*kusaybira*
89.	Salsola jordanicola	*gathgâth*
	Plantago boisseri	*ribla*
	Ifloga sp.	*zunayma*
	Helianthemum lippii	*arja*
	Medicago laciniata	*nifal*
90.	Launea sp.	*huwwa'*
	Hyocyamus sp.	*hîshî*
	Calendula sp.	*henwa*

(The slopes have more or less continual cover of)

	Stipa capensis	*sama'*
	Schismus barbatus	*khafûr*
	Plantago sp.	*ribla*
	Picris babylonica	*hawthân*
	Plantago ciliata	*gutayna*
	or Filago desertorum	

Location 28. (*Ghadir* al Hifna)

	(On slopes, stands of)	
	Stipa capensis	*sama'*
	Artemisia sieberi	*shîh*
	Astragalus spinosus	*shitâde*
	Papaver rhoeas	*deydehân*
	Picris babylonica	*hawthân*
	(In ghadir, between boulders)	
91.	? Salvia spinosa	?
	Artemisia sieberi	*shîh*
	Achillea fragrantissima	*gaysûma*
	Astragalus spinosus	*shitâde*
	Gynandris sp.	*sa'id*
	Picris babylonica	*hawthân*
	Papaver rhoeas	*deydehân*
	Plantago ciliata	*gutayna*
	Trigonella sp.	*nifal*
92.	Thymus sp.	*za'tar* (drunk as tea)
93.	? Euphorbia densa or granulata	?
	(On slopes going south)	

	Stipa capensis	*sama'*
	Plantago ciliata	*gutayna*
	Picris babylonica	*hawthân*
	Schismus barbatus	*khafûr*
	Trigonella hamosa	*nifal*
	Centaurea pseudosinaica	*mrâr*
	Gymnarrhena micrantha	*gutayna*
	Launea ? capitata	*huwwa'*
	Erodium laciniata	*ba<u>kh</u>atrî*
	Astragalus annularis	*gafâ'a*
94.	Farsetia aegyptica	*jurayba* (mangebush).
	Anthemis melampodina	*'arbayyân*
	Neotorularia sp.	*slîh, <u>kh</u>u<u>sh</u><u>sh</u>ayn*
	Aizöon hispanica	*mulayh*
	Diplotaxis acris	*<u>sh</u>iggâra*
	Ifloga sp.	*zunayma*
	Farsetia burtonae (first season)	?
95.	Lallemantia royleana	?

Location 29. (Next *sha'îb* going south)

	Haloxylon salicornica	*rimth*
	Brassica tournefortii	*<u>kh</u>af<u>sh</u>*
	Farsetia aegyptica	*jurayba*
	Astragalus spinosus	*<u>sh</u>itâde*
	Picris babylonica	*hawthân*
	Erodium laciniatum	*ba<u>kh</u>atrî*
	Ifloga sp.	*zunayma*
	Schismus barbatus	*<u>kh</u>afûr*
	Artemisia sieberi	*<u>sh</u>îh*
	Roemeria hybrida	*deydehân*
	? Rostraria pumila	*sajil*
	Pulicaria undulata	*jathjâth*
96.	Anisosciadum sp.	?
	Carduus pycnocephalus	*<u>sh</u>add al-jamâl*
	Gynandris sp.	*sa'id*
	Trigonella hamosa	*nifal*
97.	Astragalus sieberi	*mi<u>sh</u>t a<u>dh</u>-<u>dh</u>îb* (wolf's comb), *gafâ'a*
	Papaver rhoeas	*deydehân*

Appendix 411

    Echinops sp.   *shuwwaykh, kharshûf,*
                                                     kharshîr
    ? Onopordum sp.
    Stands of Stipa sp. going south.

Location 30. (By *ghadîr*, north of the Dumaithât)

	Farsetia aegyptica	*jurayba*
	Astragalus spinosus	*shitâde*
	Anisosciadum sp.	?
	Picris babylonica	*hawthân*
	Schismus barbatus	*khafûr*
	Erodium ?ciconium	*tumayr*
	Lepidum sativum	*rashâd*
	Trigonella hamosa	*nifal*
98.	? Astericus pygmaeus	?
	Carduus pycnocephalus	*shadd al-jamâl,*
		shuwwaykh
	Ifloga sp.	*zunayma*

99.	Zilla spinosoa	*shibrum, shibrîq*
	Plantago ciliata	*gutayna*
	or Filago desertorum	
	A quaking grass.	
	Hordeum sp.?	*shu'ayyira*
	Calendula sp.	*henwa*
	Anthemis melampodina	*'arbayyân*
	Astragalus annularis	*gafâ'a*
	Herniaria hirsuta	?
	Calendula sp.	*henwa*
	Centaurea pseudosinaica	*mrâr*
	Papver rhoeas	*deydehân*
	Roemeria hybrida	*deydehân*
	Achillea fragrantissima	*gaysûma*
	Astragalus haurensii	*gafâ'a*

100.	Plantago ovata	*gurayta*
	A bulb	
	(On slopes)	
	Anisosciadum sp.	?
	Schismus barbatus	*khafûr*
	Filago desertorum	*gutayna*
	or Plantago ciliata	

	Farsetia aegyptica	jurayba
	Anthemis melampodina	ʿarbayyân
	Four grasses	
	Small pink centaurea	

Location 31. (At head of northernmost Dumaitha on tiny sand ridge above flinty gravel)

		Ifloga sp.	zunayma
		Picris babylonica	hawthân
		Zilla spinosa	<u>sh</u>ubrum, <u>sh</u>ibrîq
101.		Plantago amplexicaulis	ribla
		Brassica tournefortii	<u>kh</u>af<u>sh</u>
		Astragalus spinosus	<u>sh</u>itâde

Location 32. (Wadi Dumaitha)

	Plantago amplexicaulis	ribla
	Plantago boisseri	ribla
	Zilla spinosa	<u>sh</u>ubrum, <u>sh</u>ibrîq
	Several grasses	
	Farsetia aegyptica	jurayba
	Ifloga sp.	zunayma
	Brassica tournefortii	<u>kh</u>af<u>sh</u>

102.	Salsola vermiculata	rûte

Location 32. (Between mouth of W.Dumaitha and Tell Hibr)

	Brassica tournefortii	<u>kh</u>af<u>sh</u>
	Salsola vermiculata	rûte
	Plantago boisseri	ribla
	Plantago amplexicaulis	ribla
	Plantago coronopus	gurayta
	Erodium laciniatum	ba<u>kh</u>atrî

103.	Savigna parviflora	kulayzân
104.	Cleome arabica	dhurrat an-naʿam
	Astragalus annularis	gafâʾa
	Zilla spinosa	<u>sh</u>ubrum, <u>sh</u>ubrîq

Appendix 413

Location 33. (North of Hibr, east of Jaythum al Hamad)

    Savigna parviflora   *kulayzân* (dominant)
    Artemisia sieberi   *shîh*
    Salsola vermiculata   *rûte*
    Brassica tournefortii   *khafsh*

Location 34. (Going west)

    Ifloga sp.   *zunayma*
    Salsola vermiculata   *rûte*

Location 35. (Turn to Umm al Rujm at Wisad)

    Ifloga sp.   *zunayma*
    Savigna parviflora   *kulayzân*
    Salsola vermiculata   *rûte*
    Gymnarrhena micrantha   *gutayna*
    Brassica tournefortii   *khafsh*

Location 36. (West of Wisad)

        Brassica tournefortii   *khafsh* (dominant)
        Papaver rhoeas   *deydehân*
        Picris babylonica   *hawthân*
        Launea sp.   *huwwa, adhîd*
105. Scorzonera pappos   *rubahla* (eaten)
        Plantago boisseri   *ribla*
        Cleome arabica   *dhurrat an-na'am*
        Reichardia tingitana   *halawla*
        Roemeria hybrida   *deydehân*
        Hippocrepis bicontorta   *grayna, umm al-grayn*

        Gynandris sp.   *sa'id*
        Allium sp.   *kharît*
        Ifloga sp.   *zunayma*
        Erodium glaucophyllum   *dabgha*
        Schismus barbatus   *khafûr*
        Anisosciadum sp.   ?
        Salsola vermiculata   *rûte*

106. Artemisia ?monosperma   *adhir*

107.   ? Rhanterium epapposum   arfaj

Location 37. (West of *Ga'* al Wutaidât)

    Brassica tournefortii   *khafsh*
    Haloxylon salicornica or Anabasis
    lachnantha   *ajram* or *rimth*
    Anthemis melampodina   *'arbayyân*
    Plantago ciliata   *gutayna*

Location 38. (Between Wutaidat and Marmariyya)

    Anabasis lachnantha   *ajram*
     or haloxylon salicornia   or *rimth*
    Plantago boisseri   *ribla*
    Erodium laciniatum   *bakhatrî*
    Diplotaxis harra   *khafsh*
    Schismus barbatus   *khafûr*
    Carduus pycnocephalus   *shadd al-jamâl, shuwwaykh*
    Picris babylonica   *hawthân*

Location 39. (East of Wadi Qattâfi)

    Brassica tournefortii   *khafsh*
    Senecio glaucus   *rijlet al-ghurab*
    Hordeum sp.   *shu'ayyira*
    Schismus barbatus   *khafûr*
    Salsola vermiculata   *rûte*
    Salsola sp.?   *hamdh*
    Lepidum aucheri   *slîh, rashâd*

Location 40 (Big *sha'îb* in Qattâfi basin)

    Salsola vermiculata   *rûte*
    Hordeum sp.   *shu'ayyira*
    Senecio glaucus   *rijlet al-ghurab*
    Trigonella hamosa   *nifal*
    Schismus barbatus   *khafûr*
    Plantago boissieri   *ribla*
    Artemisia sieberi   *shîh*

108.   Emex spinosa   *hambayz*

109.	Gagea lutea	s̲h̲ahhûm (fatweed; eaten)
	Aizöon hispanica	mulayh
	Plantago ciliata	
	or Filago desertorum	gutayna
	Gynandris sp.	sa'id
	Stipa capensis	sama'
	Carduus pycnocephalus	s̲h̲add al-jamâl, s̲h̲uwway<u>kh</u>
	Calendula sp.	henwa

Location 41. (West of Qattâfi, gravelly little hills)

Stipa capensis	sama'
Salsola vermiculata	rûte
Aizöon hispanica	mulayh
Several different grasses	
Astragalus spinosus	s̲h̲itâde
Anabasis sp.	
or Haloxylon sp.	ajram or rimth

Location 42. (Wadi Rajil)

110. Raetam raetam   ratam

Location 43. (Wadi Salma)

Medicago laciniata	nifal
Erodium laciniata	ba<u>kh</u>atrî
Lepidum aucheri	slîh, ras̲h̲âd
Malva parviflora	<u>kh</u>ubayza
Caylusea sp.	thanaybât
Horwoodia sp.	<u>kh</u>uzayma
Diplotaxis acris	s̲h̲iggâra
Picris babylonica	hawthân
Schismus barbatus	<u>kh</u>afûr
Anthemis melampodina	'arbayyân
Brassica tournefortii	<u>kh</u>af<u>sh</u>
Senecio glaucus	rijlet al-g̲h̲urab
Gypsophila sp.	slîh (Rwala), na'îma (Ahl al Jabal).

Aizöon hispanica   mulayh
Rumex sp.   hambasis
Gynandris sp.   sa'id, 'ansalsal

BIBLIOGRAPHY

Abbas, I. 1979 Khair ad-Din ar-Ramli's Fatawa; a new light on life in Palestine in the 11th/17th century. *Die Islamische Welt Zwischen Mittelalter und Neuzeit. Festschrift für Hans Robert Roemer zum 65 Geburtstagt*, U. Haarman & P. Bachmann, 1–19. Beirut.

Abdel Nour, A. 1984 Traits et Conflits du Monde Rural Syrien au XVIII siècle d'apres les Fatwa de Hamid al-Imadi. *Melanges de l'Université Saint-Joseph*, vol. **50**, pt 1, 71–84.

Abli, M.A.A. 1996 *The relationship between Jalwa (exile) and social change in a residential Bedouin society*, Unpublished M.A. thesis, Yarmouk University, Jordan. (in Arabic).

Abujabr, R. 1989 *Pioneers over Jordan*, IB Tauris, London.

Abu Lughod, J.L. 1989 *Before European Hegemony: The World System, AD 1250–1350*. Oxford.

Abu Rabia, A. 1994 *The Negev Bedouin & Livestock Rearing*, Mediterranean Series, Berg. Oxford and Providence.

Agnew, C.T. & Anderson, E. 1993 (unpublished report). *Water Resources in the Badia: Initial Study*. British Institute at Amman for Archaeology and History, Amman.

Agnew, C.T. & Anderson, E. *et al.* 1995 Mahafir: a Water Harvesting System in the Eastern Jordan (Badia) Desert. *Geojournal* **37**(1): 69–80.

Al-Azmeh, A. 1986 Wahhabite Policy. *Arabia and the Gulf: From Traditional Society to Modern States*, ed. IR Netton. 75–90. Tatawa, New Jersey.

Allen, M. & Smith, G.R. 1975 Some Notes on Hunting Techniques and Practices in the Arabian Peninsula. *Arabian Studies* II, Middle East Centre, Cambridge.

Al Rasheed, M. 1989 Pouvoir et Économie Caravanière dans une oasis de l'Arabie du Nord: l'exemple de Hail", *Le Nomade, l'Oasis et la Ville*, ed. J Bisson. 225–235. URBAMA 20, Tours.

Al Rasheed M. 1991. *Politics in an Arbian Oasis*. IB Tauris, London.

Altorki S & Cole, D.P. 1989. *Arabian Oasis City: the Transformation of 'Unayza*. University of Texas Press, Austin.

Amadouny V. 1994. "Infrastructural Development under the British Mandate", *Village, Steppe and State: The Social Origins of Modern Jordan*, eds. E Rogan & T Tell. 128–161. British Academic Press, London.

Amawi AM. 1994. "The Consolidation of the Merchant Class in Transjordan during the Second World War", *Village, Steppe and State: the Social Origins of Modern Jordan*, eds. E. Rogan & T. Tell. 162-186. British Academic Press, London.

Antoun, R. 1979 *Low-Key Politics: Local Level Leadership and Change in the Middle East*. State University of New York Press, Albany.

Antoun, R. 1989 *Muslim Preacher in the Modern World: a Jordanian Case Study in Comparative Perspective*. Princeton University Press, Princeton, New Jersey.

Antoun, R. 1991 Ethnicity, Clientship and Class: their changing meaning. *Syria, Culture and Polity*, (eds.) R. Antoun & D. Quataert, 1–27. State University of New York Press, Albany.

Ayalon, D. 1957–8 The System of Payment in Mamluk Military Society parts i–ii. *Journal of the Social and Economic History of the Orient* I: 37–65, 257–296.

Ayalon, D. 1993 Some Remarks on the Economic Decline of the Mamluk Sultanate. *Jerusalem Studies in Arabic & Islam, vol,* **16**, 108–124.

Al-Azzawi, A.S., Salim, A.R. & Rajjal, Y.I. 1995 The Historical Development of the Jordanian Rural House and its effects on traditional and modern buildings. *Studies of the History and Archaeology of Jordan* V, (eds.) K.'Amr *et al.* 325–332 Amman.

Bakhit, M.A. 1982 *The Ottoman Province of Damascus in the 16th Century,* Beirut.

Bakhit, M.A. & Hmud, N.R. 1989 *The Detailed Defter of Liwa Ajlun No. 970, Istanbul; 1538.* Amman.

Bakhit, M.A. & Hmud, N.R. 1991 *The Detailed Defter of Liwa Ajlun, No. 185, Ankara; 1005AH/1596AD.* Amman.

Banning, E.B. 1986 Peasants, Pastoralists and Pax Romana: Mutualism in the Southern Highlands of Jordan, *Bulletin of the American Society for Oriental Research* **261**, 25–50.

Baram, A. 1997 Neo-tribalism in Iraq: Saddam Hussein's tribal policies 1991–6, *International Journal of Middle East Studies* **29**, 1–31.

Barbir, K. 1980 *Ottoman Rule in Damascus, 1708-1758.* Princeton University Press, Princeton, New Jersey.

Behrnauer, W. 1860–61 Memoire sur les institutions de police chez les Arabes, les Persans et les Turcs, *Journal Asiatique,* series 5: xvi, 347–392: xvii, 5–76.

Betts, A.V.G. 1992 Tell el-Hibr: a Rock Shelter Occupation Site of the 4th mill. BC in the Jordanian Badiya, *Bulletin of the American Society of Oriental Research* **287**, 5–24

Betts, A.V.G. 1993 The Burqu'/Ruwayshid Project: Preliminary report on the 1991 season, *Levant* XXV, 1–12.

Betts, A.V.G & S.W. Helms (Eds.) 1991 *Excavations at Jawa 1972-1986.* Edinburgh University Press, Edinburgh.

Betts, A.V.G. & S.W. Helms (Eds). 1992 *Excavations at Tell Umm Hammad 1982–1984* Edinburgh University Press, Edinburgh.

Betts, A.V.G *et al.* 1994 Early Cotton in North Arabia, *Journal of Archaeological Science* **21**, 489–499.

Bianquis, A.-M. 1977 Le problème de l'eau à Damas et dans sa ghouta, *Revue de Geographie de Lyon,* **1**, 1–36.

Bianquis, A.-M. 1978 Le Marche en Gros des fruits et legumes a Damas, *Revue de Geographie de Lyon,* **3**.

Bianquis, T. 1986 and 1989 *Damas et Syrie sous la domination fatimide.* Institut Français de Damas. Damascus.

Bianquis, T. 1991a L'anier de village, le chevalier de la steppe, le cavalier de la citadelle, trois personnages de la transition en Syrie, *Bilad al-Sham During the Abbasid Period (132AH/750AD – 451 AH/1059AD),* eds. MA Bakhit & R Schick, Amman.

Bianquis, T. 1991b Le Hawran de l'avènement de l'Islam à la conquète ottomane, *Le Djebel al-'Arab,* eds. J-M Dentzer & J. Dentzer-Feydy. 89–100. Direction Generale des Antiquities et des Musées de la République Arabe Syrienne and ERA 20 du Centre de Recherches Archaéologiques (CRNS) Paris.

Bienkowski, P. 1985 New caves for old: Bedouin architecture in Petra, *World Archaeology* **17**,2; 149–60.

Bienkowski, P. 1995 The architecture of Edom, *Studies in the History and Archaeology of Jordan* V, eds. K.'Amr et al. 135–144, Amman.

Biewers, M. 1992 Occupation de l'espace dans le village traditionnel de 'Aima, *Studies in the History and Archaeology of Jordan* IV, ed. M. Zaghlul. 397–402. Amman.

Biewers, M. 1993 *L'Habitat traditionnel du sud de la Jordanie*. Amman.

Bisheh, G. 1985 Qasr al-Hallabat: an Umayyad desert retreat or farmland?, *Studies in the History and Archaeology of Jordan* II, ed. A Hadidi. 263–5. Amman.

Bisheh, G. 1987 Qasr al Mshatta in the light of a recently discovered inscription, *Studies in the History and Archaeology of Jordan* III, ed. A Hadidi. 193–7. Amman and London.

Bisheh, G. 1989 Qasr Mshash & Qasr 'Ayn al-Sil: Two Umayyad Sites in Jordan, *The Fourth International Conference on the History of the Bilad al-Sham during the Umayyad Period*, eds. MA Bakhit & R Schick. 81–93. Amman.

Bisheh, G. 1992 The Umayyad monuments between Muwaqqar & Azraq: palatial residences or caravanserais?, *The Near East in Antiquity* III, Ed. S. Kerner. 35–42. Goethe-Institut, German Protestant Institute for Archaeolgy, Al Kutba Publishers, Amman.

Blunt Lady, A. [1881] 1968 *A Pilgrimage to Nejd*. 2 vols. Frank Cass and Co. Ltd, London.

Bocco, R. 1985 La Notion de Dirah chez les Tribus Bedouines en Jordanie: Le Cas des Bani Sakhr, *Terroirs et sociétes*, ed. B. Cannon, 2, 193–214 Maison de l'Orient, Lyons.

Bocco, R. 1986 Petites villes et citadinité en Jordanie: quelques pistes de reflexion sur l'urbanisation en zone pastorale, *Petites Villes et villes moyennes dans le monde Arabe*, ed. P. Signoles. 167–183. URBAMA. Tours.

Bocco, R. 1989a Espaces étatique et espaces tribaux dans le sud Jordanien: Legislation fonciere et redefinition des liens sociaux, *Maghreb/Mashrek* **123**; 144–162.

Bocco, R. 1989b L'état producteur d'identités locales: lois electorales et tribus bedouines en Jordanie, *Le Nomade, l'Oasis et la Ville*. ed. J. Bisson. URBAMA. Tours.

Bocco, R. 1990 Ingénieurs agronomes et politiques de developpement dans les steppes du sud jordanien, *Batisseurs et bureaucrates: Ingénieurs et sociétes au Maghreb et au Moyen-Orient*. 255–280. Collection Etudes sur le Monde Arabe **4**, Lyons.

Bocco, R. & Tell, T. 1994 Pax Britannica in the Steppe. British Policy and the Transjordanian Bedouin, *Village, Steppe and State*, eds. R. Bocco & T.Tell. 108–127. British Academic Press. London.

Bounni, A. 1985 Palmyra: the Caravan city, *Ebla to Damascus*, ed. H. Weiss, Smithsonian Institute, Washington, DC.

Bowersock, G.W. 1983 *Roman Arabia*. Harvard University Press, Cambridge Mass. and London.

Braemer, F. 1988 Prospections archaéologiques dans le Hawran; ii, les reseaux de l'eau, *Syria* LXV,110–131.

Braemer, F. 1990 Formes d'irrigation dans le Hawran (Syrie du Sud). *Techniques et Pratiques Hydro-Agricoles Traditionelles en Domaine Irrigue*, ed. B Geyer. 453–474. Institut français d'archaeologie du Proche Orient; Beirut, Damas, Amman. Librairie Orientaliste Paul Guenther, Paris.

Braemer, F. 1993 Prospections archaéologiques dans le Hawran (Syrie), *Syria* LXX, 117–170.

Bresenham, M.F. 1985 Descriptive and Experimental Study of Contemporary and Ancient Pottery Techniques at Busra, *Berytus* xxxviii, 89–101.
Briant, P. 1982 *État et pasteurs au Moyen-Orient ancien*. Cambridge University Press/Editions de la Maison des sciences de l'homme, Paris.
Bruins, H.J. 1986 *Desert Environment and Agriculture in the Central Negev and Kadesh-Barnea during Historical Times*. MIDBAR Foundation, Nijkerk, Netherlands.
Bruins, H.J. 1994 Comparative chronology of climate and human history in the southern Levant from the late Chalcolithic to the early Arab period, *Late Quaternary Chronology and Paleoclimates of the Eastern Mediterranean*, eds. O Bar-Yosef & RS Kra, Radiocarbon, 301–314.
Buckingham, J.S. 1825 *Travels among the Arab tribes inhabiting the countries east of Syria and Palestine*. London.
Burckhardt, J.L. [1822] 1992 *Travels in Syria and the Holy Land*. Darf Publishers Ltd, London.
Burckhardt, J.L. [1829] 1993 *Travels in Arabia*. Darf Publishers, London.
Burckhardt, J.L. [1831] 1967 2 vols. *Notes on the Bedouins & Wahabys*, Colburn & Bentley, London and Johnson Reprint Company, London and New York.
Cahen, C. 1979 Ikta', *Encyclopedia of Islam*, 2nd edition, vol.III. Brill, Leiden.
Canaan, T. [1927] no date. *Mohammadan Saints & Sanctuaries in Palestine*. Ariel, Jerusalem.
Canaan, T. 1932 and 1933 The Palestinian Arab House, *Journal of the Palestine Oriental Society* **12**, 223–247: **13**, 1–83.
Carlier, P & F Morin. 1986 Qastal. Un site umayyade complet, *Archiv für Orientforschung* **33**, 187–206.
Carswell, J. 1996 Review of Necipoglu 1995, *The Art Newspaper*, June 26, 1996.
Carruthers, D. 1935 *Arabian Adventure: to the Great Nefud in search of the Oryx*. Witherby, London.
Caskel, W. 1954 The Bedouinisation of Arabia, *Studies in Islamic Cultural History*, ed. GE von Grunebaum, 36–46, **2**: American Anthropologist **56**, No 2, Part 2, Memoir 76.
Chamberlain, M. 1994 *Knowledge and Social Practice in Medieval Damascus, 1190–1350*. Cambridge University Press, Cambridge.
Chatelus, M. 1987 Rentier or producer economy in the Middle East? The Jordanian response, *The Economic Development of Jordan*, eds. B Khader & A Badran. 204–220. London.
Chatelus, M. 1990 Policies for Development: Attitudes towards Industry and Services, *The Arab State*, ed G Luciani. 99–128. Routledge, London.
Chatty, D. 1986 *From Camel to Truck: The Bedouin in the Modern World*. Vantage Press, New York.
Cohen, A. 1973 *Palestine in the 18th Century*. Jerusalem.
Cohen, A & B Lewis. 1978 *Population and Revenue in the Towns of Palestine in the Sixteenth Century*. Princeton University Press, Princeton.
Cole, D.P. 1985 Bedouin and Social Change in Saudi Arabia, *Arab Society*, ed. NS Hopkins & SE Ibrahim. 286–305. American University in Cairo Press, Cairo.
Colledge, S. 1994 *Plant exploitation on Epipalaeolithic and early Neolithic sites in the Levant*. PhD dissertation, Department of Archaeology and Prehistory, University of Sheffield.
Conder, C.E. 1883a *Heth & Moab*. London.
Conder, C.E. 1883b The *Survey of Eastern Palestine*. London.

Costa, P. 1991 *Musandam: Architecture and Material Culture of a Little Known Region of Oman*. Immel, London.
Costa, P. & Wilkinson,T.J. 1987 The Hinterland & of Sohar: Archaeological Surveys and Excavations within the region of an Omani Seafaring City, *Journal of Oman Studies* **9**, 1–238.
Crone, P. 1987 *Meccan Trade and the Rise of Islam*. Princeton University Press, Princeton.
Crowfoot, G. 1932 Pots Ancient and Modern, *Palestine Exploration Quarterly*, 179ff.
Crowfoot, G. 1945 The tent beautiful: a study of pattern weaving in Trans-Jordan, *Palestine Exploration Fund Quarterly*, 34–47.
Dakar, N. 1984 Contribution a l'étude de l'évolution de l'habitat bedouin en Syrie, *Nomades et Sedentaires: perspectives ethno-archeologiques*, ed. O. Aurenche. 51–79. Editions Recherche sur les civilisations, Paris.
Dalman, G. 1932 *Arbeit und Sitte in Palästina, band II: Der Ackerbau*. Bertelsman, Gutersloh.
d'Arvieux, Chevalier. 1735 *Mémoires*, ed. J.-P. Labat. Delespine le Fils, Paris.
de Boucheman, A. 1939 *Une Petite Cité Caravanière: Sukhné*. Documents d'Études Orientales. Institut Française de Damas. Damascus.
de Jong, F. 1984 Islamic Mysticism in Palestine: Observations and Notes Concerning Mystical Brotherhoods in Modern Times, *3rd International Conference on the Bilad al-Sham, 1980*, vol 2, *Geography and History of Palestine*. University of Jordan and Yarmouk University, Amman and Irbid.
Dentzer, J.-M. & Dentzer-Feydy, J. 1991 *Le djebel al-Arab*. Direction Générale des Antiquitiés et des Musées de la République Arabe Syrienne: ERA No. 20 du Centre de Recherches Archaéologiques (CRNS), Paris.
d'Hont, O. 1990 Evolution recente dans l'utilisation des espaces de la moyenne vallée de l'Euphrate, *Techniques et Pratiques Hydro-Agricoles Traditionelles en Domaine Irrigué*, ed. B Geyer. 239–248. Institut française d'archaéologie du proche orient. Librairie Orientaliste Paul Geuthner, Paris.
d'Hont, O. 1992 Production Pastorale et Consommation Citadine: Lait et laitages en Syrie Orientale, *Le Nomade, l'Oasis et la Ville*. ed. J. Bisson. 213–221. URBAMA 20, Tours.
d'Hont, O. 1994 *Vie Quotidienne des 'Agedat; Techniques et occupation de l'espace sur le Moyen-Euphrate*. Institut Français de Damas, Damascus.
Dickson, H.R.P. 1949 *The Arab of the Desert*. Allen & Unwin, London.
Dissard, J. 1905 Les Migrations et les Vicissitudes de la Tribu d'Amer. *Revue Biblique* **14**, 410–424.
Donner, F. McG. 1981 *The Early Islamic Conquests*. Princeton University Press, Princeton, New Jersey.
Dostal, W. 1959 The Evolution of Bedouin Life, *L'antica societa bedouina*, ed. F Gabrieli, 11–34, Studi Semitici 2, Universita di Roma, Rome.
Doughty, C. [1888] 1936 *Travels in Arabia Deserta*. vol. 2 Jonathan Cape, London.
Doumani, B. 1995 *Rediscovering Palestine: Merchants & Peasants in Jabal Nablus, 1700–1900*. University of California Press, Berkeley, Los Angeles.
Doumani, B. forthcoming, 1998 Writing family: Waqf, property and gender in Greater Syria, 1800–1860, *Comparative Studies in Society and History*.
Dresch, P. 1989 *Tribes, Government and History in Yemen*. Clarendon Press, Oxford.
Dussaud, R. 1927 *Topographie Historique de la Syrie Antique et Mediévale*. Paris.

Ducos, P. 1993 "Proto-élevage et élevage an Levant sud an vlle millénaire BC. Les données de la Damascène." *Paléorient* **19**, 153–74.

Eadie, J.W. 1989 Strategies of Economic Development in the Roman East: the Red Sea Trade Revisited, *The Eastern Frontier of the Roman Empire*, eds. DH French and CS Lightfoot, **2** vols. 113–120, BAR International series 553, Oxford.

Eickelman, D. 1989 2nd edition. *The Middle East, an anthropological approach*. Prentice-Hall, Englewood Cliffs, New Jersey.

Eph'al, I. 1982 *The Ancient Arabs*. Magnis Press, Jerusalem.

Fabietti, U. 1982 Sedentarisation as a means of detribalisation. Some policies of the Sa'udi Arabian government towards the nomads, *State, society and economy in Saudi Arabia*, ed. T. Niblock. 186–197. Croom Helm, London.

Fabietti, U. 1990 Between Two Myths: Underproductivity and Development of the Bedouin Domestic Group, *Sociétés Pastorales et Développement*, eds. E. Bernus & F. Pouillon. 237–253. Cahiers des Sciences Humaines **26**. Paris.

Fabietti, U. 1993 Politiques étatiques et adaptations bedouines: l'Arabie du nord (1900–1980), *Steppes d'Arabies. États, pasteurs, agriculteurs et commercants: le devenir des zones séches*, eds. R. Bocco, R. Jaubert & F. Métral.135–146. Presses Universitaires de France and Cahiers de l'Institut Universitaire d'Etudes de Developpement, Paris and Geneva.

Faroqhi, S. 1994 *Pilgrims and Sultans: the Hajj under the Ottomans*. IB Tauris, London.

Farrag, A. 1977 The wastah among Jordanian villagers, *Patrons & Clients in Mediterranean Societies*, eds. E. Gellner & J. Waterbury. 225–238. Duckworth, London.

Fernea, R. 1987 Technological innovation and class development among the Bedouin of Hail, Saudi Arabia, *Terroirs et Sociétés au Maghreb et au Moyen-Orient*, ed. B Cannon. 389–405 Cahiers Études sur le Monde Arabe, Maison de l'Orient, Lyons.

Field, M. 1984 *The Merchants: the big business families of Arabia*. John Murray, London.

Finn, J. 1878 *Stirring Times, or Records from the Jerusalem Consular Chronicles of 1853–56*. **2** London.

Firestone, Y. 1975 Production and Trade in an Islamic Context; Sharika contracts in the transitional economy of northern Samaria 1853–1943. *International Journal of Middle Eastern Studies* **6**, 185–209.

Fischbach, M.R. 1994 British Land Policy in Transjordan, *Village, Steppe and State: The Social Origins of Modern Jordan*, eds. E. Rogan and T. Tell. 80–107. British Academic Press, London.

Gardiner, M & A McQuitty. 1987 A Water Mill in the Wadi el Arab, North Jordan and Water Mill Development, *Palestine Exploration Quarterly*, **119**, 24–32.

Garrard, A., Colledge. S, & Martin. L, 1996 The emergence of crop cultivation and caprine herding in the Marginal Zone of the southern Levant, *The origins and spread of agriculture and pastoralism in Eurasia* ed. DR Harris, UCL Press, London.

Garrard, A. & Gebel, H.-G. (Eds.) 1988. *The prehistory of Jordan*. BAR International Series 396 i and ii, Oxford.

Gaube, H. 1979 Die Syrischen Wüstenschlosser: Einige wirtschaftliche und politische Gesichtspunkte zu ihrer Entstehung, *Zeitschrift des Deutschen Palastina-Vereins* **95**, 182–209.

Gaudefroy-Demombynes, M. 1923. *La Syrie a l'époque des Mamlouks*. Paris.

Gawlikowski, M. 1994 Palmyra as a Trading Centre, *Iraq*, LVI, 27–35.
Geyer, B. 1990 Aménagements hydrauliques et terroir agricole dans la moyenne vallée de l'Euphrate, *Techniques et Pratiques Hydro-Agricoles Traditionelles en Domaine Irrigué*, ed. B. Geyer. 63–86. Institut Française d'archaéologie du Proche Orient. Librairie Orientaliste Paul Geuthner, Paris.
Ghawanmeh, Y. 1982 Al-Qaryah fî Junûb ash-Shâm (al-Urdun wa-Filastîn) fî al-Asr al-Mamlîkî, *Studies in the History and Archaeology of Jordan* I, ed. A Hadidi, 363–371. Amman and London.
Ghazzal, Z. 1993 *L'économie politique de Damas durant le XIXième siecle*. Institut Française de Damas, Damascus.
Gilsenan, M. 1977 Against patron-client relations, *Patrons and Clients in Mediterranean Societies*, eds. E. Gellner & J. Waterbury. 167–183. Duckworth, London.
Ginat, J. 1983 Meshamas – the Outcast in Bedouin Society, *Nomadic Peoples* **12**, 26–47.
Gingrich, A. & Heiss, J. 1986 *Beitrage zur Ethnographie der Provinz Sa'da (Nordjemen)*. Verlag der Osterreichischen Akademie der Wissenschaften, Vienna.
Glubb, J.B. 1938 The Economic Situation of the Jordanian Tribes, *Journal of the Royal Central Asian Society*, **25**, 448–459.
Glubb, J.B. 1959 *Britain and the Arabs. A Study of 50 Years 1908–58*. London.
Goblot, H. 1979 *Les Qanats, une technique d'acquistion de l'eau*. Mouton, Paris.
Goitein, S.D. 1967 *A Mediterranean Society. Vol I, Economic Foundations*. University of California Press, Los Angeles.
Grabar, O. *et al.* 1978 *City in the Desert. Qasr al-Hayr East*. Cambridge, Mass.
Graf, D. 1989a. Rome and the Saracens: Reassessing the Nomadic Menace, *Colloque International sur l'Arabie préislamique et son environnement historique et culturel*, Strasbourg, June 1987, ed. T. Fahd, 341–400. Brill, Leiden.
Graf, D. 1989b Zenobia and the Arabs, *The Eastern Frontier of the Roman Empire*, eds. DH French and CS Lightfoot, 143–167, **2** vols. BAR International Series **553**, Oxford.
Grant, C. 1937 *The Syrian Desert: Caravans,Travels, Explorations*. A and C Black, London.
Groom, N. 1981 *Frankincense & Myrrh. A Study of the Arabian Incense Trade*. Longman: Librairie du Liban, London and New York.
Guarmani, C. [1917] 1938 *Northern Najd*. The Argonaut Press, London.
Gubser, P. 1973 *Politics and Change in al-Karak, Jordan*. Oxford University Press, Oxford.
Haj Ibrahim, M. 1990 Les techniques de captage et de conduite de l'eau aux fins d'irrigation et les modes de culture irrigate dans les ghoutas de l'Ouadi el Majarr (Kalamoun-nord): l'exemple de Deir Attiyeh, *Techniques et pratiques hydro-agricoles traditionelles en domaine irrigué*, ed. B Geyer. **2** vols. 295–312. Institut Français d'Archaéologie du Proche-Orient, Beirut, Damas, Amman. Librairie Orientaliste Paul Guenther, Paris.
Hamza, H.H. 1982 *Public Land Distribution in Saudi Arabia*. London.
Hannoyer, J. & Thieck, J.-P. 1984 Observations sur l'élevage et le commerce du mouton dans la region de Raqqa en Syrie , *Production Pastorale et Societe* **14**, 47–65.
al-Harawi Abu, H.A. 1957 *Guides des lieux du pèlerinage*, trans. J. Sourdel-Thoumine. Damascus.
Harlan, J.R. 1995 *The Living Fields*. Cambridge University Press, Cambridge.

Harrison, S.G., Masefield, G.B. & Wallis, M. 1969. *The Oxford Book of Food Plants*. Oxford University Press, Oxford.

Hartmann, R. [1907] 1993 Die Geographischen Nachrichten über Palästina und Syrien from Khalil az-Zahiri's Zubdat Kashf al-Mamalik, ed. F Sezgin. *Islamic Geography*, **79**, 53–154. Publications of the Institute of Arabic-Islamic Science, Johann Goethe University, Frankfort.

Hasselquist, F. 1766 *Voyages and Travels in the Levant in the years 1749–1752*. London.

Haut Commissariat de la République Francaise. 1930 *Les Tribus Nomades et Semi-Nomades des États du Levant placés sur Mandat Française*. Beirut.

Havemann, A. 1991 Non-urban Rebels in Urban Society – The Case of Fatimid Damascus, eds. *Bilad al-Sham during the Abbasid Period (137AH/750 AD–451 AH/1059 AD)* M.A. Bakhit & R. Schick, 81–90. Amman.

Helms, S.W. 1981 *Jawa: lost city of the Black Desert*. London and New York.

Helms, S.W. 1990 *Early Islamic Architecture of the Desert*. Edinburgh University Press.

Herzfeld, M. 1987 *Anthropology through the Looking-Glass*. Cambridge University Press, Cambridge.

Hill, G. 1896 A Journey East of the Jordan and the Dead Sea, 1895, *Palestine Exploration Fund Quarterly*, 24–46.

Hillman, G.C., Colledge, S. & D.R . Harris. 1989 Plant-food economy during the Epipaleolithic period at Tell Abu Hureyra, Syria: dietary diversity, seasonality, and modes of exploitation in *Foraging and Farming*, eds. D.R. Harris & G.C. Hillman, Unwin Hyman, London.

Hinnebusch, R.A. 1991 Class and State in Ba'thist Syria, *Syria: Society, Culture and Polity*, eds. R. Antoun & D. Quataert. 29–48. State University of New York Press, Albany.

Hirschfield, Y. 1992 *The Judean Desert Monasteries in the Byzantine period*. Yale University Press: New Haven and London.

Hiyari, M.A. 1975 The Origins and Development of the Amirate of the Arabs during the 7th/13th and 8th/14th centuries. *Bulletin of the School of Oriental and African Studies* XXXVIII, 509–524.

Hobbs, J. 1992 *Bedouin Life in the Egyptian Wilderness*. University of Texas Press, Austin.

Hopkins, D.C. 1985 *The Highlands of Canaan: agricultural life in the Early Iron Age*. The Social World of Biblical Antiquity Series 5: Almond Press, Decatur.

Hutteroth, W.-D. and Abdulfattah, K. 1977 *The Historical Geography of Palestine, Transjordan and Southern Syria in the late 16th Century*. Erlangen.

Ingold, T. 1980 *Hunters, pastoralists & ranchers*. Cambridge University Press, Cambridge.

Irby, C. & Mangles, J. 1823 *Travels in Egypt and Nubia, Syria and Asia Minor during the years 1817 and 1818*. London.

Irwin, R. 1986 *The Middle East in the Middle Ages: the Early Mamluk Sultanate 1250–1382*. Beckenham.

Isaac, B. 1990 *The Limits of Empire: the Roman Army in the East*. Clarendon Press, Oxford.

Isaac, B. 1994 Tax Collection in Roman Arabia: A New Interpretation of the Evidence from the Babatha Archive, *Mediterranean Historical Review* **9**, 256–266.

Issawi, C. 1988 *The Fertile Crescent 1800–1914: a documentary economic history*. Oxford University Press, New York and Oxford.

Jabbur, J.S. (trans. L. Conrad). 1995 *The Bedouins and the Desert*. State University of New York Press, Albany.

al-Jaludi, 'A. & Bakhit, M.A. 1992 *Qada' 'Ajlūn fī asr al-tanzimât al-'uthmânî*. Amman.

Jaussen Le Pere, A. 1927 *Coutumes Palestiniennes: Naplouse et son district*. Librairie Paul Geuthner, Paris.

Jaussen Le Pere, A. 1948 (orig. 1907). *Coutumes des Arabes au Pays de Moab*. Maison Adrienneuve, Paris.

Jaussen Le Pere, A. & Le Pere R. Savignac. [1914] 1920 Coutumes des Fuqara, Supplement to vol. ii of *Mission Archeologique en Arabie*. Librairie Paul Geuthner, Paris.

Johns, J. 1992 Islamic Settlement in Ard al-Karak. *Studies in the History and Archaeology of Jordan* IV, eds. M Zaghloul & K 'Amr. 363–368. Amman & Lyons.

Johns, J. 1994 The Longue Durée: State and Settlement Strategies in Southern Transjordan Across the Islamic Centuries. *Village, Steppe and State. The Social Origins of Modern Jordan*, eds. E. Rogan & T. Tell, British Academic Press, London and New York.

Kana'an, R. 1993 *Patronage and Style in Mercantile Residential Architecture of Ottoman Bilad al-Sham*. Unpublished M. Phil. thesis, Oxford.

Kana'an, R. & McQuitty, A. 1994 The Architecture of al-Qasr on the Karak Platea: an essay in the chronology of vernacular architecture. *Palestine Exploration Quarterly* 126, 127–150.

Kennedy, D. 1995 Water Supply and Use in the Southern Hauran, Jordan, *Journal of Field Archaeology*, **22**, 275–290.

Kennedy, H. 1986 *The Prophet and the Age of the Caliphates: the Islamic Near East from the sixth to the eleventh centuries*. Longman, London and New York.

Kennedy, H. 1992 The Impact of Muslim Rule on the Pattern of Rural Settlement in Syria, *La Syrie de Byzance à L'Islam VII–VIII siècles*, eds. P. Canivet & J.-P. Rey-Coquais. 291–297. Institut Francaise, Damas. Damascus.

Khalaf, S. 1991 Land Reform and Class Structure in Rural Syria. *Syria: Society, Culture, and Polity*, eds. R. Antoun & D. Quataert. 63–78. State University of New York Press, Albany.

Khalidi, T. 1984 Tribal Settlement and Patterns of Land Tenure in Early Medieval Palestine. *Land Tenure and Social Transformation in the Middle East*, ed. T Khalidi, 181–188, American University of Beirut, Beirut.

Khalidi, T. 1994 *Arabic Historical Thought in the Classical Period*. Cambridge University Press, Cambridge.

Khammash, A. 1986 *Notes on Village Architecture in Jordan*. Lafayette, Louisiana.

Killick, A.C. 1987 *Udhruh: Caravan City and Desert Oasis*. Basingstoke.

King, G.R.D. 1987 The distribution of sites and routes in the Jordanian and Syrian deserts in the early Islamic period. *Proceedings of the Arabian Seminar* 17, 91–105.

King, G.R.D. 1989 The Umayyad Qusur and related settlements in Jordan. *Fourth International Conference of the History of the Bilad al-Sham during the Umayyad period*, eds. MA Bakhit & R Schick. vol. **2**, 73–80. Amman.

King, G.R.D. 1992 Settlement Patterns in Islamic Jordan: the Umayyads and their Use of the Land. *Studies in the History and Archaeology of Jordan* IV, eds. M. Zaghloul & K. 'Amr. 369–376. Amman & Lyons.

King, G.R.D. 1994 "Settlement in Western and Central Arabia and the Gulf in the sixth – eighth centuries A.D.," in The Byzantine and Early Islamic Near

East: Land Use and Settlement Patterns, eds. G.R.D. King and Averil Cameron. 181–212. Darwin Press, Princeton, New Jersey.

Kingston, P.W.T. 1994 Breaking the Patterns of Mandate: Economic Nationalism and State Formation in Jordan 1951–57. *Village, Steppe and State. The Social Origins of Modern Jordan*, eds. E. Rogan and T. Tell. 187–216. British Academic Press, London.

Kohler-Rollefson, I. 1996 The one-humped camel in Asia: origin, utilisation and mechanisms of dispersal. *The origins and spread of agriculture and pastoralism in Eurasia*, ed. D.R. Harris, 282–294, UCL Press, London.

Kostiner, J. 1988 Britain and the Northern Frontier of the Saudi State, 1922–1925. *The Great Powers in the Middle East 1919–1939*, ed. U. Dann. 29–48. Holmes and Meier, New York/London.

Kostiner, J. 1991 Transforming Dualities: Tribe and State in Saudi Arabia. *Tribes and State Formation in the Middle East*, eds. PS Khoury & J Kostiner. 226–251. IB Tauris, London.

Kostiner, J. 1993 *The Making of Saudi Arabia 1916–1936*. Oxford University Press, Oxford and New York.

Koszinowski, T. 1981 The Arabian peninsula in the 19th and 20th centuries; Saudi Arabia. *The Muslim World: a Historical Survey, pt iv, Modern Times, fasc.i*, 199–209. Brill, Leiden.

Kressel, G. 1984 Changes in employment and social accomodations of Bedouin settling in an Israeli town. *The Changing Bedouin*, eds. E. Marx & A. Shmueli. 125–154. Transaction Books, New Brunswick, New Jersey.

Kressel, G. & Ben David, J. forthcoming, The Bedouin Market: Corner Stone for the Founding of Beersheba. *Nomadic Peoples*.

Lancaster, W. 1981 *The Rwala Bedouin Today*. Cambridge University Press, Cambridge.

Lancaster, W. & Lancaster, F. 1986 The Concept of Territoriality among the Rwala Bedouin. *Nomadic Peoples* 20.

Lancaster, W. & Lancaster, F. 1988 Thoughts on the Bedouinisation of Arabia, *Proceedings of the Arabian Seminar*, **18**, 51–62.

Lancaster, W. & Lancaster, F. 1990 Desert Devices: The Pastoral System of the Rwala Bedu, *The World of Pastoralism: Herding Systems in Comparative Perspective*. eds. J.G. Galaty & D.L. Johnson. 177–194 Guildford Press, New York and Belhaven Press, London.

Lancaster, W. & Lancaster, F. 1991 Limitations on Sheep and Goat Herding in the Eastern Badia of Jordan: an Ethno-archaeological enquiry, *Levant* XXIII, 125–138.

Lancaster, W. & Lancaster, F. 1992a. Tribal Formations in the Arabian Peninsula, *Arabian Archaeology and Epigraphy* 3, 145–172.

Lancaster, W. & Lancaster, F. 1992b. Tribe, Community and the Concept of Accesss to Resources: Territorial Behaviour in South-East Ja'alan, *Mobility and Territoriality*, eds. M.J. Casimir & A. Rao. 343–364 Berg, New York and Oxford.

Lancaster, W. & Lancaster, F. 1993a. Sécheresse et stratégies de reconversion économique chez les bedouins de Jordanie, *Steppes d'Arabie. États, pasteurs, agriculteurs et commercants: le devenir des zones séches.* 223–246. Presses Universitaires de France and Cahiers de l'Institut universitaire d'études de developpement, Paris and Geneva.

Lancaster, W. & Lancaster, F. 1993b. Graves and funerary monuments of the Ahl al-Jabal, Jordan, *Arabian Archaeology and Epigraphy* **4**, 151–169.
Lancaster, W. & Lancaster, F. 1995 Land Use and Population in the Area North of Karak, *Levant* XXVII, 103–124.
Lancaster, W. & Lancaster, F. 1996 Some Comments on Peasant and Tribal Pastoral Societies of the Arabian Peninsula (with Particular Reference to the Karak Plateau of Jordan), *The Anthropology of Tribal and Peasant Societies*, eds. U Fabietti & P. C. Salzman. 389–401. Ibis, Como and Collegio Ghislieri, Pavia.
Lancaster, W. & Lancaster, F. 1997a. Rowton's theses of Dimorphic Structure and Enclosed Nomadism: a reconsideration, *Proceedings of the International Symposium on Syria and the Ancient Near East 3000–300BC*, ed. F. Ismail, Aleppo University Publications, Aleppo.
Lancaster, W. & Lancaster, F. 1997b. Sulayb, *Encyclopedia of Islam*, second edition, Volume IX.
Lancaster, W. & Lancaster, F. 1997c. Indigenous resource management systems in the Badia of the Bilad ash-Sham, *Journal of Arid Environments* 35, 367–378.
Lancaster, W. & Lancaster, F. 1997d. Jordanian Village Houses in their Contexts: Growth, Decay and Rebuilding, *Palestine Exploration Quarterly*, 38–53.
Landry, P. 1990 Eaux souterraines et qanats d'après un livre arabe du Xie siècle, *Techniques et pratiques hydro-agricoles traditionelles en domaine irrigue*, ed. B. Geyer. 271–284 Institut Français d'Archeologie du Proche Orient. Librairie Orientaliste Paul Guenther, Paris.
Lane, E.W. [1863] 1984 2 vols. *Arabic-English lexicon*. London and Edinburgh.
Lapidus, I. 1967 *Muslim cities in the later Middle Ages*. Cambridge University Press, Cambridge.
Layne, L. 1987a. Village-Bedouin: Patterns of Change from Mobility to Sedentism in Jordan, *Method and Theory for Activity Area Research; an Ethno archaeological Approach*, ed. S Kent. 345–373. Columbia University Press, New York.
Layne, L. 1987b. Tribesmen as Citizens: Primordial Ties and Democracy in Rural Jordan, *Middle East Elections: Implications of Recent Trends*, ed. L Layne. 113–151. Westview Press, Boulder.
Layne, L. 1994 *Home and Homeland: the dialogics of tribal and national identities in Jordan*. Princeton University Press, Princeton, New Jersey.
Lenzen, C.J. 1992 Irbid & Beit Ras: Interconnected Settlements c. AD 100–900, *Studies in the History and Archaeology of Jordan* IV, eds. M. Zaghloul & K.'Amr. 299–302, Amman.
Le Strange, G. 1890 *Palestine under the Moslems*. Cambridge.
Lewis, N. 1987 *Nomads and Settlers in Syria and Jordan, 1800–1950*. Cambridge University Press, Cambridge.
Lightfoot, D. 1996a. Qanats in the Levant: Hydraulic Technology at the periphery of Early Empires, *Technology and Culture* **38** (2), 432–451.
Lightfoot, D. 1996b. Syrian Qanat Romani: History, Ecology, Abandonment, *Journal of Arid Environments* 33 (3), 321–273.
Lipschitz, N. *et al.* 1991 The beginning of Olive (Olea europaea) Cultivation in the Old World: a Reassessment, *Journal of Archaeological Science* 18, 441–453.
Longenesse, E. 1980 L'industrialisation et sa significance sociale, *La Syrie d'Aujourd'hui*, ed. A Raymond. 327–358. Editions du Centre de la Recherche Scientifique, Paris.
Longrigg, S.H. 1954 *Oil in the Middle East*. Oxford University Press, Oxford.

MacAdam, H.I. 1984 Some Aspects of Land Tenure and Social Development in the Roman Near East: Arabia, Phoenicia and Syria, *Land Tenure and Social Transformation in the Middle East* ed. T Khalidi, 45–62. American University of Beirut, Beirut.

MacAdam, H.I. 1986 *Studies in the History of the Roman Province of Arabia*. BAR International Series 295. Oxford.

McClellan, T.L. 1997 Irrigation in Dry-farming Syria: Agricultural Intensification in the Bronze Age, *Proceedings of the International Symposium on Syria and the Ancient Near East 3,000–300BC*, ed. F. Ismail. 52–76. Aleppo Univerity Publications. Aleppo.

Macdonald, M.C.A.1991 Was the Nabatean Kingdom a 'Bedouin State'? *Zeitschrift des Deutsches Palästina-Vereins* **107**, 102–119.

Macdonald, M.C.A. 1993 Nomads and the Hawran in the late Hellenistic and Roman periods: a reassessment of the epigraphic evidence, *Syria*, LXX, 3/4, 303–413.

Macdonald, M.C.A. 1995a. Herodian Echoes in the Syrian Desert, *Trade, Contact and the Movement of Peoples in the Eastern Mediterranean: Papers in Honour of JB Hennessy*, eds. S. Bourke, J.-P Descoudres, A. Walmsley; Mediterranean Archaeology, supplementary volume **2**. Sydney.

Macdonald, M.C.A. 1995b. North Arabia in the First Millenium BCE, *Civilisations of the Ancient Near East*, ed. JM Sasson, 1355–1369, vol.ii., Simon & Schuster Macmillan, New York.

McQuitty, A. 1986 Architectural Study of Bait Ras, *Archiv für Orientforschung* **33**, 153–155.

McQuitty, A. 1995 Water-mills in Jordan: Technology, Typology, Dating and Development, *Studies in the History and Archaeology of Jordan* V, ed. K.'Amr et al. 745–752. Amman.

McQuitty, A. 1997 Ovens in Town and Country, *Berytus* XLI, 53–76.

McQuitty, A. & Falkner R. 1993 A Preliminary Report on the Khirbet Faris Project: the 1989, 1990 and 1991 Seasons, *Levant* **25**, 37–61.

Mandaville, J. 1990 *Flora of Eastern Arabia*. Kegan Paul International, London and New York jointly with the National Commission for Wildlife Conservation and Development, Riyadh.

Mantran, R. & Sauvaget, J. 1951 *Règlements Fiscaux Ottomans: Les Provinces Syriennes*. Institut Français de Damas, Adrien-Maisonneuve, Paris.

Marsot, A.L. as-S. 1984 *Egypt in the Reign of Muhammad Ali*. Cambridge University Press, Cambridge.

Marx, E. 1984 Economic change among pastoral nomads in the Middle East and Changing employment patterns of bedouin in south Sinai, *The Changing Bedouin*, eds. E Marx & A. Shmueli. 1–16, 173–186. Transaction Inc. New Brunswick, New Jersey.

Marx E. 1996. Are there Pastoral Nomads in the Arab Middle East?, *The Anthropology of Tribal and Peasant Societies*, eds. U Fabietti & PC Salzman. 101–115. Ibis, Como and Collegio Ghislieri, Pavia.

Merrill, S. 1881 *East of the Jordan*. London.

Mershen, B. 1985 Recent Hand Made Pottery from Northern Jordan, *Berytus*, xxxiii, 75–87.

Mershen, B. 1992 Settlement History and Village Space in Late Ottoman Northern Jordan, *Studies in the History and Archaeology of Jordan* IV, ed. M. Zaghlul & K. 'Amr. 409–416. Amman.

Métral, F. 1982 Le droit de l'eau dans le code civil ottoman de 1869 et la notion de domaine public, *L'homme et l'éau en Mediterranée et au Proche Orient II*,

Aménagements hydrauliques, État et legislation. 125–142. Travaux de la Maison de l'Orient, Lyons.

Métral, F. 1984a. Transferts de Technologie dans l'Agriculture Irriguée en Syrie: Stratégies Familales et Travail Feminin, *Terroirs et Societes au Maghreb et au Moyen Orient*, ed. B Cannon. 331–368. Études sur le Monde Arabe **2**, Maison de l'Orient, Lyons.

Métral, F. 1984b. State and Peasants in Syria: a Local View of Government Irrigation Project, *Peasant Studies* **11**, 2, 69–90.

Métral, F. 1991 Entre Palmyre et l'Euphrate: Oasis et Agriculture dans la Region de Sukhné, *Rites et Rythmes Agraires*, ed. M.C. Cauvin. 87–108. Travaux de la Maison de l'Orient **20**, Lyons.

Métral, F. 1993 Élevage et agriculture dans l'oasis de Sukhne (Syrie): gestion de risques par les commerçants-entrepreneurs, *Steppes d'Arabies: États, pasteurs, agricultuers et commerçants: le devenir des zones séches*. eds. R. Bocco, R. Jaubert & F. Métral. 195–222. Presses Universitaires de France and Cahiers de l'Institut Universitaire d'études de développement, Paris and Geneva.

Métral, F. & Métral, J. 1989 Une ville dans la steppe, la tribu dans la ville: Sukhné (Syrie), *Le Nomade, L'Oasis et la Ville*, 153–171. ed. J. Bisson. URBAMA **20**, Tours.

Miller, A.G. & Morris, M. 1988 *Plants of Dhofar. The southern region of Oman - traditional, economic and medicinal uses*. The Office of the Adviser for Conservation of the Environment, Diwan of the Royal Court, Sultanate of Oman.

Miller, J.M. 1991 *Archaeological Survey of the Karak Plateau*. Atlanta.

Milne, J. 1971 *Problem of patrilateral parallel cousin marriage with special reference to a southern Jordanian village, el-Ji*. Unpublished MA thesis, University of Edinburgh.

Miquel, A. 1980 *La géographie humaine du monde musulman jusqu'au milieu du Xie siècle*. vol.3. Mouton.

Mount, F. 1992 *The British Constitution Now*. Heineman, London.

Mundy, M. 1992 Share-holders and the State: Representing the Village in the late 19th century Land Registers of the Southern Hauran, *The Syrian Lands in the 18th and 19th centuries*, ed. T. Philipp. 217–238. vol. **5**, Berliner Islamstudien, Berlin.

Mundy, M. 1994 Village Land and Individual Title: Musha' & Ottoman Land Registration in the Ajlun District, *Village, Steppe and State: the Social Origins of Modern Jordan*, eds. E. Rogan & T. Tell. 58–79. British Academic Press, London.

Mundy, M. 1995 *Domestic Government: Kinship, Community and Polity in North Yemen*. IB Tauris, London.

Mundy, M. & Smith, R.S. (eds). 1991 *Part-time Farmers*. Publications of the Institute of Archaeology and Anthropolgy, University of Yarmouk, Jordan.

Musil, A. 1908 *Arabia Petraea*. 3 vols. Vienna.

Musil, A. 1927 *Arabia Deserta*. American Geographical Society Oriental Explorations and Studies, No. **2**. New York.

Musil, A. 1928a. *The Manners and Customs of the Rwala Bedouins*. American Geographical Society Oriental Explorations and Studies, No. **6**. New York.

Musil, A. 1928b. *Palmyrena*. American Geographical Society Explorations and Studies, No. **4** New York.

Muzzolini, A. 1989 La 'Neolithisation' du nord de l'Afrique et ses causes, *Neolithisations*, eds. O. Aurenche et J. Cauvin, BAR International Series, 145–185, Oxford.

Nasif, A.A. 1988 *Al-'Ula: an historical and archaeological survey with special reference to its irrigation system*. King Sa'ud University, Riyadh.
'Naturalist, A.' 1885 A Journey to Sinai, Petra and South Palestine, *Palestine Exploration Fund Quarterly*, 17, 231–286.
Noca, L. 1985 *Smakieh: un village de Jordanie*. Lyons.
Northedge, A., Bamber, A. & Roaf, M (eds). 1988 *Excavations at 'Ana*. Iraq Archaeological reports 1. Published for the British School of Archaeology in Iraq and the Directorate of Antiquities of Iraq by Aris and Phillips Ltd., Warminster, England.
Ochsenwald, W. 1980 *The Hijaz Railroad*. Charlottesville, Virginia.
Oleson, J.P. 1995 The origins and designs of Nabatean water-systems, *Studies in the History and Archaeology of Jordan* V, eds. K.'Amr et al. 707–20. Amman.
Owen, R. 1981 *The Middle East in the World Economy 1800–1914* Methuen, London.
Palmer, C. forthcoming 1998 'Following the plough' - the agricultural environment of North Jordan. *Levant* XXX.
Palmer, C. & K.W. Russell. 1993 Traditional ards of Jordan. *Annual of the Department of Antiquities of Jordan* XXXVII, 37–53. Amman.
Palmer, E.H. & Tyrwhitt Drake. 1871 The Desert of Tih and the Country of Moab, *Palestine Exploration Fund Quarterly*, 3–73.
Parker, S.T. 1986 *Romans and Saracens, a History of the Arabian Frontier*. American Society for Oriental Research, Dissertation Series, Chicago.
Parr, P.J. 1989 Aspects of the Archaeology of Northwest Arabia, *L'Arabie Pré-Islamique et son environnement historique et culturel*, ed. T. Fahd, 39–66. Actes du Colloque de Strasbourg Juin 1987; Brill, Leiden.
Pascual, J.-P. 1991 La Montagne du Hawran du XVI siècle á nos jours, *Le Djebel al-'Arab*, eds. J-M Dentzer & J. Dentzer-Feydy. 101–108. Direction Generale des Antiquities et des Musées de la République Arabe Syrienne et ERA 20 du Centre de Recherches Archaéologiques (CRNS). Paris.
Peake, F. 1958 *The History of Jordan and its Tribes*. Coral Gables, Florida.
Perthus, V. 1992 The Syrian private industrial and commercial sectors and the state, *International Journal of Middle East Studies* 24, 207–230.
Perthus, V. 1995 *The Political Economy of Syria under Asad*. IB Tauris, London.
Petersen, A.D. 1995 The Fortification of the Pilgrimage Route during the first three centuries of Ottoman Rule (1516–1757), *Studies in the History and Archaeology of Jordan* V, ed. K.'Amr et al. 299–306. Amman.
Pitard, W.T. 1987 *Ancient Damascus*. Eisenbrauns, Winona Lake, Indiana.
Politis, K.D. 1995 An Ethnoarchaeological Study on the Technology and Use of Adobe in the Jordan Rift Valley, *Studies in the History and Archaeology of Jordan* V, eds. K.'Amr et al. 321–324 Amman.
Porter Rev, J.L. 1855 *Five Years in Damascus*. 2 vols. London.
Postgate, J.N. 1992 *Early Mesopotamia: Society and Economy at the Dawn of History*. Routledge, London.
Postgate, J.N. 1994 In Search of the First Empires, *Bulletin of the American Society for Oriental Research* 293, 1–13.
Postgate, J.N. & S Payne. 1975, Some Old Babylonian Shepherds and their Flocks, *Journal of Semitic Studies* 20, 1–21.
Potts, D.T. 1990 *The Arabian Gulf in Antiquity*. 2vols. Clarendon Press, Oxford.

Potts, D.T. 1994a. Contributions to the agrarian history of Eastern Arabia 1: Implements and cultivation techniques, *Arabian Archaeology and Epigraphy*, vol **5**, no.**3**, 158–168.

Potts, D.T. 1994b. Contributions to the agrarian history of Eastern Arabia II: The Cultivars, *Arabian Archaeolgy and Epiography*, vol.**5**, no.**4**, 236–275.

Rafeq, A.-K. 1981 Economic Relations between Damascus and the Dependent Countryside. 1743–1771, *The Islamic Middle East, 700–1900*, ed. AL Udovitch, Princeton University Press, Princeton, New Jersey.

Rafeq, A.-K. 1987 New Light on the Transportation of the Damascene Pilgrimage during the Ottoman Period. *Islamic and Middle Eastern Studies*, ed. R. Olson. 127–136. Amana Books, Brattleborough, Vermont.

Rafeq, A.-K. 1992 City and Countryside in a Traditional Setting: the Case of Damascus in the First Quarter of the 18th Century. *The Syrian Lands in the 18th–19th Centuries*, ed. T. Philipp. 295–332. Berliner Islamstudien, Band 5. Franz Steiner Verlag. Stuttgart.

Reilly, J.A. 1989 Status Groups and Property Holding in the Damascus Hinterland, 1828–1880, *International Journal of Middle Eastern Studies* **21**, 517–539.

Reilly, J.A. 1990 Properties Around Damascus in the 19th Century, *Arabica* **37**, 91–114.

Ripinsky, M. 1975 The camel in ancient Arabia, *Antiquity* **49**, 295–8.

Robinson, E. & Smith, E. 1841 *Biblical Researches in Palestine, Mount Sinai and Arabia Petraea*. 3 vols. London.

Rogan, E. 1991 *Incorporating the Periphery: the Ottoman Extension of Direct Rule over Southeastern Syria*. Unpublished Ph.D. thesis, Harvard University.

Rogan, E. 1992 Moneylending and Capital Flows from Nablus, Damascus & Jerusalem to Qada' al-Salt in the last Decades of Ottoman Rule. *The Syrian Lands in the 18th and 19th Centuries*, ed. T Philipp. 239–260. Vol. 5, Berliner Islamstudien, Berlin.

Rogan, E. 1994 Bringing the State Back: the limits of Ottoman rule in Jordan 1840–1910. *Village, Steppe and State: the Origins of Modern Jordan*, eds. E. Rogan & T. Tell. 32–57. British Academic Press, London.

Rogan, E. 1995 Reconstructing Water Mills in Late Ottoman Jordan. *Studies in the History and Archaeology of Jordan* V, eds. K.'Amr *et al*. 753–756. Amman.

Ron, Z. 1989 Qanats & Spring Flow Tunnels in the Holy Land. *Qanat, Kariz and Khattara*, eds. P. Beaumont, M Bonine & K. McLachlan. 211–236. Middle East and North African Studies Press Ltd., Wisbech.

Rowton, M.B. 1973a. Autonomy and Nomadism in Western Asia. *Orientalia* **42**.

Rowton, M.B. 1973b. Urban Autonomy in a Nomadic Environment. *Journal of the Near Eastern Society*, **32**.

Rowton, M.B. 1974 Enclosed Nomadism. *Journal of the Economic and Social History of the Orient* XVII, 1–30.

Russell, K. 1988 *After Eden*. BAR International Series **391**, Oxford.

Russell, K. 1995 Traditional Bedouin Agriculture at Petra: Ethnoarchaeological insights into the evolution of food production. *Studies in the History and Archaeology of Jordan* V, eds. K.'Amr *et al*. 693–706. Amman.

Safadi, C. 1990 La foggara, syteme hydraulique antique, serait-elle toujours concevable dans la mise en valeur des eaux souterraines en Syrie?. *Techniques et Pratiques Hydro-Agricoles Traditionelles en Domaine Irrigué*, ed. B Geyer. 285–294 2 vols. Institut française d'archaéologie du proche orient. Librairie Orientaliste Paul Geunther. Paris.

Salameh, E. & Bannayan, H. 1993 *Water Resources of Jordan: Present Status and Future Potentials*. Friedrich Ebert Stiftung and Royal Society for the Conservation of Nature, Amman.

Sartre, M. 1982 *Trois études sur l'Arabie romaine et byzantine*. Collection Latomus **178**, Bruxelles.

Sauer, J. 1995 Artistic and Faunal Evidence for the Influence of the Domestication of Donkeys and Camels on the Archaeological History of Jordan and Arabia in *Studies in the History and Archaeology of Jordan* **V**, eds. K.'Amr et al. 39–48, Amman.

Sauer, J. & Blakely, J.A. 1988 Archaeology Along the Spice Route of Yemen, *Araby the Blest*, ed. D.T. Potts, 90–115, Carsten Niebuhr Institute, Copenhagen.

Schilcher, L.S. 1981 The Hawran Conflicts of the 1860s: A Chapter in the Rural History of Modern Syria. *International Journal of Middle East Studies* **13**, 159–179.

Schilcher, L.S.1991a. Violence in Rural Syria in the 1880s and 1890s: State Centralization, Rural Integration and the World Market, *Peasants and Politics in the Modern Middle East*, eds. F. Kazemi & J. Waterbury. 50–84 Florida International University Press, Miami.

Schilcher, L.S. 1991b. The Grain Economy of Late Ottoman Syria. *Landholding and Commercial Agriculture in the Middle East*, eds. C. Keyder & F. Tabak. 173–196. State University of New Press, Albany.

Schmidt-Neilsen, K. 1964 *Desert Animals*. Clarendon Press, Oxford.

Schumacher, G. 1886 *Across the Jordan*. London.

Schumacher, G. 1889a. Der Arabische Pflug. *Zeitschrift für Deutsches Palästein-Verein* **12**, 157–166.

Schumacher, G. 1889b. *The Jaulan*. London.

Seale, P. 1991 Asad: Between Institutions and Autocracy. *Syria: Society, Culture, and Polity*, eds. R. Antoun & D. Quataert. 97–110. State University of New York Press, Albany.

Seeden, H. & Kaddour, M. 1984 Space, Structures and Land in Shams ed-Din Tannira on the Euphrates: an ethnoarchaeological perspective. *Land Tenure and Social Transformation in the Middle East*, ed. T Khalidi. 495–526. American University of Beirut, Beirut.

Seeden, H. & Wilson, J. 1989 The AUB-IFEAD Habur Village Project Preliminary Report: Rural settlement in the Syrian Gezira from prehistoric to modern times, *Damaszener Mitteilungen* **4**, 1–31.

Seetzen, U.J. 1810 *A Brief Account of the Countries Adjoining the Lake of Tiberias, the Jordan and the Dead Sea*. London.

Seetzen, U.J. 1854 *Reisen dürch Syrien, Palästina, Phonicien, die Transjordan-Länder, Arabia Petraea und Unter-Aegypten*. Berlin.

Seikaly, S.M. 1984 Land Tenure in 17th Century Palestine: the Evidence from the al-Fatawa al-Khairiyya. *Land Tenure and Social transformation in the Middle East*, ed. T. Khalidi, 397–408, American University in Beirut, Beirut.

Seurat, M. 1980 Les populations, l'état et la société, *La Syrie Aujourd'hui*, ed. A. Raymond. 87–142. Centre d'études et de recherches sur l'orient arabe contemporain, Editions du CRNS, Paris.

Shaban, M.A. 1976 *Islamic History: a new interpretation.* Vol.1 AD 600–750/ AH132. Cambridge University Press, Cambridge.

Shaban, M.A. 1978 *Islamic History: a new interpretation.* Vol. 2 AD 750–1055/ AH 132–448. Cambridge University Press, Cambridge.

Shahid, I. 1984 *Rome and the Arabs: a prolegomenon to the study of Byzantium and the Arabs.* Dumbarton Oaks, Washington.

Shami, S. 1989 19th century Circassian settlements in Jordan. *Studies in the History and Archaeology of Jordan* IV, ed. M. Zaghlul. 417–421. Amman.

Shehadeh, N. 1985 The Climate of Jordan in the Past and Present, *Studies in the History and Archaeology of Jordan* II, ed. A. Hadidi, Amman.

Shoup, J. 1990 Middle East Sheep Pastoralism and the Hima System. *The World of Pastoralism: Herding Systems in Comparative Perspective.* Eds. J.G. Galaty & D.L. Johnson. 195–215. Guildford Press, New York and Belhaven Press, London.

Shunnaq, M. forthcoming. Dairy Products and the Role of the Middleman. *Nomadic Peoples.*

Sidebotham, S.E. 1989 Ports of the Red Sea and the Arabia-India Trade. *The Eastern Frontier of the Roman Empire*, eds. D.H. French and C.S. Lightfoot, 485–513, BAR International Series 553i, Oxford.

Singer, A. 1992 *Palestinian Peasants and Ottoman Officials.* Cambridge University Press, Cambridge.

Strommenger, E. 1985. Assyrian Domination, Aramean Persistence, *Ebla to Damascus*, ed. H. Weiss, 322–329, Smithsonian Institute, Washington, DC.

al-Sudairi, Amir 'A. al-R. 1995 *The Desert Frontier of Arabia: al-Jawf through the Ages.* Stacey International, London.

al-Tarawnah, M.S. 1992 *Tarîkh mantiqa al Balqa' wa Ma'ân wa al-Karak 1864–1918.* Amman.

Thoumin, R. 1935 Notes sur l'aménagement et la distribution des eaux à Damas et dans sa Ghouta, *Bulletin d'Etudes Orientales,* IV, 1–26.

Thoumin, R. 1936 *Géographie Humaine de la Syrie Centrale.* Librairie Ernest Leroux, Paris.

Tresse, R. 1929 L'irrigation dans la Ghouta de Damas, *Revue des Études Islamiques,* 459–473.

Tristram, H.B. 1873 The Land of Moab. London.

Uerpmann, H.-P. 1989 Problems of Archaeo-zoological research in Eastern Arabia, *Oman Studies,* eds. PM Costa & M. Tosi. 163–168. Serie Orientale Roma LXIII, ISMEO, Rome.

Van Zeist, W. & Bakker-Heeres, J.A.H. 1979 Some economic and ecological aspects of the plant husbandry of Tell Aswad, *Paleorient* 5, 161–168.

Velud, C. 1995 Syriie: tribus, mouvement national et Etat mandataire (1920–1936), *Tribus, tribalismes et Etats au Moyen-Orient,* eds. R. Bocco & C. Velud. 48–71. Maghreb-Machrek, Special Number 147. Paris.

Vidal, F.S. 1955 *The Oasis of al-Hasa.* ARAMCO, Dhahran.

Villeneuve, F. 1986 Contribution de l'Archaéologie a l'Hawran a l'Histoire Économique et Sociale des villages du Hawran, iv ème - vii ème siècles ap.J-C, *IVth International Conference on the Bilad ash-Sham during the Byzantine period,* eds. MA Bakhit & M Asfour, 108–119, Amman.

Villeneuve, F. 1991 L'économie et les villages, de la fin de l'époque hellenistique a la fin de l'époque byzantine, *Le djebel al-'Arab,* eds. J-M Dentzer & J.

Dentzer-Feydy. 37–43. Direction Générale des Antiquites et des Musées de la République Arabe Syrienne and ERA 20 du Centre de Recherches Archaéologiques (CRNS), Paris.

Wahlin, L. 1982 *Education as Something New*. Kulturgeographiskt Seminarium 3/82. Stockholm University, Stockholm.

Wahlin, L. 1993a. *Tribal Society in Northern al-Balqa', Jordan: an historical geographical survey*. Kulturgeografiskt Seminarium 3/93, Stockholm University, Stockholm.

Wahlin, L. 1993b. *Villages north of as-Salt, Jordan: an historical geographical survey*. Kulturgeografiskt Seminarium 4/93, Stockholm University, Stockholm.

Wahlin, L. 1994 How long has land been privately held in Northern al-Balqa', Jordan?, *Geografiska Annaler*, vol. **76b**, I, 3–19.

Wallin, G.A. 1854 Narrative of a journey from Cairo to Medina and Mecca, by Suez, Araba, Tawila, al-Jauf, Jubbe, Hail and Nejd in 1845, *Journal of the Royal Geographical Society*, **24**, 115–207.

Walmsley, A.G. 1992 Fihl (Pella) and the Cities of North Jordan during the Umayyad and Abbasid periods, *Studies in the History and Archaeology of Jordan* IV, eds. M. Zaghloul & K. 'Amr. 377–384, Amman and Lyon.

Watson, A. 1983 *Agricultural Innovation in the early Islamic World*. Cambridge University Press, Cambridge.

Weir, S. 1970 *Spinning and Weaving in Palestine*. British Museum, London.

Weiss, H. 1985 Protohistoric Syria and the Origins of Cities and Civilisation, *Ebla to Damascus*, ed. H. Weiss, 77–83, Smithsonian Institute, Washington DC.

Wetzstein, J.G. 1857 Der Markt in Damaskus, *Zeitschrift der Deutsches Morgenlands Gesellschaft* **11**, 475–507.

Wetzstein, J.G. 1860 *Reisebericht über Hauran und die Trachonen*. Berlin.

Weuleresse, J. 1946 *Paysans de Syrie et du Proche-Orient*. Gallimard, Paris.

Wilkinson, J.C. 1977 *Water and Tribal Settlement in South-East Arabia*. Clarendon Press, Oxford.

Wilkinson, J.C. 1983 Traditional concepts of territory in Southeast Arabia, *Geographical Journal* **149/3**.

Will, E. 1957 Marchands et chefs de caravanes à Palmyre, *Syria* **34**, 262–77.

Wustenfeld, F. [1868] 1993 Die Wohnsitze und Wanderungen der Arabischen Stamme, from the foreword of al-Bakri to his geographical dictionary. *Texts and Studies on the Historical Geography and Topography of Central and South Arabia*, ed. F. Sezgin. 147–232. Islamic Geography, vol.91. Institute for the History of Arabic- Islamic Science, Johann Wolfgang Goethe University, Frankfort am Main.

Yedid, H. 1984 Crise et regression du systeme pastoral bedouine nomade des hauts plateaux du nord-est de la ville de Hama (Syrie), *Nomads et Sedentaires: perspectives ethnoarchaéologiques*. ed. O. Aurenche. 19–50. Editions Recherche sur les civilisations, Paris.

Yoffee, N. 1995 Political Economy in Early Mesopotamian States, *Annual Review of Anthropology*, **24**, 281–311.

Zahrins, J. 1989 Pastoralism in southwest Asia: the second millenium BC, *The Walking Larder*, ed. J. Clutton-Brock,125–155. Unwin-Hyman, London.

Ze'evi D. 1995 Women in 17th century Jerusalem: western and indigenous perspectives, *International Journal of Middle East Studies* **27**, 157–173.

Ze'evi, D. 1996 *An Ottoman Century: the District of Jerusalem in the 1600s*. State University of New York Press, Albany.

Ziadeh, N.A. 1970 *Urban Life in Syria under the Mamluks.* Greenwood Press, Westport, Connecticut.

Zohary, D. & Hopf, M. 1988 *Domestication of plants in the Old World. The origin and spread of cultivated plants in West Asia, Europe and the Nile Valley.* Oxford University Press, Oxford.

INDEX

Abbad 70, 73, 390
Acacia sp. 115, 118, 172, 173, 183, 232, 241
Access to resources 10, 27, 31, 33, 34, 39, 44, 48, 50, 60, 62, 70, 71, 76, 79, 303, 342, 367, 380
Accountability 365–375
Accountant 311, 318, 324, 356
Acorns 172, 175, 176
Adhra 37
Agents 184, 199, 288, 289–90, 293–4, 296, 297–8, 300–1, 316, 321, 352
  forwarding 294
  government 36, 37, 49, 50, 289–90, 384–6
  labour 311
  official agencies 314–5, 345, 367
  state 344–5, 348, 375–6, 384–6
  wakîl 78, 184
Ageyl – see Traders
Agriculture 19–20, 25, 27–30, 31–34, 115, 117–8, 120, 123, 146–151, 176–202, 337–42, 361–2, 378–80
  arable 20, 23, 25, 37–9, 68–9, 101–2, 114–20, 192–8, 198–201, 289–90, 297–8, 302–3, 304, 307–9, 314–5, 319, 338, 341–2, 370–1, 378
  changes 20, 31–33, 41–2, 109, 115, 123, 146–7, 147–9, 152–5, 178–88, 191–2, 194–6, 199–201, 304–5, 306, 309–10, 379–80
  contracting 22, 35, 43, 62, 76, 291, 292, 293, 296, 297, 310, 322
  crop failures 27, 39, 44, 50, 68, 178, 180, 183, 184, 187, 189, 193, 199, 200, 209, 211, 218, 223

development 31–2, 34–5, 37–9, 41, 178–88, 191–2, 194–5, 199, 201, 291–2, 311, 361–2, 379–80
implements 33, 152, 178–82, 186–7, 189, 191, 195–7, 199
investment 27, 39, 45, 78, 226, 227, 277, 287, 292, 301, 302, 303, 314, 315, 317, 319, 324, 353, 379
irrigated 20, 32, 33, 99, 100, 119, 131, 147–150, 152–4, 158, 162–6, 176–192, 289, 322, 379–80
land ownership 68–72, 115, 118–27, 130–1, 158, 161, 177, 179, 185–6, 188–90, 198, 289, 291, 304–12
machinery 179–81, 184, 186–9, 191, 195–7, 199, 203
Ministry of 369–71
oasis 176–82
opportunistic 30, 101–2, 146–7, 198–200, 217–8
policies 34–40, 297–8, 308, 311, 361–2, 379
production 26–7, 33–40, 178–88, 191, 193, 289, 297, 299, 302–12
rainfed 20, 68–9, 98, 100, 114, 117–8, 192–201, 290, 297, 302–3
riverain 23, 31, 32, 147, 184–5, 201
vegetables 32–3, 176–7, 179, 182–92, 198, 200–1, 294–5, 309, 311, 328–32, 378
Ahl al-Jabal 66, 68, 73, 75, 101–2, 106, 126, 132, 134, 147–8, 169, 213, 215, 217–9, 226–7, 228, 230, 250–2, 253–4, 281, 287, 306, 359, 373, 391–2
Ahl al-Karak 68, 121

437

Ahl Tibna 64, 68, 73, 75
Aid 348, 352, 357, 368, 369
Air-conditioning 259
Airline pilot 311
Ajlun 28, 38, 39, 48, 98, 102, 116, 147, 172, 174, 175–6, 203, 284, 286, 287, 304, 338
Alaf (bought feed) 207, 208, 209–10, 212, 214, 216, 217–20, 222, 225–7, 229–30, 236, 250, 288, 318, 323
Almonds (amygdalus) 20, 99, 100, 118, 176, 330
Alternatives
 courses of action 79
 to tribal identity 12
Altorki, S and D Cole 41, 55, 58, 59, 63, 70, 71, 74, 76, 163, 164, 294, 296, 317, 356, 382, 387
Amarin (al-Hishe) 119, 154, 184, 282
Amarin (Ahl al-Karak) 192, 308
Amman 24, 39, 128, 144, 165, 185, 186, 187, 204, 231, 262, 288, 298, 308, 364, 366, 375, 381
Anabasis (ajram) 103, 118, 174, 229
Aneze 75, 205, 212
 Muwahib 30
Annual plants 103–5, 215, 219, 227–9
Anti-Lebanon 13, 98–9
Antiquities 300, 376, 385–6
 Department of
Antoun, R. 57, 61, 63, 64, 65, 72, 73, n15 75, 76, 128, 198, 326, 350, 358, 368, 388, 389, 390
Apples 20, 189, 191, 198, 370
Apricots 177, 180, 189, 191, 201, 276
 presseries 276
Aqaba 46, 183, 288
Aquifers 131, 134, 141, 157, 165
Arches 262, 265, 266
Armed forces 306, 308, 310, 319, 327, 350, 353, 355, 380, 382

Aromatics 23, 24, 26, 172, 173
Artemisia (shih) 20, 99, 102–3, 108, 116, 118, 171, 174, 229, 231, 249, 391
Ashîra 64–8, 306
'*Asîl* 73
 families 368
Asparagus 32
Assets
 management 325, 336, 348, 358, 379
Assymetry 17–18
Ata'ata 119–20, 242, 319
Aubergine 32, 179, 180, 185, 198, 330
Auction 293, 295
Autonomy
 business 322
 individual 17, 63, 290–1, 300–1
Azazma 120, 175, 186, 192, 203, 207, 231–2, 236–7, 243, 301, 330, 359
Azraq 20, 100, 102, 103, 109, 110, 133, 134, 146, 165–6, 217, 218, 280–1

Badia 14, 16, 98, 103–110, 117, 128, 132–42, 174, 206, 212–31, 314–6, 319, 329, 337, 347, 350, 353, 355, 357, 372–5, 381, 392
 arable crops in 146–8, 217–8
 structures in 239–54, 283–4, 2 287–8
Badia Police 217, 253
al-Balqa 25, 30, 38, 39, 48, 98, 142, 187, 239, 264, 333, 358, 387
Bamia 32, 189
Bananas 32, 115, 185
Barley 19, 30, 37, 44, 176, 177, 179, 181, 183, 184–5, 186, 187, 192–7, 198–9, 200, 203, 210, 218
 for livestock 112, 178, 181, 183, 196–7, 198–9, 203, 207, 209, 218

varieties 193–4
volunteer 181, 194
Barqa 107
Barsîm 180–1, 189, 205
Basalt 14, 98, 100, 102, 126, 141, 243, 252, 255, 271
  hand-mills 276
  mortars 274, 276
  as temper 279
Beans 32
  broad 32, 182, 189, 198, 200
  green 32, 173, 185, 331
Bedding 247, 374
Bedouin 19, 41, 46, 58, 59, 60, 68, 71, 212, 213, 233, 259, 280, 300, 307, 313, 315, 336, 346
  bedu 55, 56, 58, 61, 72, 75, 305, 355
  domestic group 64
Beni 'Amr 30, 58, 65–6, 73, 122, 124, 168, 185, 187, 210–11, 262, 264, 265, 299, 309–10, 313
Beni Attiya 116, 119, 151, 211
Beni Hassan 39, 109, 87
Beni Hamida 119, 122, 169, 173, 175, 202–3, 264, 265–6, 274, 299, 302, 310, 312
Beni Kalb 30, 253
Beni Karim 30
Beni Khalid 70, 83, 218, 373
Beni Kilab 36
Beni Mahdi 30
Beni Sakhr 30, 38, 49, 50, 70, 73, 110, 116, 152, 169, 211–2, 219, 236, 252, 264, 275, 290, 303, 358, 373
Beni Zaid 30
Bianquis, T. 31, 36, 127, 157
Birds, migrant 110, 167
Bilâd ash-Shâm 8, 12, 13–4, 20, 27–8, 33, 34, 36, 42–3, 47, 49, 53, 61, 97, 98, 375, 387
Birka 133, 142, 144
Bitter vetch(*kersena*) 195, 196, 198, 203
Bocco, R. 48, 63, 298, 361, 368, 390

Bocco, R. and T. Tell 49, 50, 358
Borders 49, 146, 217, 218, 220, 222–4, 225, 229, 281, 283, 288, 298, 318–9, 358, 360, 362–3, 365, 375
Bowers 247
Braemer, F. 100, 147, 201
Brassica sp. *(khafsh)* 103 and appendix
Bread 174, 186, 197, 270, 328–332
  dhurra 200
  flour 197
  masliyya 329
  saj 174, 175, 247, 329
  shop 328
  tabûn 174, 271, 303
  with herbs 331
British Mandate 39, 49, 283, 304–6, 341, 347, 358, 360
Brucellosis 210
Buckingham, J.S. 206, 262, 277
Buildings
  decayed 262, 271, 272, 274
  inheritance of 267, 271–3
  land 309, 317
  renovations 271–3
Burckhardt, J.L. 28, 29, 41, 46, 70, 74, 79, 147, 153, 168, 173, 174, 182, 184, 194, 200, 202, 206, 212, 213, 262, 270, 274, 280, 282, 296, 307, 338
Bureaucracy 366–7, 382, 383
Burqu' 100, 107, 132, 134, 217, 218, 219, 372
Busaira 24, 370
Busaita 98, 165, 181
Business enterprise 291–2, 298, 322–4
  small 316
Bustard 167
Butm see *Pistacia*
Butter *(samn)* 70, 203, 206, 207, 209, 210, 213, 216, 219, 220, 231, 232, 236, 248, 296, 300, 310, 317, 318, 329, 330
  fresh 180, 203, 205, 206, 207, 208–9, 216, 329

Buying
  land 37, 149, 209, 271, 273, 288, 300, 302–3, 305, 307, 308, 309, 311, 312, 319
  speculative 302
  trees 177
  water 130, 144, 149, 158, 159, 163, 211

Cabbage 189, 198
Cairns 125, 126, 252, 283, 284
Calligonum (abal) 109
Calligonum (arta) 118
Camel 21, 22, 23, 24, 30, 40–1, 44, 50, 59, 60, 70–1, 105–6, 112–3, 133, 135, 162–4, 174, 177, 202, 233–7, 248, 281, 285, 290, 296, 303, 315, 318, 319, 326, 338, 342, 354
  hair 241, 296
  herders 22, 30, 41, 60, 70–1, 105–6, 108, 234–7, 241, 290, 354
  herding, decline of 29, 233–5, 342, 343, 354
  management 105–6, 233–7
  prices 237, 303
  saddle 24
  water-carrying 133, 214
Camp-sites 125, 214–7, 243–250
Candying fruit 189
Capital 314, 322, 325
Caravans 24, 40, 41, 43, 46, 59, 281
Carrots 32, 201
Catering services 328
Cattle 20, 25, 144, 162–3, 166, 178, 201, 202–5, 211, 262, 281
  house cows 203
  housing 203, 204, 262
Cauliflowers 32, 191
Caves 247, 248, 249, 250, 262, 263, 303
Cement 143, 150, 251, 176, 254, 259, 265–7, 269, 275, 355
  factory 370
  houses 251, 254, 259, 265–7, 269, 287, 326

Cemeteries 251–2, 273–4
Change 9, 11, 30, 50, 260, 266, 331, 341–2, 350–2
also Transformations 341–2
  in attitude to land 379–80
  in population 226–7
Charcoal 26, 174, 280
Chechen (Shishan) 53, 109
Cheetah 110
Cheese 205, 207, 216, 219, 231, 293, 295, 329
Chickens 180, 262
Chickpeas 176, 192, 195, 196, 198, 329
Children herding 216–7, 220, 221, 231, 236, 337, 353, 355
Christians 25, 53, 61, 150, 263, 264, 265, 267, 273, 279, 303, 317, 323, 336
Circassians 53, 267
Cisterns 25, 120, 130, 131, 138, 140, 142–6, 151, 152, 254, 263, 265, 266, 269, 270, 273, 278, 285, 287
Citizens 348, 349, 352, 353, 357
Citrus 115
City – see Towns
City-states 23
Civil Service 353
Classification of plants, animals 104, 110–114, 194
Clay 279
Climate
  change 21
Climatic factors – see Landscapes
Clinics 331, 357
Cloth – see textiles
Clothes 231, 296, 317, 320, 323, 326, 328, 332–4, 337
Clubs 365, 382–3
Coffee 174, 247, 256, 262, 328
  apparatus 259, 332, 335
  pots 247, 252, 259, 260, 295, 329
Colocasia 32, 185
Compensation 291, 302, 312, 337, 360, 381–2

Computers 321, 375
Consumer goods 351
Consumption 328–37
Context 17, 18, 53, 56, 60, 72, 111, 301
Contracts 17, 43, 62, 76, 216, 282, 291, 292, 300, 314, 337, 372
Cooking 174–5, 255, 262, 269, 270, 328–32
  dung 174–5
  equipment 247, 260, 329
  also household utensils
  gas stove 174, 351
  primus 174
  wood 174–5, 233, 369
Co-operatives 158, 229, 231, 361–2, 373
Coot 110
Coppicing 173–4, 241, 371
Corbelling 145, 255
Corrals *(sayra)* 215, 248–9
Cosmetics 104, 296, 317
Cotton 20, 26, 32, 37, 41, 59, 200, 243, 295, 313
Courts 292, 381
Courtyards 258, 259, 265, 287
Cousa 32, 179, 185, 188, 189, 192, 330
Crafts 9, 27, 71, 77, 173–4
  wood for 173–4
Craftspeople 35, 54, 57, 58, 177
Cranes 110
Cratagus 99, 100, 172
Credit 43, 45, 229, 231, 294, 301, 302, 312–4, 318, 319, 324
Cucumbers 32, 179, 182, 184, 189, 198, 200, 201, 329
Currencies 223, 224–5, 271, 314
Curtains 335
Cushions 32, 335
Customs 365
  duties 298
  see borders
Cypress 118, 173
Dairies 203–5, 236, 317
  see milk products
Damascus 13, 14, 16, 21, 23, 26, 27, 32, 37, 42, 45, 47, 76, 102, 127, 130, 157, 168, 173, 174, 202, 212, 222, 241, 256, 282, 283, 294, 300, 313, 316, 317, 338, 339, 340, 384, 391
Dana 117, 119–0, 145, 154, 182, 188–91, 239, 263, 270–3, 284, 307, 370
Date palm 184, 229, 256, 258
Dates 20, 24, 26, 44, 70, 166, 176–9, 258, 295–6, 311, 357
  feral/wild 115, 118, 172
Dams 128, 132, 146, 147, 150, 374–5
Dead Sea 13, 114, 115, 208, 280–1, 358
Dealers
  see traders
de Boucheman, H 46, 59, 172, 173, 174, 296, 318
Debt 43, 45, 301, 302, 313, 318, 319, 320, 325
Deficit 9, 222, 301, 372
Democracy 366–8
Desk-top publishing 320
Descent 16, 21, 65
  tribal 58, 68
Development 164
  aims of 348
  of assets 321–5
  projects 134, 146, 373–5, 376–8, 380
Dew 98, 105, 178, 215
Dhibon 24
d'Hont, O. 29, 33, 120, 138, 166, 221, 230, 261, 280
Dhulail 203–5, 219
Dibs 44, 329
Diesel 152, 166, 200, 278
Diseases
  animal 180, 205, 210, 248
  human 25, 28, 30, 110, 177, 264, 328, 390–4
  plant 177, 193
Dispute 305–7, 308, 344
  management 366, 368, 381–2
  water 149, 150–2
Dira 69–70
Doctor 311, 356

Documents 15–6
Dogs 113, 172, 221
Domestication
  animals 19–21, 111–4
  plants 19–21
Donkeys 22, 44, 114, 168, 182, 187, 189, 196, 197, 202, 206, 211, 214, 221, 222, 227, 285
Doughty, C. 27, 29, 30, 31, 46, 71, 75, 113, 160, 163, 167, 168, 173, 275, 280, 296
Doumani, B. 27, 39, 42, 48, 297, 299, 300, 304, 313, 314, 316, 318, 323, 338, 339, 340, 346
Drainage systems
  see Wadis
Dress-making 317
Drought 39, 50, 211, 266, 287, 312, 371
Druze 53, 165–6, 191, 203, 281, 333–4, 341
Drying fruit/veg. 189
Duck 110
Dung 174
  burning 174
  fertiliser 180, 192, 195, 205

Eagles 110, 113
Education 211, 225, 234, 259, 267, 286, 311, 327, 328, 335, 351, 353–6, 366, 376, 377, 378, 386–7
  Christian 387
  Quranic 387
  state 387
  tribal 387–8
Eggs 180, 329
Elections-Jordan 368, 390
Electricity 128, 164–5, 179, 203, 259, 266, 270, 275, 298, 314, 323, 331, 334, 357, 379
Employment 178, 182, 185, 187, 205, 208, 211, 213–5, 226, 234, 254, 287, 310, 311, 315, 323, 327, 345, 351, 352, 354, 356, 360, 364, 374–5, 376–7, 385, 395

  on dev. projects 374–5, 376–7, 385
  govt. service 345, 348, 352–3, 356, 385
  graduate 355, 356
  of women 311–2, 317–8, 323–4, 355–6, 386
Engineers 260–1, 323–4, 355, 375
Escarpments *(jal)* 108
Euphrates River 14, 31, 32, 166, 201, 230
Exports
  cotton 41, 338
  fruit, vegetables 184, 191, 362, 378
  grain 25, 26, 41, 197, 338
  gums & aromatics 23, 42, 171, 172–3
  kilw 37, 42, 171, 173, 296, 313, 338
  oil 342, 345
  sheep 212, 224, 225, 294, 296
  sugar 41

Factories 309, 314
Fainan 24, 110, 117, 119, 120, 121, 123, 154, 170, 172, 175, 182–3, 329, 331–2, 335, 359, 385
Falaj 155–66, 277, 303
Fallow 192, 193, 194–6, 201, 236
Family
  hâdhr 61
  learned 56
  sayyid/sada 75
  waqfs 299
Faqîrs 75, 392
Falcons 110, 112, 167, 169, 170–1, 373
Farming
  see agriculture
Feasts 255, 328, 329, 330, 331, 335
Fed'an 18, 98, 349
Feddan 196
Fellah 55, 58, 68
Fertiliser 180, 181, 187, 191–2, 193, 194–5, 232, 328

Field systems 284–6
Figs 20, 44, 176, 183, 185, 189, 192, 201, 287
Finches 110
Finders 318
Fire 193
Firewood 116, 174–5, 233, 243, 371
Flies 110, 249
Flint 98, 102, 126
Flour 171, 172, 186, 197, 271, 276, 277, 278, 319, 330
Fodder
 -green 177, 178, 181, 189, 192, 194, 198, 218
Fog 101
Foggara
 see falaj
Food (all types) 171–3, 328–32
 additives 328
 wild 104, 171–3, 331
Forests 115, 211, 368–72
 department of 369–72
Foxes 110
French Mandate 49, 296, 345, 346–7, 349
Fuel 104, 116, 174, 243
Fugara 152, 185–7, 264, 266, 274, 307
Furnishings 334–5
Furniture business 316
Fuware 106, 218, 287

Garages *(mahatta)* 288, 297, 318
Garaigara 154–5, 183–4
Gardening Calendar 189
Gardens 176–7, 182, 183, 186, 188–190, 203, 269, 270, 305, 309, 310, 316, 317
Garlic 198, 201
Gas 128, 134, 174, 270, 283
Gathering 9, 42, 167, 171–6, 280–1
Gazelle 110, 112, 167, 168–9, 170, 372, 373
Geese 110
Genealogical idiom 16
Geological configs. 98–102, 114–5

Ghadîr 100, 105, 128
Ghawarna 115, 116, 185, 186, 202, 203, 301, 308
Ghayyath 73, 106, 217, 218, 219, 220–1, 228, 229, 231, 250
Ghor/Aghwar 102, 114–5, 147, 149, 202, 215, 236–7, 277, 282, 283, 361, 381
Ghutas 98, 99, 131, 157–9, 160, 173–4, 200–1, 241
Gipsies 54, 56, 63
Glubb Pasha 49–50, 222, 306, 369
Goats 20, 30, 59, 101, 106, 112, 114, 115, 119, 120, 135, 144, 174, 175, 182, 183, 190, 194, 201, 202, 203, 208, 212, 213, 226, 230–33, 240–241
Government 31, 33, 35–8, 47–9, 50, 72–3, 352–3, 380–1
 functions of 343–4, 377–9
 hukuma 343–4, 384
 irrigated farms 53
 state/daulat 49, 74, 98, 128, 283, 302, 311, 314, 344–50, 360, 366, 367–7, 383–5, 387
 taxation 345
Grain 38–40, 41, 42, 44, 45, 47, 186, 187, 267, 271, 275, 302
 Government centre 194, 197, 238–9
 market 290, 33, 316
 milling 276–8
 processing 246
 production 30, 158, 338
 requirements 197
 sales 197, 267, 323, 340–1
 storage 199, 250, 253–4, 263–7, 271, 273,
 subsidies 179, 181, 267, 380
 trading 267, 312, 338
 yields 193
Graffiti 250
Grapes 20, 44, 166, 176, 178, 179, 180, 182, 183, 184, 185, 186, 187–8, 189, 191, 192, 198, 201, 295

Gravel 8, 75, 281, 283
Graves 157, 158, 233–4, 226, 244
Grazing 78, 83, 85, 184, 187, 189–90, 203–5, 203–4, 204–5, 370, 372, 374–5
Growing season 156, 157, 166, 175
Guarantees 11, 55, 56, 60, 62, 67
  trade 65, 67
Guests 227, 228, 237
Guest-room 55, 233, 318, 330
Gulf War 283
Gunpowder 31, 251
Gymnasiums 290
Gypsies 47, 56

Ha'il 63
Hair styles 306
Halawa 302
Hallabat 157
Haloxyletum (ghada) 80, 86, 95, 147, 208, 211, 213
Haloxyletum (rimth) 80, 85, 86
Hamad 8, 57, 66, 70, 74, 76, 77, 79–85, 88, 91, 104, 105, 143, 144, 149, 187, 189, 192–6, 204, 214, 216, 220, 222
  development plan 110, 112, 121
  water in 110, 111, 113, 115, 121
  wells, modern 111–2
  vegetation 79–81
'Handing on' 292–9, 321
Hares/rabbits 110, 167
Harra 14, 19, 46, 97, 98, 100–3, 106–7, 109, 132–8, 140, 141–2, 146–8, 169, 199–200, 192–6, 217–9, 220, 225, 230–1, 241, 243, 248–50, 252, 373
  rainfall 100
  vegetation 102–3
  water in 106–7, 109, 132–8, 140, 141–2, 146–8
Harvests 178, 181, 185, 186, 189, 191, 195, 196, 197, 199, 200, 209, 217, 218, 306, 308, 309, 310
  –ing by hand 191, 195, 196
grain 181, 185, 186, 196, 199, 200, 209, 217, 218, 306, 308, 309
Hauran 25, 47, 48, 70, 98, 99–100, 102, 105, 147, 200, 201, 213, 230, 263, 277, 279, 281, 307, 333, 340–1, 346, 349
Hazm (ridge) 107
Healers 75, 318, 391–3
Health 274, 326, 328, 351, 357, 377, 386, 390–394
Health centres 288, 386
Hebron (Khalil) 27, 173, 282, 313
Hedgehog 110
Herbicides 181, 196
Herding 12, 20, 21, 22
  and see grazing 23, 25, 26, 27, 29, 30, 37, 40, 60, 69, 76, 101, 105–6, 108, 111–2, 120, 132–40, 142–5, 201–37, 240–1, 243–7, 250, 292, 314–5, 320, 322–3, 334, 337, 338, 351, 353–4, 355–358, 365, 370–2, 379
  part-time 355
Hijaz 14, 46, 47, 59, 117, 155, 163, 282, 299, 313
  railway 46, 47, 176
  Hired labourers 37, 69, 181, 185, 187, 191, 205, 208, 214, 292, 297, 316
Hishe
  see Shera
Hisma 25, 98
Historical truth 16
History 13–6, 19
  Arab writings 14–15
  local histories 15
  local traditions 13–15, 18
  mediaeval Islamic 15
Hobbs, J. 111, 114, 166, 169, 170, 172, 175
Honour 17, 56, 72, 77
  and shame 17
  defended 17, 63, 123

Horses 40, 117, 167, 168, 187, 189, 202, 210, 278
Hospitality 8, 44, 77, 247, 255, 259, 260, 300, 301, 322, 327, 335, 345
Hotels 294, 315
Household goods 32, 247, 256, 259, 260, 276, 279–80, 318, 330–2, 334–5, 350–1
Houses 25, 29, 209, 243–4, 248, 253, 254–71, 287–8, 326, 328, 335, 336
  badia 253–6, 287–8
  block 259–60
  Dana 270–1
  decoration 261, 308
  design 260–1
  al-Jauf 256–9
  Karak 262–70
  as-Salt 262
  villas 259–60, 287–8
Humidity 96
Hunting 112, 113, 167–71, 373
Hutaim 56, 70–1, 72, 167, 168
Huwaitat 49, 110, 111, 119, 212, 219, 236, 243, 386
Hyaena 110, 113, 140

Ibex 110, 112, 167, 169–70
Ibn 'amm group 62, 64, 65, 67, 186, 189, 213, 306, 308, 336
Identity/ies 10–11, 12, 54, 56, 60–1, 66, 79, 388, 395
Idioms of
  changes in 61, 122
  claims 98
  closeness 18, 111
  formal/informal 18
  genealogical 16, 53
  official and local 18
  public/private 18, 123
Indigo 147, 182, 184
Industries
  fertiliser 153
  food processing 24, 26, 36, 44, 147, 173, 180, 184, 204–5, 276, 287, 293, 302, 314–5, 365
  glass 42

  small scale 173, 293, 295, 296–7, 299, 309, 310, 311, 315, 316, 317, 320, 321, 365
  soap 33, 37, 42, 173, 302, 313, 342
  tanning 26, 40, 173
  textile, clothing 26, 32, 35, 41, 42, 147, 208, 230, 240, 293, 302, 313, 320, 342
Inheritance 190, 291–2, 309, 310, 311, 316, 336
Inscriptions 24, 250, 252
  rock carvings 110, 113, 126
  modern 250, 252
  Safaïtic 24, 252
  Thamudic 24
Investment 123, 184, 292, 302–4, 312–7, 319, 324–5, 378
IPC 50, 133–4, 253, 288, 315
Irrigation 32, 33, 99, 115, 119, 130, 131, 146–7, 149–51, 152, 153–4, 155–8, 161–4, 166, 177, 178, 179–80, 184, 187, 191, 198, 199, 201
  modern 131, 150, 152, 153, 164, 165, 178, 179, 180, 184, 186, 199, 201
  traditional 131, 146–7, 149–50, 153, 155–64, 187, 198, 201
Islam 15, 26, 31, 127, 129, 387, 388–9

Jabal Ajlun
  see Ajlun
Jabal al-'Arab 25, 30, 33, 98, 99, 100, 146–7
(J. Druze, J.Hauran) 121, 145, 147, 172, 174, 199–200, 203, 281
Jabal Nablus
  see Nablus 75, 271, 277, 286, 313
Jabal ash-Shaikh 99
Jam 189, 329
Jama'a 62, 68, 149, 321, 322, 335, 338, 385–6
al-Jauf 14, 70, 108, 109, 159, 168, 171, 176–8, 202, 256,

258, 259, 175, 280, 286,
  295–6, 299, 335
Jaulan 102, 105, 200
Jaussen, A. 62, 73, 211, 265,
  274, 278, 388, 391
Jawa 102, 132, 140–2, 146–7,
  199–200
Jerboa 110
Jerusalem 27, 34, 173, 206, 277,
  282–3, 304, 323, 338
Jewellery 296, 320, 337
al-Jibal 25
Jird 110
Jordan Valley 13, 19, 20, 39, 169
  JVA 123, 151, 153, 361–2
al-Jûba 98, 105, 108–10, 133,
  179–82, 259–60, 311–12,
  379–80
Jumlan 106, 218
Juniper 118, 173, 241, 271–2

Kafîla 61, 282, 300, 337, 363–5
Kafr al-Ma' 63–5, 76, 128, 198,
  388–9, 390
Karak 25, 28, 30, 48, 50, 62,
  65–6, 67–8, 79, 98, 114–7,
  121–2, 124, 143–4, 147–53,
  155, 172, 173, 175, 176,
  185–8, 192–9, 201, 203,
  206–12, 236, 239–47, 249,
  262–70, 273, 274, 276–9,
  282–3, 284–6, 290, 292, 295,
  305, 307–10, 313, 317, 320,
  331, 333, 340, 345, 350, 358,
  380–1, 302, 385–6, 387, 390
  plateau 114–7, 143–4, 186,
    192–8, 206–12, 290
  vegetation 116
Karbala 105
Khabra 105, 108, 131–2
Khaibar 27, 30, 70, 71, 160
Khana 267–70, 273, 299
Khuwa 62, 75–6, 77, 233–4,
  300, 337, 347, 372
Kilw 37, 42, 171, 173, 296, 313,
  338
al-Labbah 108, 138, 252
Lambs 114, 206, 207–10, 212,
  213, 215–6, 219–22

Land
  abandoned 28–30, 305–6, 307
  access to 28–9, 68–9, 120,
    198, 299, 304
  for agriculture 119–20, 198,
    264, 286–7, 299, 323, 379–80
  buying 283, 307–8, 312
  claims 188, 198, 272
  control 69
  defence of 121, 99, 101
  disputes 122–3, 278–9
  exchange 150
  for herding 39, 49, 69–70,
    120, 290
  grants 35, 39, 290, 339
  inheritance of 37, 280–2
  kharaj 342, 344, 345, 368,
    370
  management 291, 302, 378–8
  market 312
  miri 128, 304, 368, 370
  mortgages 39
  mulk 30, 304
  musha' 39, 304–5
  ownership 34, 37, 39, 69,
    118–125, 183–4, 286–7, 289,
    299, 304, 309, 310, 311–2,
    321, 325, 339
  owning/using/living 21, 27,
    30–1, 118–28, 289, 315, 347,
    359, 371
  registration 38, 152, 179, 199,
    254, 264–5 286–7, 305,
    306–7, 333, 358, 359
  rights of usufruct 37, 38, 69,
    123, 289, 371
  sales 37, 285
  State owned 37, 38, 39, 69,
    128, 201, 259, 264, 290, 298,
    339, 345, 361
  transfers 301–2, 308–9, 312–3
  use 13, 19, 25, 198, 235–8
  waqf 31

Lapwing 110
Layne, L. 60, 70, 73, 75, 127,
  261, 361, 390, 395
Law
  courts 34, 292, 366, 368, 371

Islamic   68–9, 129–30, 313, 368, 371
State   69, 368, 371, 381
tribal, customary   62, 292, 296, 371, 372, 381
Legumes   20, 26
Lemons   26, 176, 184, 158
Lentils   176, 192, 193, 194, 195, 196, 198, 200, 329, 332
Leopard   110
Licences   174, 180, 297, 310, 365, 381
Lime   143, 278–9
Lions   110
Livelihoods   9, 11, 12, 13, 64, 77, 97, 115, 119, 120, 121, 124–5, 128, 289, 290–2, 301, 303, 307, 312, 321, 324–5, 327, 337, 344, 353, 358
 rights to   301, 367, 371
Livestock rearing   19, 202–238, 292, 299, 314–5, 319, 322, 341
Lizard (*dhubb*)   168
Loans, government   302
Local
 analyses   14–5, 16–8, 31, 62, 76, 121, 122, 123, 125, 126, 190–1, 203, 208–9, 223–4, 225–6, 227–8, 229–32, 266, 281, 285, 309, 324–5, 351–2, 384, 394–5
 interests   50–1, 54, 122–4, 126, 128, 338–40, 341–52, 359, 361–8, 368–5, 386, 394–5
 knowledge   14, 16, 97–8, 101, 102–5, 106–8, 109, 117, 118, 119, 120, 121, 126, 134–42, 143–6, 158–9, 164, 166, 176–82, 184–190, 192–8, 278–9, 375–6
 perception of
  animals & plants   102–5, 110–3, 115–6, 118, 120, 128, 175–6, 228–9, 232, 371–3
 practice   18, 50, 53, 56–7, 58, 69, 71–2, 75, 79–80, 111–2, 121, 122, 126, 149–54, 159, 161–2, 198, 201, 204–5, 210–11, 213–22, 235–7, 251–3, 260, 298
 premises   7, 8, 51, 61–2, 72, 79–80, 198, 224, 232, 234–5, 252–3, 300, 354–5, 357–8, 382
Local government   275, 380–1,
 and offices   384–6
Lorries
 also pickups   211, 217, 223, 227, 230, 237, 243, 249, 283, 295, 296, 318–9

Ma'aita   185, 186
Ma'an   25, 70, 76, 157, 171, 315
Mafraq   165, 191, 215, 218, 223, 231, 249, 283, 286, 293, 295, 305, 318, 366, 381
Magazines   331, 334
Mahfûr   101, 107–8, 132, 135–8
Maize   32, 185, 188, 189, 198
Majali   48, 66, 73, 79, 122, 144, 185, 186, 192, 208–9, 211, 243, 263, 264, 265–6, 273, 283, 301, 307–8, 382
Marj   105
Markets   26, 27, 30, 40, 42, 43, 44, 73, 77, 179, 180, 182, 183, 184, 185, 188, 235, 258–9, 275, 295, 322, 338, 346, 347, 352, 358, 359, 362
 grain   39, 313, 343
 livestock   206, 222, 224, 225, 230, 232, 233, 293, 343
 retail   275, 294
 seasonal   288
 urban   76, 172, 173, 205, 206, 313, 331, 343
 vegetable/fruit   32, 294–5, 311, 331
 wholesale   187, 275, 294, 362
Marriage   16–17, 56, 64, 65, 66, 67, 68, 231, 250, 309, 312, 335–6, 350
Matthiola (shiggara)   103, 228
Mattresses   32
Mediators (*wasît*)   344, 363–4, 366, 371
Medical bills   335

Medicine
  modern   315, 393–4
  traditional   386, 390–3
Melons   200
  sweet   32, 123, 179, 182–3,
    184, 186, 187, 198, 295
  watermelon   32, 146, 184, 189,
    191, 196, 198, 199, 201, 282,
    283, 295
Menaja'a   119, 170, 183
Merchants
  see also traders   37, 38, 39, 40,
    41, 43, 45, 46, 49, 50, 54, 57,
    58, 64, 168, 184, 188, 200,
    208, 213, 240, 277, 294, 295,
    296, 297, 300, 313, 314–5,
    316, 318, 341, 343, 346, 351
Merrill, S.   32, 147, n25 255,
    277, 338
Metral, F.   27, 48, 59, 64, 129,
    201, 213, 226, 296, 314, 317,
    350
Metral, F. and J. Metral   62, 70,
    n15 75
Mice   110, 189
Micro-climates   101, 102
Migrant workers   50, 54, 185,
    205, 288, 316, 363–4
Milk   20, 177, 211
  camel's   234, 235, 236, 328,
    331
  cheese   219, 268, 295
  contracts for   219, 292, 317
  cow's   202–5, 317, 331
  goat's   220, 230, 231, 233, 236
  processing   221, 225, 249, 331
  processors   225, 249, 265
  production   14, 178–80, 230
  products   20, 221, 222, 230,
    310, 317, 332
  sheep's   206–7, 210–11, 213,
    216, 220, 221, 225, 230, 314,
    315, 317, 329, 331
Millet (dhurra)   32, 146, 176,
    183, 184, 185, 192–4, 196,
    198, 199–200, 209
Mills
  diesel   186, 197, 278
  hand   276

  steam   278
  water   150, 154, 158, 197,
    276–8, 297, 302
Millstones   277
Ministry of Interior   381, 382, 385
Ministry of Tourism   369
Mlosi wells   105, 133
Mobility   9 13, 20, 24, 30, 44,
    58–9, 64, 69, 198, 200, 222,
    395
Models   8, 17, 30, 72
Money-changing   315
Money-lending   37, 38, 39, 312,
    313, 317
Moral order   11, 15, 124
Mosques   252–3, 273, 388, 389
Mosquitoes   110, 116
Mount Carmel   332
Mountains   13–14, 64, 97,
    98–100, 102, 110, 117–21,
    145, 153–4, 157, 169–70,
    183, 188, 202, 213, 231–2,
    270, 371
Movement   9, 13, 29, 44, 105–6,
    175, 216, 358, 360
  seasonal   108, 115, 120, 182,
    270, 358, 370, 372
Mukhtar   366–7, 368, 380–1
Mules   187, 196, 202, 276
Multi-resource   9, 12, 13, 19
  economics   289, 299, 322,
    350
Musil, A.   32, 41, 46, 59, 62, 65,
    71, 73, 74, 76, 77, 79, 104,
    113, 115, 116, 122, 125, 126,
    138, 157, 160, 167, 168, 172,
    176, 182, 184, 193, 194, 196,
    202, 212, 213, 235, 241, 247,
    251, 252, 253, 256, 258, 270,
    276, 279, 280, 282, 302, 304,
    313, 341, 346, 387
Muwaqqar   70, 199

Nablus-Jabal   16, 27, 36, 39
  Nablus   48, 173, 275,
    299–300, 304, 313–4, 316–7,
    323, 338, 339–40, 346,
    391–2, 393
Na'imi   205

Najaf 105
Nation state 13, 259, 298, 348, 358
Nationality 358–61
  local views on 332–4
Nectarines 191
Nefud 14, 98, 102, 108–9, 114, 140–115
Negev 19, 21, 119, 213
Negotiation 8, 11, 301, 346
  local/official 346, 371–2, 375
  by locals 381, 385, 350–1
Nejd 97, 171, 212
Networks 20, 44–5, 54, 62, 67, 117, 186, 213, 222, 233, 300, 307, 326, 327, 376, 382
NGOs 305, 368–372, 378, 383
NRA 369
Nuria 166

Oak *(ballût, sindian)* 116, 118, 172, 173, 175–6, 211, 271, 279
  woodland 19–20, 100, 285
Oases 27, 30, 70, 73, 105, 108, 109, 110, 157, 162, 176–9, 182, 201–2
Oats 181, 184
Oil (petroleum) 50
  companies 253, 259, 283, 288, 311, 351, 380
  economy 62, 164, 343, 345, 348–9, 352
  industry 342, 343
  wealth 259, 348, 352, 379
Oils 26, 390, 391
Oleander *(difla)* 115, 116, 118, 241, 271
Olives 20, 44, 166, 170, 180, 183, 185, 186, 192, 198, 309, 310, 311, 312, 329, 335
  oil 36, 39, 172, 176, 180, 183, 276, 310, 313, 332, 338
  presseries 180, 276, 310, 365
  processing 276
Oman 58–9, 61, 64, 67, 68
Onagers 110
Onions 32, 177, 179, 188, 189, 190, 198, 200, 294, 329, 330

Oranges 32, 176, 178, 184, 198
Orchard trees 20, 120, 175, 187, 310, 316
Orders for goods 295, 296, 298, 318–9
Oryx 110, 112, 167, 169, 372
Ostrich 110, 112, 168, 372
Ottomans 14, 36–9, 45–8, 122, 128, 282, 297, 323, 339–40, 341, 345, 346, 387
  officials 307, 346
  reforms 264, 340
  tax registers 14, 122
Over-grazing denied 103–4
Owls 110

Palestine 26, 31, 36, 37, 46, 50, 65, 119, 197, 275, 277, 278, 325, 331, 346, 358, 359, 387, 388
Palestinians 38, 109, 117, 203, 204, 266, 267, 288, 293, 297, 308, 323, 351, 352, 359
Palmyra 13, 23, 24, 213, 257, 280
Palmyrena 40, 171
Panthers 110
Paper 32
Paraffin (kerosene) 174, 295
Partridge 167
Patrons/clients 299–301, 345
Peaches 176, 178, 191
Peanuts 184
Pears 176, 198
Peasants also *Fellah* 10, 38, 39, 48, 55, 58, 64–5, 68, 300, 304, 307, 313, 315, 340–1, 346, 384
Pedlars 320
Pensions 60, 310, 311, 316, 322, 325, 326–7, 353, 380
Peppers 33, 187
Pesticides 181, 187
Pests, insect 193
Petra 23, 371
Petrol (gasoline) 297
Pickups –
  see Lorries
Pigeons 167, 170

Pigs 20
Pilgrimage 26, 27, 274
    Mecca 37, 45–6, 70, 74, 76, 182, 206, 282
    local 245, 274–5, 392
Pistacia (butm) 20, 99, 100, 102, 116, 118, 172–3, 176, 211, 232, 275–6, 370
    flour 276
    processing 275–6
Plains and plateaux 98, 99, 102, 108, 114, 115, 116, 117–8, 143, 144, 145–6, 198
Plants
    annual 103, 104, 105, 111, 171, 227–9, 231
    ashes
    see *kilw*
    categories (local) 104
    collecting wild 171–4, 331, 370
    diseases 167
    food 20, 171–3, 176, 331, 370
    gums 26, 171–3, 176
    industrial 23, 26, 171, 172–4, 176
    medicinal 171, 176, 187
    nursery 165
    repertoire 19, 21
    resilience 103, 396
    seasonal grazing 102, 103–4, 108, 109, 110, 118, 227–9, 231, 232
Plaster 26, 143, 251, 278
Plastic 26, 279, 361
    pipes 116, 123, 152, 153, 154, 155, 164, 165, 186, 286, 295, 361
    sheeting 116, 152, 153, 164, 186, 286
    tunnels 179, 380
Ploughing 33, 178, 186, 187, 188, 189, 192, 193, 195–6, 283, 290
Plums 176, 189, 201
Point 4 132, 146
Police 224, 288, 311, 348, 354, 366, 381, 382
Police posts 128, 217, 253, 265, 266, 363, 381

Pomegranates 20, 44, 154, 166, 178, 183, 201
Poplar, white 173, 201, 241
Poplar, black 173
Popular resistance 36, 39, 42, 48, 49, 51, 307, 340–1, 346, 369
Porcupines 110, 189
Potatoes 32, 181, 184, 198, 200, 201, 294, 365
    chips 330, 365
Pottery 276, 279–80
Pounding holes 275–6, 370
Poverty 71, 75, 78, 326–7
Power
    local ideas of 16, 17
    relations of 326
    'over' 17, 18, 375–6
    'to' 17, 18, 375–6
Practice, social 12, 15, 53, 61, 63, 79–80, 289, 344, 361, 372
    constructed 10, 56–8, 78–9
Presseries
    Apricot 276
    Grape 158
    Olive – see olives
Primus
    cooking 174
Productivity 289–328
    grain 193–4, 277
Professions 311, 318, 323, 353, 356, 382, 386
Profitability
    arable crops 179, 184, 193–5, 197, 206, 338
    businesses 292, 325
    camels 234, 236–7, 290
    chickens 180
    cow dairies 203, 205, 206
    date gardens 178
    fruit 184, 185, 187, 191, 295, 378
    goats 212, 231–2, 233
    land 225, 302, 378
    olive gardens 310
    sheep 181, 206, 208–9, 212, 224–6, 234, 290, 310, 315, 338, 379
    small businesses 293, 295, 310, 315, 316, 317

trading 319, 320, 323
  vegetables 179, 183, 184, 185, 191, 295, 378
Protection 17, 76, 213, 266, 282, 287, 312, 337, 381–2
Pruning 188, 189
Pumpkins 32

Qadisiyya 117, 120, 145, 175, 236, 270, 272, 316, 319, 370–1
Quail 167
Qalamoun 30, 31, 98, 99, 102, 131, 157, 159, 160, 170, 172, 173, 174, 198, 200–1, 213, 240, 241, 277, 278, 280, 299, 303, 314, 332
Qanat 150, 154, 155–7, 303
Qara 61, 178–9, 256–8, 259
Qariatain 168, 172, 318
al-Qasr (place) 263, 265–7, 278, 283, 307, 310, 380–1, 382
Qasr, Qusûr 35, 125, 132,177, 249–50, 251, 258
Quarries 281
Quarters 258
Quilts 32, 247, 335, 374
Quince 189

Rabbit
  see Hares
Radishes 189
Raetam (rattam) 102
Rafîq 62, 282, 300, 312, 337
Raids 22, 49–50, 62, 73, 300, 362, 372
  on pilgrims 46
Rains 203–4
Rainfall 14, 27, 98, 99, 100, 101, 103, 104–5, 105, 108, 109, 117, 118, 129, 131, 132, 138, 142, 144, 153, 157, 175, 192–3, 195, 197, 198, 199–200, 214, 215, 217–9, 227–9, 235, 241, 243, 374–5
  local terms for 104–5
  variability 100, 105
  winter 98, 104, 105
Raisins 117, 287, 330, 365

Rashaiyida 119, 122–3, 154, 175, 183, 212, 385
Ravens 110, 113
Red Sea 13, 23, 70
Refrigerators 331, 350
Regeneration of
  grazing 103–4, 227–8, 371–2, 372–3, 375
Religion 26
  local practice 388
  official 388–90
Remittances 352
Renting
  businesses 34
  camels 41, 296
  gardens 177, 299
  land 25, 33, 308, 309
  shops 315, 316
  trees 177, 299
  waqfs 37
Reputation 16, 53, 63, 66, 72, 74, 78, 319, 321, 322, 325, 327, 363
Reserves 87, 342, 343, 344–9
  forage 369, 370
  forest 368–71
  grazing; *hîma* 40, 229, 370, 372, 375
  mahmiyya 110, 369, 372–3
Resilience
  in production 77, 342
  of social practice 60, 125, 395
  of vegetation 103, 227, 396
Responsibility 36, 363, 365, 367, 382, 394
  individual 300, 367
Restaurants 46, 288, 316, 328, 331
Restitution 61, 300, 364
Retailers 294, 299
Rhantherium (arfaj) 108
Rice 32, 71, 328, 329, 330, 331, 332
Rifles
  see weapons
ar-Rîshas 70, 107, 125, 128, 134, 146, 198–9, 217, 250–1, 252–3, 287–8, 316
Risk 292, 315, 356, 362

Roads   128, 251, 259, 266, 267–9, 270, 281, 282–4, 297, 331
  building   128, 283
  routes   282–4
Rock carvings   110, 126, 135, 169
Rogan, E.   38, 47, 48, 302, 313
Roofs
  basalt slabs   255
  metal beams   254
  weakening   271–2, 242
  wooden beams   253, 256, 258, 262, 267, 271–2
RSCN   369–71, 372
Rugs   247, 317, 335, 374
Ruins, including
  khirba   253, 254, 255, 262, 263, 272, 280
Ruling   57, 58, 71–4, 106, 122, 343–4, 261
  hukm (rule)   72–4, 83, 343
Rushes   116, 178, 271
Ruwaishid   128, 217, 218, 219, 251, 288, 295, 315–6, 317, 318, 331
Rwala   55, 59–60, 62, 65, 66, 67, 69–70, 71, 72–3, 74, 76, 101, 105–7, 109, 110, 113, 125, 126, 133, 134, 135, 138, 142, 169, 172, 213, 215, 217–9, 223, 228, 229, 234–6, 241, 247, 250, 251–3, 259–60, 287–8, 298, 300–1, 302, 311, 315, 316, 317, 321–2, 326, 330, 333, 334, 336, 340, 346–7, 356, 364–5, 373, 379–80

Sab'a Biyar   105, 133
Safi   115, 153, 182, 184–5, 282
Sakaka   70, 140, 142, 165, 176–79, 258–9, 275, 294–5, 311, 321, 331, 357, 387
Salad   32, 328, 330
Salkhad   28
Salsola (ruth)   103, 229
Salt   26, 102, 280–1, 329, 332
as-Salt (place)   27, 32, 39, 48, 61, 172, 173, 205, 206, 262, 263, 274–5, 312, 313, 323, 338, 345
Saltpetre   174, 280
Samn
  see butter
Sand   102, 107, 108, 109, 118, 138, 139, 140, 171, 181
  dunes   108–9
  sandstone, see soils
Sandgrouse   167
Sardiyya   46, 106, 110, 165, 191, 218, 254–6, 359, 373
Sawwân   14, 98
Sayl –   118, 131, 141
  flood   147, 149, 152, 154, 231
Sa'idiyin   70, 75, 119, 154, 360, 370
Sba'a   106, 296, 298, 300
Schilcher, L.   39, 47, 48, 297, 340, 341, 346
Schools   259, 266, 272, 273, 287, 288, 378, 385, 387, 389
Scorpions   178
Seeps   114, 117, 118, 141, 153, 157
Semh
  mesembryanthemum forskalei   118, 171
Sesame   198
Settlements   27, 30, 31, 38, 263, 265–7, 287
  in *badia*   70, 125, 251
  expansion/decline   22, 23, 24, 25, 286, 287–8
  of nomads   25, 70, 259
Shaikhs/tribal leaders   18, 25, 35–6, 39, 45, 46, 72–5, 121–2, 125, 133, 251, 252, 264, 275, 277, 298, 299–300, 339, 340, 346–7, 348, 349, 364–5, 374–5, 387, 395
Shammar   64, 68
Shararat   70, 71, 110, 111, 113, 169, 217, 373
Shares   27, 34, 45, 291, 292
  buda'a   315
  farming   29, 30, 37, 38, 69, 177, 185, 186, 191, 197–8,

210, 211, 292, 299, 305, 308, 309, 313, 317
  partnerships 24, 29, 37, 69, 190, 204, 208, 214, 292, 305, 313, 314, 315, 319
  selling of 34, 45, 204
  sub-contracting 204, 252
  three-sided 315
  transfer of 315, 319
  urban enterprises 292, 313, 315–6, 320
Shaubak 28, 119, 120, 123, 145, 212
Shawabke 119, 122–3
Sheep 20, 21, 22, 23, 37, 40, 60, 101, 105, 178, 181, 205–30, 255, 256, 290, 310, 311, 314, 317
  breeds 205–6
  calendar 207–8, 214–6, 220
  flock size 181, 206, 208, 209, 210, 217, 218, 220, 226–7
  herding
    see herding
  housing 248–9, 256, 262, 271
  ownership 206, 211, 212–4, 314
  trading 212, 222–3, 224–5, 293–4, 296, 297
Sheepskin 293, 294, 334
Shelters, animal 203, 204, 233, 248–9, 256, 262, 271
Shera' 28, 36, 98, 117–21, 145–6, 153–5, 170, 172, 173, 176, 231–2, 240, 247–8
  vegetation 118
Shih
  see *artemsia*
Shops 203, 250, 269, 270, 275, 288, 293, 294, 295, 309, 310, 315, 316, 328, 329, 331
Shrines 252, 253, 264, 274–5, 392
Shwaya 59–60, 212
Sidr
  see zizyphus
Silk 26, 32
Sinai 19, 46, 60, 171, 172
Sirhan 110, 373

Skins 26, 168, 263, 293, 294
Slaughterhouses 293
Slaves *(abd)* 56, 58
Smuggling 39, 225, 287, 296, 298, 300, 331, 337, 362
Snakes 178
Snow 98, 99, 101, 105, 192, 199, 218
Social class 349–50
Sorghum
  see millet/*dhurra*
Soil types 98, 216
  clay silts 102, 105, 133, 138, 140, 164, 181
  cultivable 101, 102, 105, 108, 111, 115, 116, 117, 179
  gravels 102, 115, 118, 133, 139, 171
  limestone 102, 114, 144, 145
  moisture levels 101, 105, 109, 131, 146, 199, 200
  properties 101, 105, 131–2, 146
  red volcanic 99, 102, 144
  sand 102, 108–9, 118, 138, 139, 140, 171, 196
  sandstone 108–9, 117, 145, 160, 164
  shi'bân 101, 105, 126, 216, 228
Sources 15–16
Sowing rates 192–3, 195–6, 201
Spinach 182, 189
Sponsors
  see *khafila*
  state as 298, 364
Sponsorship 259, 282, 337, 363–5
Springs 76, 114, 117, 145, 147, 150–3, 157–9
Star calendar 104
State
  attitudes to tribes 298, 305, 321, 339, 341, 344, 347–8, 395
  capitalism 342
  control by 341, 344, 347, 349, 350, 358–60, 395

control of resources 297,
  338–9, 357, 358–60, 366,
  379, 388
daulat 343–4
income 46, 297–8, 318, 339,
  341, 349
institutions 340, 357–8, 367,
  368, 383–5
land 39, 106, 304–5, 368–372,
  379–80
land reforms 361–2
as partner 298, 345, 384
pricing policy 179, 181, 197,
  267, 379, 380
provide security 339, 342
redefine themselves 344, 348–9
rentier 341–2, 348
restrictions 41, 331–6
subsidies 197
Stipa sp. (sam'a) 103
Stone clearance 249, 284
Storage 249–50, 251, 253–56,
  258, 262–4, 265, 266, 267,
  269, 270–1, 274, 275
Storks 110
Stratification 56, 350
Straw also *Tibben* 180, 181, 196–7,
  200, 202, 203, 207, 208, 250,
  254, 255, 267, 271, 278–9
Structural units of
  'aila/ family 64, 65, 66, 67
  ashira 64, 65, 67, 68
  fakhdh 59, 65, 66
  fara 59, 66, 68
  hamûla 59, 66
  khamsa 64, 381
  luzum 64, 68
  qabîla 64, 67
  rural society 64–68
Subayh/Subayha 254
Sub-contracting 204
Suez Canal 46, 338
Sugar 32, 41, 147, 173, 184, 295
Sukhne 27, 46, 59, 64, 68, 70,
  171, 172, 173, 201, 213, 230,
  293, 296, 314, 318
Sulaib 56, 71, 105, 139, 167,
  168, 169, 173, 280, 364–5
Sumac 173, 201

Summer crops 31, 182–3, 185,
  189, 200
Sunflowers 188, 198
Supermarkets 316
Surplus 31, 33, 77, 299, 337–9,
  343, 345, 349, 377, 378

Tabûn ovens 271, 279
Taima 23, 27, 105, 163, 168,
  169, 178
Tamarisk *(ithl/tarfa)* 103, 115,
  116, 118, 163, 227, 232, 241,
  256, 258
Tankers 191, 199, 204, 214,
  215, 226, 227, 235
Taxation 75–7, 338
  of agriculture 35, 291, 338, 341
  of animals 40, 76, 290–1, 341
  collection 34–5, 73, 339, 340,
    345
  delegated 36, 339, 345
  direct 39
  'farming' 35, 37, 39, 339
  local 381
  malikana 28
  registers 30, 36, 290
  revenues 39, 44, 338
Technologies 14, 24, 47, 128,
  131, 133, 165, 296, 343
  agricultural 164, 165, 179,
    361, 379
  computers 311, 320, 350, 9
  satellite dishes 260, 335, 350
  steamships 47
  telecommunication 311, 350
  telegraph 47
  telephone 311, 335, 350, 354
  television 259, 270, 334, 335
Tent 64, 202, 214, 217,
  239–250, 251, 252, 255, 259
  Ahl al-Jabal 217, 230
  Ata'ata' 242, 270
  Azazma 231, 243
  Beni Attiya 243
  Beni Hamida 243
  Beni Sakhr 243
  canvas, square 250
  care of 247
  cooking in 174, 247, 276

furniture 237, 247, 259
Majali 209, 243, 264
metal frames for 259, 231, 328
in mountains 239, 240, 241, 247
pegs 242, 295
poles 173, 214, 239, 241, 242, 288, 295
ropes 173, 241–2, 295
Rwala 234, 241, 247, 258–60, 287–8, 328
sewing 240, 242
siting 230–1, 234, 243–4, 250, 251
shade 247
Tentcloth 233, 240–1, 288, 295, 300, 302–3
camelhair 241
canvas 250
cotton 243
goat-hair 203, 232, 240–1
sacking 243
wool 241
Terns 110
Terraces 189, 202, 284–5
Territory 70
Textiles 20, 32, 35, 42, 247, 298, 313, 317, 327, 335
Thoumin, R. 33, 76, 157, 159, 200, 213, 241, 261, 276, 277, 278, 303, 314, 347
Threshing 196–7
Tibben
see straws
Tobacco 188, 192, 310
Tomatoes 32, 115, 179, 183, 185, 186–7, 188–9, 191, 192, 198, 201, 295, 309–10, 329, 330
dried 189, 331
paste factory 365
Towns and cities 22–3, 57–8, 258–9, 260, 266–7, 273, 351–2, 354, 378, 382
population/s 26, 28, 29, 33, 54–5, 61, 307, 319
Trade 23–4, 28, 41–6, 299, 320, 323, 352
agricultural 25–6, 38–40, 179, 213, 294–6, 315, 358
animals 40, 70, 222, 224–5, 293–4, 315, 318–9
animal products
see product
aromatics 23, 26
cross-border 298–9, 318–9, 320, 322
hunted/gathered
products 167–73
inter-regional 22, 43, 117, 171, 230, 287, 313, 338, 341, 358
intra-regional 43–5, 174, 230, 313, 341
mixed 296, 323, 352
routes 28, 42–3, 70, 282, 342
at shrines 274–5
smuggling as 362
Traders 24, 38–40, 119, 275, 316, 351
big 293, 294
from Kubayza 59, 296
from Ramtha 287, 315
small 179, 204, 231, 232, 233–2, 288, 296, 320
from Sukhne 52, 171, 172, 230, 296
from Syria 168, 172, 288, 296
Uqayl/Ageyl 41, 42, 74, 296 299
women traders 317–8
Tranj (citrus fruit) 176
Transformations 11–12, 54, 61–2
also Change 233–4, 341–2, 378
Transport 22, 292, 295, 296, 297, 311, 319, 321, 323
companies 323,
Transporters 295, 296, 300, 307, 315
Travel 282, 359
Trees 99, 100, 109, 115, 116, 118, 232, 241, 247, 285, 370–1
as crops 171–4, 183
destruction of 116, 128, 179, 370
importance of 111, 128, 175, 232, 345
products 171–4, 272

456 *People, Land and Water*

Tribal leaders
  – see Shaikhs
Tristram, HB   30, n10 54, 142,
  147, 149, 169, 194, 202, 203,
  277
Turaif   107, 253, 298, 311, 318,
  380
Turnips   181, 201

Uqayl–see Traders
al-Ula   160–2, 256, 275
'Umur   62, 106, 172, 213, 218,
  230, 287, 288, 364
'Unayza   43, 58, 59, 61, 70, 71,
  296, 317, 318, 356, 382–3
Urbanisation   351
'Usayfat   119

Valleys   98, 102, 247
  rijla   102
  risha   102
  see wadi
Vegetables
  see individual names
Vengeance   64–6, 381–2
Vicia (bicia)   195
Villas
  see Houses
Villages   22–3, 25, 27–8, 29–31,
  36–7, 48, 55, 58, 63, 101,
  102, 109, 119, 124, 169, 184,
  198, 211, 212–3, 218, 231,
  253, 258–9, 262–3, 266, 270,
  272–3, 283, 286–7, 304, 343
Volcanic cones   100, 108

Wadis   99, 100, 106–7,
  114–5, 125, 129, 131, 133,
  139–40, 141, 145, 147,
  151–2, 216, 227, 232, 287
  'Araba   8, 117–8, 145,146–8,
    153–5, 171, 172, 277, 323,
    327, 331, 333, 358
  Ibn Hammad   114, 116, 143,
    150, 185–8, 274, 278
  Sirhan   109–10, 116, 132–3,
    145,149, 171, 176, 182, 280,
    282, 284, 253, 311, 331, 334,
    358, 361, 380

Wallin, G.A.   27, 46, 70, 76, 140,
  159, 162, 171, 176, 182, 202,
  256, 258
Walnuts   188, 190, 201
Waqf   36, 37, 304, 339
Watan   69–70
Water
  aquifers   131, 134, 157, 164,
    165
  barrels   215, 243, 248, 295
  birka
    see *birka*
  channels
    see foggara/qanat
  cisterns
    see cisterns
  collection of   131–50, 152–66
  customary law   129–31, 142,
    149
  dams
    see dams
  disputes
    see disputes, water
  flows   106–8
  free   129, 131
  ghadîr   133, 135, 139, 140
  groundwater   135, 155, 157
  hamad
    see hamad
  harra
    see harra
  household   142, 144, 145
  khabra
    see khabra
  legal position   129–131
  mahfûr
    see *mahfur*
  ownership
    see owning
  -ing places   131–3, 134–42,
    143, 145
  piped/pumped   143, 145, 165,
    166, 254, 258, 269, 270
  rainpools
    see khabra
  rock basins   132–3
  run-off
    see sayl/runoff
  sayl

see sayl/runoff
-sheds 100, 107
snowmelt
see snow
springs
see springs/ seeps
storage
see water conservation
summer 115, 145
transporting 41, 214, 227
wells
see wells
Wealth
agricultural 33, 341
in countryside 302
ideas of 324–8
increases in 341–2, 350, 352
merchant 74, 314
in past 25, 337–9, 341–2
sources of 43, 290, 341, 348, 352
too much 356–7
transfer of 301, 335–7
of tribespeople 25, 74, 321
Weddings 336
Weeding 191, 193, 196
Weeds 189
Welî
and see shrines 252, 253
Wells 40, 49, 101, 105, 108, 112, 120, 125, 130, 131, 138, 139–42, 135–6, 137, 157, 158, 159, 163, 164, 166, 177, 179–80, 184, 189, 192–4, 234
deep 106, 109, 133, 162, 163, 164, 165, 191, 379
government 106, 133, 142, 155, 379
types of 138, 140–2
Wetzstein, J.G. 32, 33, 174, 202, 251, 252, 253, 340
Weuleresse, J. 27, 171, 252, 280, 281
Wheat 20, 32, 37, 44, 153, 162, 176–8, 178, 180–1, 184, 187, 192–4, 196, 198, 199, 210
einkorn 19, 194
emmer 19, 194

burghul 163, 192, 197, 172, 276, 329, 330, 335
fariqa 175, 192, 197, 276
spelt 194
varieties of 19, 193–4
Wheelbarrows 295
Wholesalers 294, 299
'Wild' (category) 104, 111
Wild animals 110, 370, 372
Wild boar 110
Wild cat 110
Willow 116, 118, 173, 241
Winds 98, 101, 108
problems with 108, 191, 193
Winter 106
cold 101, 200, 216, 370
rains 98, 105, 192–3, 214, 217–9
shelter 101, 108, 115, 120, 215, 243, 248–9, 370
warmth 115
Wisad 100, 132, 133
Witnesses 17, 45, 62, 66, 292
Wolves 110, 113, 167, 373
Women 17, 34, 67, 252, 276, 291, 295, 327, 328, 331, 332–4, 337, 365, 382, 390
education 311–2, 318, 323, 353, 354, 355–6, 387
employment of 163–4, 311–2, 318, 324, 360, 386
enterprises 208, 210, 214, 225, 231, 291, 316–8
fieldworkers 184, 186, 191
gardening 186, 188–9, 319
herding 175, 203, 207, 208, 213–4, 220–1, 225, 231, 233, 248–9
inheritance 34, 272, 309, 316
liwan 256
mourning 273–4
ownership 34, 189, 213–4, 309, 316–7, 325
pottery 279
privacy of 255, 259, 269, 287
property 284–5, 290–1, 309
relations through 67
and tents 233, 240–7, 248
traders 317–8

in villas 255–6, 259–60, 261, 269, 270, 350
Wood 26, 174–4, 176, 227, 254, 276, 279, 316, 371
Wool 20, 26, 32, 41, 206, 207, 208, 209, 210, 213, 220, 241, 296, 314, 322–3
Workshops 258, 275, 288, 296, 297, 316
Workmen 254, 260, 264, 267, 269, 284

al-Wudiyan 108
Wusta 363–4

Yabrud 99, 157, 159, 200, 201, 240, 241

Zarqa 204, 250, 293
Zirb 243, 247
Zizyphus (sidr) 20, 115, 118, 172, 173, 183, 186, 232
Zubaid 101, 106, 217, 220

54. Transformation of the God-Image: Jung's Answer to Job
Edward F. Edinger (Los Angeles). ISBN 0-919123-55-4. 144 pp. $18

55. Getting to Know You: The Inside Out of Relationship
Daryl Sharp (Toronto). ISBN 0-919123-56-2. 128 pp. $18

56. A Strategy for a Loss of Faith: Jung's Proposal
John P. Dourley (Ottawa). ISBN 0-919123-57-0. 144 pp. $18

58. Conscious Femininity: Interviews with Marion Woodman
Introduction by Marion Woodman (Toronto). ISBN 0-919123-59-7. 160 pp. $18

59. The Middle Passage: From Misery to Meaning in Midlife
James Hollis (Houston). ISBN 0-919123-60-0. 128 pp. $18

60. The Living Room Mysteries: Patterns of Male Intimacy, Book 2
Graham Jackson (Toronto). ISBN 0-919123-61-9. 144 pp. $18

61. Chicken Little: The Inside Story *(A Jungian Romance)*
Daryl Sharp (Toronto). ISBN 0-919123-62-7. 128 pp. $18

62. Coming To Age: The Croning Years and Late-Life Transformation
Jane R. Prétat (Providence, RI). ISBN 0-919123-63-5. 144 pp. $18

63. Under Saturn's Shadow: The Wounding and Healing of Men
James Hollis (Houston). ISBN 0-919123-64-3. 144 pp. $18

65. The Mystery of the Coniunctio: Alchemical Image of Individuation
Edward F. Edinger (Los Angeles). ISBN 0-919123-67-8. 112 pp. $18

66. The Mysterium Lectures: Journey through Jung's *Mysterium Coniunctionis*
Edward F. Edinger (Los Angeles). ISBN 0-919123-66-X. 352 pp. $30

83. The Cat: A Tale of Feminine Redemption
Marie-Louise von Franz (Zurich). ISBN 0-919123-84-8. 128 pp. $18

87. The Problem of the Puer Aeternus
Marie-Louise von Franz (Zurich). ISBN 0-919123-88-0. 288 pp. $25

95. Digesting Jung: Food for the Journey
Daryl Sharp (Toronto). ISBN 0-919123-96-1. 128 pp. $18

99. The Secret World of Drawings: Healing through Art
Gregg M. Furth (New York). ISBN 1-894574-00-1. 100 illustrations. 176 pp. $25

100. Animus and Anima in Fairy Tales
Marie-Louise von Franz (Zurich). ISBN 1-894574-01-X. 128 pp. $18

108. The Sacred Psyche: A Psychological Approach to the Psalms
Edward F. Edinger (Los Angeles). ISBN 1-894574-60-5. 160 pp. $18

111. The Secret Garden: Temenos for Individuation
Margaret Eileen Meredith (Toronto). ISBN 1-894574-12-5. 160 pp. $18

112. Not the Big Sleep: on having fun, seriously *(A Jungian Romance)*
Daryl Sharp (Toronto). ISBN 1-894574-13-3. 128 pp. $18

113. The Use of Dreams in Couple Counseling
Renée Nell (Litchfield, CT). ISBN 1-894574-14-1. 160 pp. $18

Discounts: *any 3-5 books, 10%; 6-9 books, 20%; 10 or more, 25%*
Add Postage/Handling: 1-2 books, $6 surface ($10 air); 3-4 books, $8 surface ($12 air);
5-9 books, $15 surface ($20 air); 10 or more, $10 surface ($25 air)

Free Catalogue of over 100 titles and *Jung at Heart* newsletter

INNER CITY BOOKS, Box 1271, Station Q, Toronto, ON M4T 2P4, Canada
Tel. 416- 927-0355 / Fax 416-924-1814 / E-mail: sales@innercitybooks.net

27. **Phallos: Sacred Image of the Masculine**
Eugene Monick (Scranton, PA). ISBN 0-919123-26-0. 144 pp. $18

28. **The Christian Archetype: A Jungian Commentary on the Life of Christ**
Edward F. Edinger (Los Angeles). ISBN 0-919123-27-9. 144 pp. $18

30. **Touching: Body Therapy and Depth Psychology**
Deldon Anne McNeely (Lynchburg, VA). ISBN 0-919123-29-5. 128 pp. $18

31. **Personality Types: Jung's Model of Typology**
Daryl Sharp (Toronto). ISBN 0-919123-30-9. 128 pp. $18

32. **The Sacred Prostitute: Eternal Aspect of the Feminine**
Nancy Qualls-Corbett (Birmingham). ISBN 0-919123-31-7. 176 pp. $20

33. **When the Spirits Come Back**
Janet O. Dallett (Seal Harbor, WA). ISBN 0-919123-32-5. 160 pp. $18

34. **The Mother: Archetypal Image in Fairy Tales**
Sibylle Birkhäuser-Oeri (Zurich). ISBN 0-919123-33-3. 176 pp. $20

35. **The Survival Papers: Anatomy of a Midlife Crisis**
Daryl Sharp (Toronto). ISBN 0-919123-34-1. 160 pp. $18

37. **Dear Gladys: The Survival Papers, Book 2**
Daryl Sharp (Toronto). ISBN 0-919123-36-8. 144 pp. $18

39. **Acrobats of the Gods: Dance and Transformation**
Joan Dexter Blackmer (Wilmot Flat, NH). ISBN 0-919123-38-4. 128 pp. $18

40. **Eros and Pathos: Shades of Love and Suffering**
Aldo Carotenuto (Rome). ISBN 0-919123-39-2. 160 pp. $18

41. **The Ravaged Bridegroom: Masculinity in Women**
Marion Woodman (Toronto). ISBN 0-919123-42-2. 224 pp. $22

43. **Goethe's *Faust*: Notes for a Jungian Commentary**
Edward F. Edinger (Los Angeles). ISBN 0-919123-44-9. 112 pp. $18

44. **The Dream Story**
Donald Broadribb (Baker's Hill, Australia). ISBN 0-919123-45-7. 256 pp. $24

45. **The Rainbow Serpent: Bridge to Consciousness**
Robert L. Gardner (Toronto). ISBN 0-919123-46-5. 128 pp. $18

46. **Circle of Care: Clinical Issues in Jungian Therapy**
Warren Steinberg (New York). ISBN 0-919123-47-3. 160 pp. $18

47. **Jung Lexicon: A Primer of Terms & Concepts**
Daryl Sharp (Toronto). ISBN 0-919123-48-1. 160 pp. $18

48. **Body and Soul: The Other Side of Illness**
Albert Kreinheder (Los Angeles). ISBN 0-919123-49-X. 112 pp. $18

49. **Animus Aeternus: Exploring the Inner Masculine**
Deldon Anne McNeely (Lynchburg, VA). ISBN 0-919123-50-3. 192 pp. $20

50. **Castration and Male Rage: The Phallic Wound**
Eugene Monick (Scranton, PA). ISBN 0-919123-51-1. 144 pp. $18

51. **Saturday's Child: Encounters with the Dark Gods**
Janet O. Dallett (Seal Harbor, WA). ISBN 0-919123-52-X. 128 pp. $16

52. **The Secret Lore of Gardening: Patterns of Male Intimacy**
Graham Jackson (Toronto). ISBN 0-919123-53-8. 160 pp. $16

53. **The Refiner's Fire: Memoirs of a German Girlhood**
Sigrid R. McPherson (Los Angeles). ISBN 0-919123-54-6. 208 pp. $18